Logic

'Paul Tomassi's book is the most accessible and user-friendly introduction to formal logic currently available to students. Semantic and syntactic approaches are nicely integrated and the organisation is excellent, with later sections building systematically on earlier ones. Tomassi anticipates all the most important traps and confusions that students are likely to fall into and provides first-rate guidance on practical matters, such as strategies for proof-construction. Never intimidating, this is a text from which even the most unmathematically minded student can learn all the basics of elementary formal logic.'

E.J. Lowe, University of Durham

Logic brings elementary logic out of the academic darkness into the light of day and makes the subject fully accessible. Paul Tomassi writes in a patient and user-friendly style which makes both the nature and value of formal logic crystal clear. The reader is encouraged to develop critical and analytical skills and to achieve a mastery of all the most successful formal methods for logical analysis.

This textbook proceeds from a frank, informal introduction to fundamental logical notions, to a system of formal logic rooted in the best of our natural deductive reasoning in daily life. As the book develops, a comprehensive set of formal methods for distinguishing good arguments from bad is defined and discussed. In each and every case, methods are clearly explained and illustrated before being stated in formal terms. Extensive exercises enable the reader to understand and exploit the full range of techniques in elementary logic.

Logic will be valuable to anyone interested in sharpening their logical and analytical skills and particularly to any undergraduate who needs a patient and comprehensible introduction to what can otherwise be a daunting subject.

Paul Tomassi is a lecturer in Philosophy at the University of Aberdeen.

Logic

Paul Tomassi

London and New York

First published 1999
by Routledge
11 New Fetter Lane, London EC4P 4EE

Simultaneously published in the USA and Canada
by Routledge
29 West 35th Street, New York, NY 10001

Typeset in Palatino and Optima by RefineCatch Limited, Bungay, Suffolk
Printed and bound in Great Britain by
TJ International Ltd, Padstow, Cornwall

British Library Cataloguing in Publication Data
A catalogue record for this book is available from the British Library.

Library of Congress Cataloguing in Publication Data
A catalogue record for this book has been requested.

ISBN 0–415–16695–0 (hbk)
ISBN 0–415–16696–9 (pbk)

To Lindsey McLean,
Tiffin and Zebedee

Contents

Figures

Preface

I felt compelled to write an introductory textbook about formal logic for a number of reasons, most of which are pedagogic. I began teaching formal logic to undergraduates at the University of Edinburgh in 1985 and have continued to teach formal logic to undergraduates ever since. Speaking frankly, I have always found teaching the subject to be a particularly rewarding pastime. That may sound odd. Formal logic is widely perceived to be a difficult subject and students can and often do experience problems with it. But the pleasure I have found in teaching the subject does not derive from the anxious moments which every student experiences to some extent when approaching a first course in formal logic. Rather, it derives from later moments when self-confidence and self-esteem take a significant hike as students (many of whom will always have found mathematics daunting) realise that they can manipulate symbols, construct logical proofs and reason effectively in formal terms. The educational value and indeed the personal pleasure which such an achievement brings to a person cannot be overestimated. Enabling students to take those steps forward in intellectual and personal development is the source of the pleasure I derive from teaching formal logic. In these terms, however, the problem with existing textbooks is that they generally make too little contribution to that end.

For example, each and every year during my time at Edinburgh the formal logic class contained a significant percentage of arts students with symbol-based anxieties. More worryingly, these often included intending honours students who had either delayed taking the compulsory logic course, failed the course in earlier years or converted to Philosophy late. Many of these students were very capable people who only needed to be taught at a gentler pace or to be given some individual attention. Moreover, even the best of those students who were not so daunted by symbols regularly got into difficulties simply through having missed classes – often for the best of reasons. Given the progressive nature of the formal logic course these students frequently just failed to catch up. As a teacher, it was immensely frustrating not to be able to refer students (particularly those in the final category) to the textbook in any really useful way. The text we used was E.J. Lemmon's *Beginning Logic* [1965]. Undoubtedly, Lemmon's is, in many ways, an excellent text but the majority of students simply did not find it sufficiently accessible to be able to teach themselves from it. In all honesty, I think that this is quite generally the case with the vast majority of introductory texts in formal logic, i.e. inaccessibility is really only a matter of degree (albeit more so in the case of some than others). And this is no mere inconvenience for students and teachers. The underlying worry is that the consequent level of fail rates in formal logic courses might ultimately contribute to a decline in the teaching of formal logic in the universities or to a significant dilution of the content of such courses. For all of these reasons, I think it essential that we have a genuinely accessible introductory text which both covers the ground and caters to the whole spectrum of intending logic students, i.e. a text which enables students to teach themselves. That is what I have tried to produce here.

Logic covers the traditional syllabus in formal logic but in a way which may significantly reduce the kind of fail rates which, without such a text, are perhaps inevitable in compulsory courses in elementary logic offered within the Faculty of Arts. In the present climate, many faculties and, indeed, many philosophy departments consider such fail rates to be wholly unacceptable. Hence, the motivation to dilute the content of courses is obvious, e.g. by wholly omitting proof-theory. Personally, I believe that this cannot be a step in the right direction. In the last analysis, such a strategy either diminishes formal logic entirely or results in an unwelcome unevenness in the distribution of formal analytical skills among graduates from different institutions. I believe that the solution is to make available to students a genuinely accessible textbook on elementary logic which even the most anxious students in the class can use to teach themselves. Thus, *Logic* is not designed to promote my own view of formal logic as such or to promote the subject in any narrow sense. Rather, it is designed to promote formal logic in the widest sense, i.e. to make a subject which is generally perceived as difficult and inaccessible open and readily accessible to the widest possible audience.

To that end, the text is deliberately written in what I hope is a clear and user-friendly style. For example, formal statements of the rules of inference are postponed until the relevant natural deduction motivation has been outlined and an informal rule-statement has been specified. The text also makes extensive use of summary boxes of key points both during and at the end of chapters. Initial uses of key terms (and some timely reminders) are given in bold and such items are further explained in the glossary. Mock examination papers are also set at regular intervals in the text by way of dress rehearsal for the real thing. Given that accessibility is a crucial consideration, the pace of *Logic* is deliberately slow and indulgent. But this need not handicap either students or teachers. The text is exercise-intensive and brighter students can simply move to more difficult exercises more quickly. Moreover, the very point of there being such a text is to enable students to teach themselves. So teachers need not move as slowly as the text, i.e. the pace of the course may very well be deliberately faster than that of the text. The point is that the text provides the necessary back-up for slower students anyway. Further, those who miss classes can plug gaps for themselves, and while I have no doubt that certain students will still have problems with formal logic the text is specifically designed to minimise the potential for anxiety attacks.

I should also add that the text is tried and tested at least in so far as a desktop version has been used successfully at the University of Aberdeen for the past three academic sessions, over which, as I write, class numbers have trebled. The success of the text is reflected as much in course evaluation responses as in the pass rate for Formal Logic 1 (only one student failed Formal Logic 1 over sessions 1994–5 and 1995–6). Further, the pass rate for the follow-on course, Formal Logic 2, was 100 per cent in the first academic session and 95 per cent in the second academic session. Despite the increase in class numbers, pass rates in both courses remain very high and the contents of course evaluation forms suitably reassuring.

A certain amount of motivation for writing *Logic* also stems from some unease not just about the style but about the content of existing textbooks. For although many excellent texts are available, there is something of an imbalance in most. For example, while a number of familiar texts are quite excellent on semantic methods these tend to be wholly devoid of (linear or Lemmon-style) proof-theory. In contrast, texts such as Lemmon, for example, show a clear bias towards proof-theory and are not as extensive in their treatment of semantic concepts and methods as they might be. Indeed, certain texts in this latter category are either devoid of semantic methods at the level of quantificational logic or devote a very limited amount of space to such topics. Yet another group of familiar texts involves rather less in the way of formal methods generally. Ultimately, I think, such texts include too little in that respect for purposes of teaching formal logic to undergraduates. Hence, there is a strong argument for an accessible textbook which strikes a fair balance between syntactic and semantic methods. To that end, *Logic* combines a comprehensive treatment of proof-theory not just with the truth-table method but also with the truth-tree method. After all, the latter method is quite mechanical throughout both propositional logic and the monadic fragment of quantificational logic. Moreover, if that method is given sufficient emphasis at an early stage students can also be enabled to apply the method beyond monadic quantificational logic. Of course, in virtue of undecidability with respect to invalidity at that level, there is no guarantee of the success of any purely mechanical application of the truth-tree method, i.e. infinite branches and infinite trees are possible. But the application of the method at that level, together with examples of infinite trees and branches, vividly illustrates the consequences of undecidability to students and goes some way towards making clear just what is meant by undecidability. Finally, given that the method is also useful at the metatheoretical level, supplementing truth-tables with truth-trees from the outset seems a sound investment. In terms of content, then, the text covers the same amount of logical ground as any other text pitched at this level and, indeed, more than many.

In summary, *Logic* is primarily intended as a successful teaching book which students can use to teach themselves and which will enable even the most anxious students to grasp something of the nature of elementary logic. It is not intended to be a text which lecturers themselves will want to spend hours studying closely. Rather, it is intended to make a subject which is generally perceived as difficult and inaccessible open and easily accessible to the widest possible audience. In short, I hope that *Logic* constitutes a solution to what I believe to be a substantive teaching problem. However, if the text does no more than make formal logic accessible, comprehensible and above all useful to anxious students for whom it would otherwise have remained a mystery, then it will have fulfilled its purpose.

Paul Tomassi

Acknowledgements

I personally owe a number of debts of gratitude here. First, to those who taught me formal logic at the University of Edinburgh, principally, Alan Weir, Barry Richards and (via his *Elementary Logic*) Benson Mates. Next, I am indebted to E.J. Lemmon (via his *Beginning Logic*), to Stephen Read and Crispin Wright (via *Read and Wright: Formal Logic, An Introduction to First Order Logic*), to Stig Rassmussen and to John Slaney. This text owes much to all those people but especially to John Slaney, who first taught me how to teach formal logic. The text also owes much to all those undergraduate students at the Universities of Edinburgh and Aberdeen who have studied formal logic with me over the years. For me at least, it has been a particular pleasure. I gratefully acknowledge the *British Medical Journal* for permission to reproduce some of the arguments and illustrations published in *Logic in Medicine*. I am also very grateful to Louise Gregory for help preparing the manuscript, to Roy Allen for the index to the text, to Stephen Priest and to Stephen Read for useful comments and even more useful encouragement at an early stage of preparation, and to Patricia Clarke for helpful discussions of Chapter 1; and I am particularly indebted to Robin Cameron for all his generous help and support with the project.

1
How to Think Logically

1
How to Think Logically

I
Validity and Soundness

To study logic is to study **argument**. Argument is the stuff of logic. Above all, a logician is someone who worries about arguments. The arguments which logicians worry about come in all shapes and sizes, from every corner of the intellectual globe, and are not confined to any one particular topic. Arguments may be drawn from mathematics, science, religion, politics, philosophy or anything else for that matter. They may be about cats and dogs, right and wrong, the price of cheese, or the meaning of life, the universe and everything. All are equally of interest to the logician. Argument itself is the subject-matter of logic.

The central problem which worries the logician is just this: how, in general, can we tell good arguments from bad arguments? Modern logicians have a solution to this problem which is incredibly successful and enormously impressive. The modern logician's solution is the subject-matter of this book.

In daily life, of course, we do all argue. We are all familiar with arguments presented by people on television, at the dinner table, on the bus and so on. These arguments might be about politics, for example, or about more important matters such as football or pop music. In these cases, the term 'argument' often refers to heated shouting matches, escalating interpersonal altercations, which can result in doors being slammed and people not speaking to each other for a few days. But the logician is not interested in these aspects of argument, only in what was actually said. It is not the shouting but the sentences which were shouted which interest the logician.

For logical purposes, an *argument* simply consists of a sentence or a small set of sentences which lead up to, and might or might not justify, some other sentence. The division between the two is usually marked by a word such as 'therefore', 'so', 'hence' or 'thus'. In logical terms, the sentence or sentences leading up to the 'therefore'-type word are called **premises**. The sentence which comes after the 'therefore' is the **conclusion**. For the logician, an *argument* is made up of premises, a 'therefore'-type word, and a conclusion – and that's all. In general, words like 'therefore', 'so', 'hence' and 'thus' usually signal that a conclusion is about to be stated, while words like 'because', 'since' and 'for' usually signal premises. Ordinarily, however, things are not always as obvious as this. Arguments in daily life are frequently rather messy, disordered affairs. Conclusions are sometimes stated before their premises, and identifying which sentences are premises and which sentence is the conclusion can take a little careful thought. However, the real problem for the logician is just how to tell whether or not the conclusion really does follow from the premises. In other words, when is the conclusion a **logical consequence** of the premises?

Again, in daily life we are all well aware that there are good, compelling, persuasive arguments which really do establish their conclusions and, in contrast, poor arguments which fail to establish their conclusions. For example, consider the following argument which purports to prove that a cheese sandwich is better than eternal happiness:

1. Nothing is better than eternal happiness.
2. But a cheese sandwich is better than nothing.

Therefore,

3. A cheese sandwich is better than eternal happiness.[1]

Is this a good argument? Plainly not. In this case, the sentences leading up to the 'therefore', numbered '1' and '2' respectively, are the premises. The sentence which comes after the 'therefore', Sentence 3, is the conclusion. Now, the premises of this argument might well be true, but the conclusion is certainly false. The falsity of the conclusion is no doubt reflected by the fact that while many would be prepared to devote a lifetime to the acquisition of eternal happiness few would be prepared to devote a lifetime to the acquisition of a cheese sandwich. What is wrong with the argument is that the term 'nothing' used in the premises seems to be being used as a name, as if it were the name of some other thing which, while better than eternal happiness, is not quite as good as a cheese sandwich. But, of course, 'nothing' isn't the name of anything.

In contrast, consider a rather different argument which I might construct in the process of selecting an album from my rather large record collection:

1. If it's a Blind Lemon Jefferson album then it's a blues album.
2. It's a Blind Lemon Jefferson album.

Therefore,

3. It's a Blues album.

Now, this argument is certainly a good argument. There is no misappropriation of terms here and the conclusion really does follow from the premises. In fact, both the premises and the conclusion are actually true; Blind Lemon Jefferson was indeed a bluesman who only ever made blues albums. Moreover, a little reflection quickly reveals that if the premises are true the conclusion must also be true. That is not to say that the conclusion is an eternal or **necessary truth**, i.e. a sentence which is always true, now and forever. But if the premises are actually true then the conclusion must also be actually true. In other words, this time, the conclusion really does follow from the premises. The conclusion is a logical consequence of the premises. Moreover, the necessity, the force of the 'must' here, belongs to the relation of consequence which holds between these sentences rather than to the conclusion which is consequent upon the premises. What we have discovered, then, is not the necessity of the consequent conclusion but the necessity of logical consequence itself.

In logical terms the Blind Lemon Jefferson argument is a **valid** argument, i.e. quite simply, if the premises are true, then the conclusion must be true, on pain of contradiction. And that is just what it means to say that an argument is valid: whenever the premises are true, the conclusion is guaranteed to be true. If an argument is valid then it is impossible that its premises be true and its conclusion false. Hence, logicians talk of validity as preserving truth, or speak of the transmission of truth from the premises to the conclusion. In a valid argument, true input guarantees true output.

Is the very first argument about eternal happiness and the cheese sandwich a valid argument? Plainly not. In that case, the premises were, indeed, true but the conclusion was obviously false. If an argument is valid then whenever the premises are true the conclusion is guaranteed to be true. Therefore, that argument is **invalid**. To show that an argument fails to preserve truth across the inference from premises to conclusion is precisely to show that the argument is invalid.

The Blind Lemon Jefferson example also illustrates the point that logic is not really concerned with particular matters of fact. Logic is not really about the way things actually are in the world. Rather, logic is about argument. So far as logic is concerned, Blind Lemon Jefferson might be a classical pianist, a punk rocker, a rapper, or a country and western artist, and the argument would still be valid. The point is simply that:

If it's true that:	If it's a Blind Lemon Jefferson album then it's a blues album.
And it's true that:	It's a Blind Lemon Jefferson album.
Then it must be true that:	It's a blues album.

However, if one or even all of the premises are false in actual fact it is still perfectly possible that the argument is valid. Remember: validity is simply the property that *if* the premises are all true *then* the conclusion must be true. Validity is certainly not synonymous with truth. So, not every valid argument is going to be a good argument. If an argument is valid but has one or more false premises then the conclusion of the argument may well be a false sentence. In contrast, valid arguments with premises, which are all actually true sentences must also have conclusions which are actually true sentences. In Logicspeak, such arguments are known as **sound** arguments. Because a sound argument is a valid argument with true premises, the conclusion of every sound argument must be a true sentence. So, we have now discovered a very important criterion for identifying good arguments, i.e. *sound arguments are good arguments*. But surely we can say something even stronger here. Can't we simply say that sound arguments are definitely, indeed, definitively good arguments? Well, this is a controversial claim. After all, there are many blatantly circular arguments which are certainly sound but which are not so certainly good.

For example, consider the following argument:

1. Bill Clinton is the current President of the United States of America.

Therefore,

2. Bill Clinton is the current President of the United States of America.

We can all agree that this argument is valid and, indeed, sound. But can we also agree that it is really a good argument? In truth, such arguments raise a number of questions some of which we will consider together later in this text and some of which lie beyond the scope of a humble introduction to what is ultimately a vast and variegated field of study. For present purposes, it is perfectly sufficient that you have a grasp of what is meant by saying that an argument is *valid* or *sound*.

To recap, sound arguments are valid arguments with true premises. A valid argument is an argument such that if the premises are true then the conclusion must be true. Hence, the conclusion of any sound argument must be true. But do note carefully that validity is not the same thing as truth. Validity is a property of arguments. Truth is a property of individual sentences. Moreover, not every valid argument is a sound argument. Remember: a valid argument is simply an argument such that if the premises are true then the conclusion must be true. It follows that arguments with one or more premises which are in fact false and conclusions which are also false might still be valid none the less. In such cases the logician still speaks of the conclusion as being validly drawn even if it is false. On false conclusions in general, one American logician, Roger C. Lyndon, prefaces his logic text with the following quotation from Shakespeare's *Twelfth Night*: 'A false conclusion; I hate it as an unfilled can.'[2] That sentiment is no doubt particularly apt as regards a false conclusion which is validly drawn. None the less, it is perfectly possible for a false conclusion to be validly drawn. For example:

1. If I do no work then I will pass my logic exam.
2. I will do no work.

Therefore,

3. I will pass my logic exam.

So, not all valid arguments are good arguments, but the important point is that even though the conclusion is false, the argument is still valid, i.e. if its premises really were true then its conclusion would also have to be true. Hence, the conclusion is validly drawn from the premises even though the conclusion is false.

Moreover, valid arguments with false premises can also have actually true conclusions. For example:

1. My uncle's cat is a reptile.
2. All reptiles are cute, furry creatures.

Therefore,

3. My uncle's cat is a cute, furry creature.

This time both premises are false but the conclusion is true. Again, the argument is valid none the less, i.e. it is still not possible for the conclusion to be false if the premises are true. Further, while we might not want to say that this particular argument is a good one, it is worth pointing out that there are ways in which we can draw conclusions from a certain kind of false sentence which leads to a whole class of arguments which are obviously good arguments. We will consider just this kind of reasoning in some detail later in Chapter 3. For now, remember that validity is not synonymous with truth and that validity itself offers no guarantee of truth. If the premises of a valid argument are true then, certainly, the conclusion of that argument must be true. But just as a valid argument may have true premises, it may just as easily have false premises or a mixture of both true and false premises. Indeed, valid arguments may have any mix of true or false premises with a true or false conclusion excepting only that combination of true premises and false conclusion. Only sound arguments need have actually true premises and actually true conclusions. Therefore, soundness of argument is the criterion which takes us closest to capturing our intuitive notion of a good argument which genuinely does establish its conclusion. Whether we can simply identify soundness of argument with that intuitive notion of good argument remains controversial. But what is surely uncontroversial is that validity and soundness of argument are integral parts of any attempt to make that intuition clear.

II
Deduction and Induction

In the ordinary business of daily life (and particularly in films about Sherlock Holmes) we generally find the term 'deduction' used in a very loose sense to describe the process of reasoning from a set of premises to a conclusion. In contrast, logicians tend to use the same term in a rather narrower sense. For the logician, **deductive argument** is valid argument, i.e. validity is the logical standard of deductive argument. Hence, you will frequently find valid arguments referred to as *deductively valid* arguments.

In Logicspeak the premises of a valid argument are said to **entail** or **imply** their conclusion and that conclusion is said to be *deducible* from those premises. But deduction is not the only kind of reasoning recognised by logicians and philosophers. Rather, deduction is one of a pair of contrasting kinds of reasoning. The contrast here is with **induction** and **inductive argument**. Traditionally, while deduction is just that kind of reasoning associated with logic, mathematics and Sherlock Holmes, induction is considered to be the hallmark of scientific reasoning, the hallmark of scientific method. For the logician deductive reasoning is valid reasoning. Therefore, if the premises of a deductive argument are true then the conclusion of that argument must be true, i.e. validity is truth-preserving. But validity is certainly not the same as truth and deduction is not really concerned with particular matters of fact or with the way things actually are in the world. In sharp contrast, and just as we might expect of scientists, induction is very much concerned with the way things actually are in the world.

We can see this point illustrated in one rather simple kind of inductive argument which involves reasoning, as we might put it, from the particular to the general. Such arguments proceed from a set of premises reporting a particular property of some specific individuals to a conclusion which ascribes that property to every individual, quite generally. Inductive arguments of this kind proceed, then, from premises which need be no more than records of personal experience, i.e. from *observation-statements*. These are **singular sentences** in the sense that they concern some particular individual, fact or event which has actually been observed. For example, suppose you were acquainted with ten enthusiastic and very industrious logic students. You might number these students 1, 2, 3 and so on and proceed to draw up a list of premises as follows:

1. Logic student #1 is very industrious.

2. Logic student #2 is very industrious.

3. Logic student #3 is very industrious.

4. Logic student #4 is very industrious.

.

.

10. Logic student #10 is very industrious.

In the light of your rather uniform experience of the industriousness of students of logic you might well now be inclined to argue thus:

Therefore,

11. Every logic student is very industrious.

Arguments of this kind are precisely inductive. From a finite list of singular observation-statements about particular individuals we go on to infer a general statement which refers to all such individuals and attributes to those individuals a certain property. For just that reason, the great American logician Charles Sanders Peirce described inductive arguments as 'ampliative arguments', i.e. the conclusion goes beyond, 'amplifies', the content of the premises. But, if that is so, isn't there a deep problem with induction? After all, isn't it perfectly possible that the conclusion is false here even if we know that the premises are true? Certainly, the industriousness of ten logic students does not guarantee the industriousness of every logic student. And, indeed, if that is so, induction is invalid, i.e. it simply does not provide the assurance of the truth of the conclusion, given the truth of the premises, which is definitive of deductive reasoning. But aren't invalid arguments always bad arguments? Certain philosophers have indeed argued that that is so.[3] On the other hand, however, couldn't we at least say that the premises of an inductive argument make their conclusion more or less likely, more or less probable? Perhaps a list of premises reporting the industriousness of a mere ten logic students does not make the conclusion that all such students are industrious highly probable. But what of a list of 100 such premises? Indeed, what of a list of 100,000 such premises? If the latter were in fact the case, might it not then be highly probable that all such students were very industrious?

Many philosophers have considerable sympathy with just such a probabilistic approach to understanding inductive inference. And despite the fact that induction can never attain the same high standard of validity that deduction reaches, some philosophers (myself included!) even go so far as to defend the claim that there are good inductive arguments none the less. We cannot pursue this fascinating debate any further here. For, if there are good inductive arguments, these have a logic all of their own. Interested parties can find my own account of the logic of scientific reasoning and a defence of the idea that there can be good inductive arguments in my paper 'Logic and Scientific Method'.[4] For present purposes, it is sufficient to appreciate that inductive reasoning is not valid reasoning.

III
The Hardness of the Logical 'Must'

In the previous section we again noted that invalid arguments fail to establish the truth of their conclusions even when the premises of such an argument are actually true. In contrast, given that the premises are true, the conclusion of any valid argument must be true. So, what is it about a valid argument with true premises which compels us to accept the conclusion of that argument? In the course of ordinary daily life, we find that many different things can compel us to accept the conclusion of an argument as a consequence of its premises: large persons of a violent disposition will often secure agreement to the conclusions of their arguments, for instance. But it is not the threat of violence that compels us to accept the logicians' conclusions. Rather, it is **logical force**, the force of reason. Again, we can appeal to the definition of validity to cash out quite what logical force comes to: valid arguments establish their conclusions conditionally upon the truth of all their premises. Consider a very clear example of valid argument:

1. All human beings are mortal.
2. Prince is a human being.

Therefore,

3. Prince is mortal.

Of course, the premises may not be true. Some human beings may be immortal. Prince may not be a human being. But if all human beings are mortal and if Prince is a human being then it must follow that Prince is mortal. So, once I have accepted the truth of the premises here I am forced to accept the truth of the conclusion. Why? Because if I do not accept the truth of the conclusion having accepted the truth of the premises then I have blatantly contradicted myself. In this case the contradiction consists in believing that all human beings are mortal and that Prince is a human being who is not mortal. It cannot be rational to believe contradictions. Therefore, I must accept the truth of the conclusion, on pain of irrationality. (Check that this is also the case in each of the valid examples given earlier.) The hardness of the logical 'must' is the hardness of reason. Logical force is the force of reason.

This point gives some insight into the traditional definition of logic as 'the science of thought', the study of the rationality of thinking. In the last analysis, we ourselves may not want to defend quite such a subjective, psychologistic definition of the subject, but supposing that we can identify the laws of logic and represent them mathematically (we shall see later that we can) we can at least make clear sense of George Boole's account of the laws of logic, in his *Mathematical Analysis of Logic* [1847]:

> The laws we have to examine are the laws of one of the most important of our mental faculties. The mathematics we have to construct are the mathematics of the human intellect.

IV
Formal Logic and Formal Validity

Earlier, I noted that logic is not really about matters of fact or particular cause and effect relations but is concerned instead with validity which is independent of any such worldly, factual or, in philosophical terms, **empirical** matters. Recall the Blind Lemon Jefferson example. Perhaps you find it unconvincing. You might think that Blind Lemon may have been a milkman rather than a bluesman. But if you substitute the name of your own favourite blues performer the argument at once appears convincing and sound.

In one sense, it really doesn't matter which particular performer's name I actually used: we can legitimately substitute the name of any performer or any band and still retain a valid argument. Indeed, it needn't even be a Blues band. What is important is not the name of the band but the pattern of argument. When you substitute the name of your favourite band for 'Blind Lemon Jefferson' something changes. But something also remains the same: the pattern or structure of the argument. In fact, the only thing that changes is the particular name used in each sentence. The type of sentence is the same and the overall structure of the argument is the same. What is in common between your favourite example and my favourite example is the **logical form** of the argument. What is important to the formal logician is not the content of the argument but its form.

Change the name of the supposed bluesman as you will, the form of argument remains exactly the same. So, logic is not really about particular matters of fact and it is not really about particular bluesmen either. Rather, formal logic is about **argument-forms** (logically enough). Most importantly, as we shall see, the formal logician can use the notion of logical form to investigate the concept of validity. For example, consider the Blind Lemon Jefferson argument. Clearly, it is a valid argument. If the premises are true the conclusion must be true. But, as we have just seen, we can change the name of the bluesman or even substitute the name of any band and still get a valid argument. Moreover, as you will see, we can in fact change the most basic sentences which make up the premises and conclusion and still produce a valid argument. What makes this possible is the fact that the validity of this argument does not depend on particular matters of fact or particular bluesmen. Rather, it is the form and structure of the sentences in the argument and the relations between those sentences which guarantee that we cannot have true premises with a false conclusion in any such argument.

It follows that any argument of that particular logical form will also be a valid argument. Thus, formal logicians use the notion of logical form to investigate the concept of validity. Indeed, many formal logicians will now encourage us to replace the intuitive definition of validity we have been working with so far in favour of the following purely **formal definition** of validity:

An argument is valid if, and only if, it is an instance of a valid logical form.

Hence, formal logic is fundamentally concerned with valid logical forms of argument. Formal logic, we might say, investigates formal validity. Further, it can be argued that the intuitive or **modal definition** is not an entirely adequate one (the term 'modal' is appropriate here because it refers to the notion of necessity, the 'must' element of our definition). For example, consider the following argument carefully:

1. Snow is white.

Therefore,

2. $1 + 1 = 2$.

This argument seems to satisfy the modal definition. But the conclusion surely is not a logical consequence of the premise. However, the argument is not an instance of any valid logical form of argument. Therefore, it is formally invalid. So, perhaps we should adopt a purely formal definition. But is validity always a formal matter? In all honesty, it is not entirely clear that it is. Some arguments are intuitively valid, i.e. valid in terms of the modal definition, even though they seem to exhibit no valid logical form. Here is an example:

1. The Statue of Liberty is green.

Therefore,

2. The Statue of Liberty is coloured.

In fact, although this particular argument is intuitively valid, it is, as we shall see, an instance of at least one obviously invalid logical form. Moreover, the problem here is a deep, intractable one for there does not seem to be any way in which we can faithfully amend the sentences composing the argument which would result in the argument becoming formally valid. For example, consider another case which might seem similar:

1. All unmarried men are unmarried men.

Therefore,

2. All bachelors are unmarried men.

Again, the problem is precisely that the argument is an instance of an invalid logical form. In this case, however, the premise is obviously a necessary or logical truth while the conclusion is not obviously so. But the terms 'bachelors' and 'unmarried men' are synonyms. And if we substitute the term 'bachelors' in the conclusion with 'unmarried men' we generate the following argument:

1. All unmarried men are unmarried men.

Therefore,

2. All unmarried men are unmarried men.

This argument is obviously circular but it is also obviously valid and sound, and, crucially, it can now be shown to be an instance of a valid logical form. So, while we cannot honestly say that the first version of the argument is an instance of a valid logical form we can say that it is an argument which will become an instance of a valid logical form after appropriate substitution of synonyms.

But now consider the example about the Statue of Liberty again. In this case, synonym substitution is not legitimate. The terms 'green/coloured' do not represent a synonym pair. Certainly all green things are coloured. But not all coloured things are green! What does this prove? In the last analysis, it may well prove that validity cannot be completely explained in purely formal terms. In truth, however, this again is a matter of some controversy.

As we shall see, the notion of logical form is not an absolute one, i.e. the same argument can be an instance of more than one form. Perhaps we have simply failed to find that valid form of which our argument about the Statue of Liberty is an instance. Perhaps not. Alternatively, we could simply adopt the formal definition and find another term to describe those arguments which seem to slip through the formal net, as it were. Be that as it may, our hand is not forced here. And so, although we should note this important controversy carefully, we will not abandon the intuitive or modal conception of validity we have been working with to date in favour of a purely formal definition.

While it is not incumbent upon us to resolve the controversy about the definition of validity here, it is crucially important to appreciate that logic is a discipline which contains many fascinating and important controversies. Indeed, within formal logic itself there is even room for disagreement about the validity of the argument-forms sanctioned by a given formal system, i.e. about the correctness of the formal system itself. Notably, it is precisely that possibility which Captain James T. Kirk regularly overlooks in the well-known television programme *Star Trek* when he accepts Mr Spock's allegations of illogicality. What Kirk fails to realise is that there exist a number of distinct, competing systems of formal logic which sanction distinct sets of argument-forms. Hence, in one sense, there is no single correct logic. It follows that a proper formal definition of validity is only fully specified for a particular set of forms and so one can only really make an informed judgement once that set has been laid out. The particular system of formal logic upon which we will focus in this text is the traditional or **classical logic** which was formulated first. In all honesty, alternative systems are best (and most easily) understood as revisions of that traditional system which arise from both formal and philosophical thinking about classical logic. So, it is the classical system to which we should devote ourselves first.

Finally, in the light of the possible limitation to the adequacy of the formal definition of validity considered above, one might wonder whether logicians should concentrate on purely formal logic. Would we do better to pursue our logical investigations informally?

This is a ticklish question. Part of the answer to it is just that formal logic embodies many of the very standards we would need to pursue our informal investigations! So, the best student of informal logic will be the one who has first mastered formal logic. Moreover, even if we do accept that the formal logician cannot completely explain validity in purely formal terms, classical formal logic captures a huge class of valid forms none the less. Therefore, formal logic remains a crucially important and highly effective means of investigating the concept of validity.

V Identifying Logical Form

In the previous section, we noted that the formal analysis of validity may be incomplete. However, we should not be too daunted by that fact. The job of the formal logician consists in unearthing valid forms of argument and the sheer extent to which the formal logician is able to do that job effectively is astonishing. But just how far does formal validity go? To provide an answer to that very question is the purpose of this book. And where better to begin than with old Blind Lemon Jefferson? In the Blind Lemon Jefferson case, the structure of the sentences and the pattern of argument are very easy to see. The first sentence is clearly an 'If . . . then --- ' sentence:

1. *If* it's a Blind Lemon Jefferson album *then* it's a blues album.
2. It's a Blind Lemon Jefferson album.

Therefore,

3. It's a blues album.

Stripped bare, as it were, this argument has the form:

1. *If . . . then ---*
2. . . .

Therefore,

3. ---

Looked at in this way, there are two gaps or *places* to be filled in the first premise, i.e. ' . . . ' (pronounced ' dot, dot, dot ') and ' ---' (pronounced 'dash, dash, dash'). So, the structure of the first premise is just: If . . . then ----. In the Blind Lemon Jefferson case, the first gap is filled in by the sentence 'It's a Blind Lemon Jefferson album' which is precisely the same sentence as the second premise. The second gap is filled by the sentence 'It's a blues album' which is precisely the same sentence as the conclusion: 'It's a blues album.' But now it is clear that as long as we stick precisely to the same form of argument we could have used any two sentences and we would still have had a valid argument. Hence, any argument of this form is bound to be valid, i.e. any argument consisting of any sentences in those relations must be valid. And that is just what it means to say that a form of argument is valid. So, not only is logic not concerned with particular matters of fact, or particular bluesmen, it is not even concerned with particular sentences.

Logically enough, formal logic is fundamentally concerned with forms of argument. Forms of argument are really **argument-frames** or *schemas*, i.e. patterns of inference with gaps which, for present purposes, can be filled using any particular sentences we choose to pick, provided only that we do complete the form exactly. Since it doesn't matter which particular sentences are involved in a given form it would be useful to have symbols which just marked the gaps, **place-markers**, for which we could substitute any sentence. This would save us writing out whole sentences, or marking gaps with ' . . . ' and ' --- '.

In algebra, mathematicians generally use the symbols '*x*' and '*y*' to stand for any numbers. Because such symbols mark a place for *any* number they are called **variables**. But the logician has no need to borrow these variables. Logicians have their own variables. In the present context, the logicians' variables mark places not for numbers but for sentences or, in more traditional logical terms, **propositions**. A proposition is thought of as identical with the meaning or sense of a sentence rather than with the actual sentence itself. So, intuitively, two different sentences which are really just two different ways of saying exactly the same thing are said to express one and the same proposition. For example, the following two sentences would be said to express only one proposition:

1. Edinburgh lies to the north of London.
2. London lies to the south of Edinburgh.

Talking of propositions rather than sentences can constitute a linguistic economy and many find the concept of a proposition both natural and intuitive. The idea is not uncontroversial and a fascinating debate has grown up around the simple questions of whether there are such things as propositions and, if so, just what kind of thing they might be. Those interested in these questions would do well to read the first chapter of W.V.O. Quine's *Philosophy of Logic*,[5] though, unfortunately, such questions lie beyond the scope of the present text.

For present purposes, we will bypass this particular debate by simply taking the lower-case letters '*p*', '*q*', '*r*' and so on as being **sentential variables**, i.e. variables whose values are simply well-formed sentences.

As schematic letters, sentential variables make it very easy to express precisely the bare pattern or logical form of an argument. For example, we can easily represent the logical form of the Blind Lemon Jefferson argument, as follows:

1. If p then q
2. p

Therefore,

3. q

When formalising a given argument, the crucial point to note is just that the same variable must mark a place for the same sentence throughout that formalisation. (The example about the logic examination also has exactly the same logical form we have just identified. This time, however, the variables 'p' and 'q' mark places for two different sentences – work out which.)

Above all, the formal logician is interested in forms of argument. Therefore, the central problem for the logician becomes: how are we to tell good forms of argument from bad forms of argument? In other words, how do we distinguish valid forms from invalid forms? According to the formal logician, a form of argument is valid if, and only if, every particular instance of that argument-form is itself valid. Thus valid argument forms are patterns of argument which, when followed faithfully, should always lead us to construct particular valid arguments as instances. For obvious reasons, this is known as the **substitutional criterion of validity**. I will offer a precise definition later but for the moment here is an analogy. Consider the following simple algebraic equation: $2x + 2x = 4x$. For every particular value of the variable x in this equation, be it apples, pears or double-decker buses, it will always be true that two of them added to another two will add up to four in total. Analogously, for any valid argument form, every particular argument which really is a **substitution-instance** of that form will itself be a valid argument, whether it concerns Blind Lemon Jefferson, passing your exams or anything else.

Unfortunately, we may have to recognise another limitation to the purely formal account later. Certain logicians have argued that the substitutional criterion is ultimately incomplete, just as it stands. These logicians allege that the criterion turns out to sanction as valid certain forms which have obviously invalid instances.[6] If that is so, we must indeed recognise another limitation to the purely formal account. This particular allegation raises a number of questions which, again, lie beyond the scope of the present text. Be that as it may, it should now be clear that formal logic is fundamentally concerned with valid forms of argument. Indeed, the traditional or classical logic which we will consider together in this text is one attempt to identify and elucidate all the valid forms of argument.

As such, logic is the study of the structure and principles of reasoning and of the nature of sound argument. But it is important to note that logicians need not always arrive at those principles of deductive inference which form the subject-matter of their field of study by collecting data about the way people actually argue. Boole's rather traditional definition might well give that impression but the relation between formal logic and actual argument is more complex. The two interact. As we have already seen, logic has traditionally been described as the science of thought. If it is a science, however, logic is a theoretical science, not an empirical science.

A good way of elucidating this distinction is with an analogy to games. Chess, in particular, is an excellent example. Logic, in the analogy, is like the rules of the game of chess, the rules of play which govern the game and define what chess is. The relation between the logical principles of deductive inference and the actual arguments people use, the inferences made by 'the person in the street' or 'the person on the Clapham omnibus', as it used to be said, is analogous to the relation between the rules of the game of chess and the actual playing of the game. The famous Austrian philosopher Ludwig Wittgenstein, in Remark 81 of his *Philosophical Investigations*, quotes a definition of logic by the mathematical logician F.P. Ramsey as a 'normative science'.[7] This is a good description which allows us to develop (and update) our definition of formal logic: formal logic constitutes a set of rules and standards, ideals of inference, or *norms*, independent of the thinking of any actual individual, in terms of which we appraise and assess the actual inferences which individuals make. So, in its concern with the ways in which people do actually argue, logic is *scientific* but in so far as logic provides standards of argument it is also *normative*.

To sum up, formal logic is fundamentally concerned with the form and structure of arguments and not, primarily, with their content. In terms of the chess analogy, it is the study of the rules of the game, not of the strategies of any particular player.

VI
Invalidity

According to the modal definition of validity an argument is valid if, and only if, whenever its premises are true its conclusion must also be true, i.e. if, and only if, it would be impossible for its premises to be true and its conclusion false. It follows logically that no valid argument can have true premises and a false conclusion. Indeed, to show that an argument is invalid is precisely to show a way in which that argument could have true premises and a false conclusion. In general then, an argument is invalid if it is such that its premises could all be true and its conclusion false.

Therefore, in order to demonstrate that a given argument is invalid it is sufficient to indicate that even if the premises are true the conclusion is actually false, or could be false, while the premises were true. For example, the former is precisely what is the case as regards the very first argument concerning the cheese sandwich which we considered at the outset of this chapter. Therefore, that argument is invalid. However, particular arguments are of interest to the formal logician only in so far as they exhibit logical forms of argument. Above all, logic is the study of forms of argument. Therefore, the fundamental question at this stage is just: how do we show that a given form of argument is invalid?

Recall the substitutional criterion: a form of argument really is valid if, and only if, every substitution-instance of that form is itself a valid argument. It follows that an argument-form is valid if, and only if, it is not the case that there is any instance of that form which has true premises and a false conclusion.

In order to demonstrate that a given form of argument is invalid, then, it is sufficient to exhibit some particular example of the form in question that could have actually true premises and a false conclusion. Any such invalid particular instance of a form is known as a **counterexample** to that form. The method of proving invalidity by means of a counterexample is known as **refutation by counterexample**. In practice, it is a devastatingly effective argumentative technique. Consider the following argument-form:

1. If p then q
2. q

Therefore,

3. p

Here is a counterexample to the form concerning my black cat, Zebedee (for the purposes of many of the examples in this book it is worth bearing in mind that I am the proud owner of two small black cats called Tiffin and Zebedee):

1 If all cats are black then Zebedee is black.

2. Zebedee is black.

Therefore,

3. All cats are black.

Now check for yourself:

1. That the argument is an instance of the logical form in question.
2. That the premises are actually true in this case.
3. That the conclusion is actually false.

Consider another argument-form:

1. If p then not q
2. Not p

Therefore,

3. q

Here is a counterexample to this form:

1. If Tiffin is a dog then it is not the case that Tiffin is an elephant.
2. Tiffin is not a dog.

Therefore,

3. Tiffin is an elephant.

Again, check for yourself:

1. That the argument really is an instance of the form in question.
2. That the premises are true.
3. That the conclusion is false.

It is important to note that I am not using any algorithm, i.e. any step-by-step, mechanical decision-procedure, to produce these counterexamples. At this stage, producing actual counterexamples requires art and imagination (and a fair bit of practice!). So, don't worry if you cannot come up with your own examples. It is sufficient that you understand the particular examples given.

VII
The Value of Formal Logic

Many students are very daunted by the prospect of a first logic course and feel extremely anxious at the outset of their course. If that is your experience, you can at least rest assured that you are not alone in your angst. In fact, many formal logicians will themselves have felt just as you do at this stage in the inquiry. So, the point is not simply that you have company but rather that you are in good company. Moreover, I feel sure that you will have found at least some of the ground we have covered together in the present chapter both accessible and intuitive. It is important to realise why that should be so. The point is a very simple one: as a matter of fact, we do all reason logically in daily life perfectly successfully and in ways which are often just as complex as those we will consider together in the present text.

As we noted earlier, formal logic is the study of the rules of the game rather than the strategy of the individual player. None the less, we should never lose sight of the fact that we do all reason logically in ordinary life. As I might put it, we all do on at least a part-time basis what the formal logician does full-time. And that fact is underwritten by a still more fundamental point: human beings are born with a natural ability to argue, to reason and to think logically. In his later work, Wittgenstein rightly made much of the simple point that many of our attitudes and abilities, ways of acting and ways of reacting, follow from the *form of life* we share just as human beings. Fortunately, the ability to argue and to reason logically is part of that natural legacy.

To realise that the study of formal logic is not really a matter of memorising and applying daunting mechanical rules but is rather a reflective study of how well we can all naturally reason at our very best is to realise the true value of the study of formal logic. The logician A.A. Luce puts this point very well when he notes that:

> the study acquires a new status and dignity when viewed as a conscious awakening of an unconscious natural endowment.[8]

As this book develops, our concern with argument will inevitably focus upon forms of argument rather than the particular arguments which we might construct day to day in a natural language such as English. But we should never lose sight of the fact that formal logic has its roots in just such natural language arguments and has enormous applicability to arguments in natural language, quite generally.

For the philosopher in particular, formal logic is a potentially devastating weapon which can and should be deployed in debate. If you lose sight of the applicability of formal logic to natural language arguments then you will miss out on a crucial aspect of the power and value of formal logic and much of its excitement. Something of the applicability of formal logic should be clear already. After all, the classical logician has provided us with some powerful tools for telling good arguments from bad, for identifying logical forms of argument, and for exposing the invalidity both of particular arguments and of argument-forms.

It is often difficult to exploit formal logic in debate but when it can be brought to bear it can be extremely effective. There is a famous story of a debate between the eminent classical logician Bertrand Russell and Father Frederick Copleston which clearly illustrates just how useful knowledge of formal logic can be. The debate in question concerned a particular argument known as the 'cosmological argument'. This argument is one of the traditional arguments (we cannot say 'proof', for that begs the question) for the existence of God. The argument moves from the premise that every event has a cause to the conclusion that there must, at some point, be a first cause and this is God. Father Copleston defended the cosmological argument in the debate. What is of interest to us here is the way in which Russell attacked the argument.

In effect, Russell represented the cosmological argument as follows:

1. Every event has a cause.

Therefore,

2. Some event is the cause of every event.

Next, Russell tried to identify the form of the argument, thought hard about the validity of that form, and then produced the following counterexample:

1. Everyone has a mother.

Therefore,

2. Someone is the mother of everyone.

What Russell attempted to show is that the cosmological argument is invalid because it is an instance of an invalid form of argument. The form of reasoning which Russell highlights is certainly an invalid one. Indeed, arguments of that form exemplify a well-known fallacy, the **quantifier switch** or **quantifier shift fallacy**. Stating the form of this particular fallacy requires more logical machinery than is available to us at this stage. But, as we will see in Chapter 5, the form of the fallacy certainly can be made explicit. However, even if Russell has shown that the argument is an instance of that invalid form this does not prove that the cosmological argument is invalid. As we noted in Section IV, a particular argument may be an instance of more than one form. So, on one level of analysis, the argument might well be shown to be an instance of an invalid form but if we are not careful we may overlook the fact that it is also an instance of a more complex valid form. Perhaps Russell is biased and has given a very simplistic account of the argument form involved. Perhaps a deeper analysis would reveal that the cosmological argument is also an instance of a more complex form that is in fact valid. Perhaps the argument is valid but not in virtue of form. Perhaps not. The question of the nature of logical form is one to which we will often return. But the question of the logical form of the cosmological argument need not worry us here. It is sufficient to note just how powerful and valuable an ally formal logic can be in debate in natural language, whatever the topic under discussion might be.

In truth, the form which Russell appeals to here is of quite a high level of complexity; as, indeed, is the very first example about eternal happiness and the cheese sandwich which we considered on p. 3. Formal logic can handle forms of this level of complexity with ease and can, in fact, handle still more complex forms of argument. (Such argument forms will be considered in detail later in Chapter 5.) Classical formal logic will prove itself to be an enormously efficient instrument for investigating the nature of argument and the concept of validity itself. To discover precisely how and why that should be the case can be genuinely exciting and will, on occasion, lead to some rather surprising results. Not all of the surprises are pleasant ones, however. Formal logic has its limits.

As we have already seen, for example, there are serious questions about whether the formal logician can ultimately account for validity in purely formal terms. Worse still, perhaps, is the fact that classical formal logic sanctions as valid some forms of argument which are rather less than intuitive. These particular limits will be considered later when we are in a better position to appreciate them for what they are. None the less, the existence of certain possible limitations to the formal logician's project detracts not one iota from the value of studying logic in general and classical formal logic in particular.

Provided that you do not lose sight of the applicability of formal logical considerations to ordinary discourse, you will quickly realise that the study of formal logic tends to produce clear-thinking, articulate individuals who can present and develop complex arguments in a rigorous way. In acquiring these communications skills you will also acquire the ability to lead discussion in a structured way and to persuade others. Further, as we have noted, formal logic provides impressive analytical machinery with which to identify the logical structure of an opponent's arguments and provides an arsenal of weaponry which may well enable one to destroy the apparent force of those arguments. All these skills are obviously valuable and useful to their possessor. Less obviously, perhaps, they are also highly coveted by many employers, particularly in the business environment.

Finally, no logic student should ever lose sight of the enormous practical value of formal logic. In 1879, while Professor of Mathematics at Jena University in Germany, Gottlob Frege [1848–1925] produced the first formal, mathematical language capable of expressing argument-forms as complex as and even more complex than those we have been considering here. The publication of Frege's *Begriffsschrift* is an event whose significance in the development of formal logic is inestimable. The publication of Frege's text certainly heralds the dawn of the modern tradition of classical formal logic with which we are concerned. Moreover, Frege's work not only constituted the first system of modern formal logic but also laid much of the foundations for the contemporary programming languages which have become such an integral part of modern daily life, from the university, college or office software package to automatic cash dispensers and bar tills. The name of the programming language PROLOG, for example, is simply shorthand for *Logic Programming*. Logic is, and always has been, an integral part of philosophy. Students of philosophy in particular should be pleased to be able to lay to rest so easily the old but still popular misconception that their subject is 'impractical' and 'unproductive'!

VIII
A Brief Note on the History of Formal Logic

In all honesty, it will be some time before you become fully aware of the value and extent of Frege's contribution to the development of formal logic. In fact, this will not really become clear until we consider the logic of **general sentences** (sentences involving terms such as 'all' and 'some', 'most' and 'many') and arguments composed of such sentences, again, in Chapter 5 of the present text. The logic of such sentences and arguments is known as **quantificational logic** and the design of the logical machinery of quantificational logic is due, above all, to Gottlob Frege. It is precisely that design which is undoubtedly the crowning glory of Frege's contribution to the development of formal logic and, perhaps, the crowning glory of formal logic itself.

As a first step towards an appreciation of the value of Frege's contribution consider the following historical sketch carefully.

The first system of logic which allowed philosophers to investigate the logic of general sentences formally was designed by Aristotle some 2,000 years before Frege. The importance of Aristotle's own role in the history of formal logic is also unique and inestimable just because formal logic itself originates in the work of that author. As the logician Benson Mates puts it:

> Aristotle, according to all available evidence, created the science of logic absolutely *ex nihilo*.[9]

Moreover, the science which Aristotle created is, as we might put it, properly formal, for it embodies the insight that the validity of certain particular arguments consists in the logical forms which they exemplify. Further, Aristotle's approach to formal logic is generally systematic, i.e. it identifies and groups together the valid forms of argument in an overall system.

Aristotle's system of logic is known as **syllogistic** just because it confines itself to a certain kind of argument known as a *syllogism*. A syllogism consists of two premises and a conclusion each of which is a general or *categorical sentence*, i.e. a sentence which makes an assertion about sets of things. Typically, such sentences will tell us that some set is or is not contained in another. So, for example, sentences such as 'All A is B' and 'Some B is C' are categorical sentences. In fact, Aristotle distinguished four different kinds of categorical sentence as foundational in his system of logic. Given a careful analysis of the place and role of the key terms in each such sentence, and of the place of each sentence in a syllogism, no fewer than 256 kinds or *moods* of syllogism can ultimately be distinguished.

As a whole, however, the system of syllogistic is a limited one which does not consider the logic of relations, for example. Moreover, certain elements of the system are, to say the least, controversial. In particular, it is not at all clear that the nature and consequences of general sentences involving 'all' are properly represented. Here we must be very careful. No sentence of the form 'All As are B' ever implies that there actually exist any As. That would be an independent claim. What does follow is just that *if* there is something which is A *then* that something is also B. To ignore this point can lead to legitimating fallacious reasoning. But it may well be the case that Aristotelian syllogistic does ignore this point. Bertrand Russell points out that just such a fallacy is involved in certain instances of one particular mood of Aristotelian syllogistic. During the Middle Ages the syllogisms were given rather exotic names such as 'Barbara' and 'Celarent' by the medieval logicians who studied them. The syllogism which Russell highlights here is known as 'Darapti' and is of the following form:

. All As are B.

. All As are C.

Therefore,

. Some Bs are C.

As you might expect, Russell offers a counterexample which, this time, concerns a mythical fire-breathing animal, the chimera:

. All chimeras are animals.

. All chimeras breathe flame.

Therefore,

. Some animals breathe flame.

Aristotle's syllogistic was developed and extended by logicians and philosophers throughout the Middle Ages and, indeed, in subsequent centuries.

Moreover, the medieval logicians conducted their own logical investigations of general sentences and made considerable progress towards a systematic theory of the logic of such sentences.[10]

The next major step forward in the development of the subject did not occur until well into the nineteenth century when the English logicians George Boole [1815–64] and Augustus De Morgan [1806–71] approached formal logic in terms of abstract algebra and, for the first time, developed algebraic logic. To this day, the algebraic perspective remains a useful and insightful one. In that mathematisation of the subject a new level of formal rigour and systematisation was achieved and, as we shall see in Chapter 4, De Morgan also contributed some extremely useful logical laws. In a sense, these logicians realised the dream of the great German philosopher and logician G.W. Leibniz [1646–1716], who had already outlined the idea of a universal calculus into which arguments could be translated and assessed. But it was not until the work of Frege in the late nineteenth century that the level of systematisation which formal logic now enjoys was achieved. As we shall see, contemporary formal logic is nothing less than a formal language into which arguments can be translated and in which they can be proved to be valid or invalid. Further, Frege not only realised Leibniz's dream but also contributed the machinery of quantificational logic, which enables the logician to dig into even the internal grammatical structure of natural language sentences. So, the particular moment in which we are studying formal logic together is one at which the subject has attained its highest level of achievement in an evolution of more than 2,000 years.

I shall say no more about the historical evolution of formal logic here. Interested parties can explore the development of logic in William and Martha Kneale's useful and accessible text *The Development of Logic* [1962]. A much briefer discussion can also be found in Chapter 12 of Benson Mates's *Elementary Logic* [1972] and, indeed, Mates's *Stoic Logic* [1953] remains a classic in the field. On Aristotelian logic in particular, I also warmly recommend both Jan Lukasiewicz's *Aristotle's Syllogistic from the Standpoint of Modern Formal Logic* [1951] and Jonathan Lear's *Aristotle and Logical Theory* [1980].

Before we turn to the first exercise in this text, Exercise 1.1, note carefully that at the end of each chapter I give a summary box of salient points, all of which are generally helpful for the exercises which follow them. (Because certain chapters cover a considerable amount of material you can also expect to find similar summary boxes at regular intervals during the course of relevant chapters.) You might also have noticed that during this chapter key words have been written in bold. Information on these items can be found in the glossary before the index at the end of the text and you will also find entries under these items in the index. Do study the contents of each summary box carefully before attempting subsequent exercises! The first summary box is Box 1.1.

BOX 1.1

- An *argument* consists of premises, a 'therefore'-type word and a conclusion.
- An argument is *valid* if, and only if, it is impossible that its premises be true *and* its conclusion false.
- An argument is *invalid* if, and only if, it *is* possible that its premises be true *and* its conclusion false.
- A *sound* argument is a valid argument with true premises.
- The claim of the formal logician is that an argument is valid purely in virtue of being a *substitution-instance* of a valid *argument form.*
- An argument form is valid if and only if *every* substitution-instance of that form is valid.
- An argument form is invalid if some substitution-instance of that form is invalid.
- A *counterexample* to a form is a substitution-instance of that form which is itself an invalid argument.

EXERCISE 1.1

1 For logical purposes, an argument is a set of sentences in which some sentences (the premises) purport to give reasons for accepting another sentence (the conclusion). But not every set of sentences constitutes an argument. Consider the following sets of sentences carefully. In each case, state whether or not that set of sentences constitutes an argument. Give reasons for your answers.

A Professor Plum was in the drawing room. Miss Scarlet was in the kitchen. The murderer used the knife and the evil act was committed in the hall.

B If Professor Plum was in the drawing room then Colonel Mustard was the murderer. Professor Plum was in the drawing room. So, Colonel Mustard was the murderer.

C Every student of logic is wise and knowledgeable. Anyone attempting this exercise is a student of logic. Therefore, anyone attempting this exercise is wise and knowledgeable.

D I am absolutely sick and tired of getting wet every time it rains. From now on I will never forget to take my umbrella with me in the morning. Even if the weather looks fine when I leave I shall certainly make a point of taking that umbrella.

2 In an argument the premises purport to give reasons for accepting the conclusion. In general, the examples considered in this chapter have listed the premises before stating the conclusion. In ordinary discourse, however, premises are not always stated before their conclusions. Study the following arguments carefully. In each case, state which sentences you consider to be premises and which the conclusion. Give reasons for your answers.

A Professor Plum was in the drawing room and Miss Scarlet was in the conservatory. If Professor Plum was in the drawing room and the murder weapon was found in the drawing room then Professor Plum is in big trouble. So, if the murder weapon was found in the drawing room then Professor Plum really is in big trouble.

B All human beings are mortal. So, it stands to reason that Socrates is mortal. After all, he is a human being.

C Very few elephants can fly. Very few elephants are pink. So, the pink flying elephant is truly a rare creature for fewer pink elephants than ordinary elephants can actually fly.

D Professor Plum was obviously the murderer in this instance. For the murderer used the knife and Professor Plum had the knife. And the murder was committed in the hall and Professor Plum was certainly in the hall earlier.

3 An argument is valid if and only if it is impossible that its premises be true and its conclusion false. Consider the following questions carefully before responding. In each case give reasons for answering as you do.

A When is an argument invalid?

B Can a valid argument have a false conclusion?

C Can a valid argument have actually true premises but a false conclusion?

D Can an argument have true premises and a true conclusion but not be valid?

E Can an argument be sound but invalid?

F Must the conclusion of a sound argument be true?

G When is an argument-form valid?

H When is an argument-form invalid?

4 Consider the following arguments carefully. In each case, indicate whether the argument is valid or invalid. If you find any to be valid indicate whether or not the argument is also sound:

A 1. If Abraham Lincoln was French then the Moon is made of green cheese.

2. Abraham Lincoln was French.

Therefore,

3. The moon is made of green cheese.

B 1. The Washington Redskins are better than the Miami Dolphins.

2. But the Miami Dolphins are better than the Buffalo Bills.

Therefore,

3. The Washington Redskins are better than the Buffalo Bills.

C 1. If all cats are black then Tiffin is black.

2. Tiffin is black.

Therefore,

3. All cats are black.

D 1. If all cats are black then Zebedee is black.

2. Some cats are not black.

Therefore,

3. Zebedee is not black.

5 (i) Using only: 'If . . . then---;', '*p*', '*q*', and 'not' exhibit the logical form of arguments C and D in 4 above.

(ii) State whether or not my natural language arguments C and D constitute counterexamples to those logical forms. Give reasons for your answers.

6 Provide a counterexample to the following argument-form:

1. If *p* then not *q*

2. Not *p*

Therefore,

3. Not *q*

7 Using only sentential variables exhibit the logical form of the following argument:

1. The team strip is red.

Therefore,

2. The team strip is coloured.

State: (i) whether the form is valid or not and (ii) whether the particular argument itself is valid or not.

For discussion:
What, in your view, do your answers imply as regards the purely formal definition of validity?

Notes

1 I am indebted to John Slaney for the kind of example involved here.
2 Lyndon, Roger C., [1966], *Notes on Logic*, Van Nostrand Mathematical Studies #6, Princeton NJ, D. Van Nostrand, preface, p. iii.
3 See, for example, Popper, K.R., [1972], *Conjectures and Refutations*, fourth edition, London and Henley, Routledge & Kegan Paul, Ch. 3.
4 In Phillips, Calbert (ed.), [1995], *Logic in Medicine*, London, British Medical Journal Publishing Group, Ch. 2. But see also Ch. 1 of this volume.
5 Quine, W.V.O., [1986], *Philosophy of Logic*, second edition, Cambridge MA and London, Harvard University Press.
6 I am indebted to Stephen Read for this point which was made in correspondence. However, the point is also well made in Read, Stephen, [1995], *Thinking about Logic: An Introduction to the Philosophy of Logic*, Oxford, Oxford University Press, Ch. 2. Interested parties will find a substantive discussion of relevant issues there and a useful guide to further reading in the area.
7 Wittgenstein, Ludwig, [1967], *Philosophical Investigations*, Oxford, Blackwell, Remark 81.
8 Luce, A.A., [1958], *Logic*, London, English Universities Press, p. 9.
9 Mates, Benson, [1972], *Elementary Logic*, second edition, New York, Oxford University Press, p. 206.
10 We cannot pursue this fascinating aspect of the development of formal logic but interested parties might profitably consult Broadie, Alexander, [1987], *Introduction to Medieval Logic*, Oxford, Clarendon Press. See also Boehner, Philotheus, [1952], *Medieval Logic: An Outline of its Development from 1250 to c. 1400*, Manchester, Manchester University Press.

2
How to Prove that You Can Argue Logically #1

2
How to Prove that You Can Argue Logically #1

I
A Formal Language for Formal Logic

Logic is the study of argument. But particular arguments in a natural language such as English are only really of interest to the formal logician as instances of logical *forms* of argument. Formal logic is the study of argument-forms; hence, *formal* logic. Classical formal logic constitutes one ambitious attempt to capture every logical form of argument in a single language. But that language is not English. It is not any natural language. Rather, it is a **formal language**, i.e. a symbolism or notation in which we can express arguments so that their forms show up clearly. As we shall see, we can go on to add certain rules to that formal language so that we can demonstrate when one sentence in the language follows logically from other sentences.

Unlike English, the simplest sentences in the vocabulary of the formal language are not actual sentences but symbols which abbreviate particular sentences and stand in their place. These new symbols which stand for specific sentences are just letters of the alphabet and so they are often called **sentence-letters**. In the last chapter we saw that we could use sentential variables such as '*p*', '*q*', '*r*' etc. to mark a gap or *place* in an argument-frame which might be filled by any sentence whatsoever. In contrast, sentence-letters stand in place of specific sentences. In more formal terms, sentence-letters are not sentential *variables* but **sentential constants**. This distinction is an important one. Variables are not the same thing as constants. Each does a distinct job for the formal logician and different symbols are used to mark that distinction. So, while sentential variables are represented by lower-case letters, '*p*', '*q*', '*r*', etc., sentential constants will be represented by upper-case capital letters, 'P', 'Q', 'R' and so on.

Before we go any further, it is instructive to make an important point about the notation we will use in this and subsequent chapters. Note carefully that when we mention rather than actually use a sentential variable or a sentential constant in an English sentence as we did at the end of the previous paragraph, for example, the symbol is placed in quotation marks. In effect, quotation marks are used to form an English name for each symbol, i.e. an expression which refers to the symbol itself, not to the referent of that symbol. In practice, though this is perfectly correct, it can also become perfectly tedious for both the reader and, indeed, the writer. Hence, from now on we will use these symbols in English language contexts as names of themselves or, as logicians say, we will use these symbols *autonymously*. More generally, we can now understand any sentence of the formal language as a name of itself, i.e. as an **autonym**. This subtle point will be clarified as we go. For now, it is enough to see that it allows us to reduce the use of quotation marks vastly and thus, hopefully, affords us a greater clarity of expression.

With that point in mind, consider a simple example. Suppose that we wanted to represent the Blind Lemon Jefferson argument in our formal language using sentence-letters, i.e. the argument that:

1. If it's a Blind Lemon Jefferson album then it's a blues album.
2. It's a Blind Lemon Jefferson album.

Therefore,

3. It's a Blues album.

First, we must study the argument closely so as to identify clearly the sentences which compose it. In the first instance, what we are looking for here are not *complex* sentences such as Premise 1 but rather the most *simple* or basic sentences such as Premise 2 and the conclusion, i.e. we are looking for the shortest possible well-formed sentences involved. Given our stock of sentence-letters we can easily represent any such sentence formally. Hence, we simply let the first sentence-letter P stand for the first such basic sentence involved, i.e. 'It's a Blind Lemon Jefferson album', and then let the second sentence-letter Q stand for the second basic sentence involved, i.e. 'It's a blues album.' Having done so, we can abbreviate the first premise to:

1. If P then Q

Now, the second premise is exactly the same sentence that we used P to stand for. So, we may rewrite Premise 2 as:

2. P

Finally, the conclusion is exactly the sentence that we used Q to stand for. So, we complete the formalisation by rewriting 3 as:

3. Q

Obviously, sentence-letters are just as easy to use as sentential variables. But do remember that sentence-letters stand for particular sentences. As such, they are constants and not variables. We could simply continue to use the more formal term 'sentential constant', but 'sentence-letter' is much more user-friendly and using that term may also serve to remind us that it is always particular sentences which are involved in this type of formalisation. So, I will stick to 'sentence-letter' in what follows.

II The Formal Language PL

Let me call the particular formal language which I will construct here **PL**; for 'propositional logic'. The simplest possible sentences in the formal language PL are just the sentence-letters P, Q, R and so on. So, consider again the Blind Lemon Jefferson argument:

1. If it's a Blind Lemon Jefferson album then it's a blues album.
2. It's a Blind Lemon Jefferson album.

Therefore,

3. It's a blues album.

Let P stand for: 'It's a Blind Lemon Jefferson album' and Q for 'It's a blues album.'

In a natural language such as English individual sentences can be combined or conjoined, using 'and', for example. Equally, the basic elements of the logicians' formal language can also be combined in certain ways, according to some simple grammatical rules, and combinations of basic elements can be separated out, again according to simple rules. In this way, ultimately, actual arguments can be represented formally, studied, and tested for validity within the formal language.

PL is a formal language and so PL should contain symbols that allow us to combine or *connect* sentence-letters into complexes. For obvious reasons, these symbols are known as **logical connectives**. PL contains five logical connectives which we read as follows:

&	:	And
v	:	Or
→	:	If . . . then
↔	:	If and only if
~	:	Not

Just as we can combine sentences in English using English language connectives so we can combine sentence-letters in PL using the logical connectives. But note that only the first four connectives need actually connect two sentence-letters; 'not' may apply to a single sentence-letter. This is perhaps easier to see if we spell out the connectives and their English language readings as follows:

. . . & ---	Both . . . and ---
. . . v ---	Either . . . or ---
. . . → ---	If . . . then ---
. . . ↔ ---	. . . if and only if ---
~ . . .	It is not the case that . . .

Clearly, while the first four connectives require a minimum of two sentence-letters, 'not' requires only one. Connectives which require at least two sentence-letters are known as **binary** connectives. Those connectives which can be applied to a single sentence-letter are known as **unary** connectives. 'Not' is the only unary connective in PL. So, just as we may conjoin two sentences in English using 'and' we may conjoin two sentence-letters in PL using '&', e.g. P & Q. As we shall see, each time we form a compound PL formula that formula should be enclosed in brackets, e.g. (P & Q). For the moment, however, we can safely omit these. The PL formula is simply read in English 'P *and* Q'. To connect P and Q using the logical connective '&' is to form the **conjunction** P & Q in PL. For present purposes, P & Q formalises the English language conjunction: 'It's a Blind Lemon Jefferson album *and* it's a blues album.' (Note that we do not now need to place the complex PL sentences mentioned in this paragraph in quotation marks to form the name of each such PL sentence as these too are autonyms, i.e. names for themselves.)

To connect P and Q using 'v' is to form the **disjunction** P v Q in PL.

To connect P and Q using '→' is to form the **conditional** P → Q in PL.

To connect P and Q using '↔' is to form the **biconditional** P ↔ Q in PL.

To connect '~' with any sentence-letter is to form the **negation** of that sentence-letter, e.g ~P, ~Q, etc., in PL.

Now run through the list yourself spelling out the English language equivalents for each as I did for the case of conjunction.

Every such compound or complex of the sentence-letters of PL formed using the connectives as described above is known as a *formula* of PL or, more solemnly, as a **well-formed formula** of PL. The individual sentence-letters are themselves well-formed formulas of PL. Each simple, individual sentence-letter is known as an **atomic formula** of PL. Any well-formed formula produced from the sentence-letters of PL using the connectives is known as a **compound formula** of PL. To date, we have only considered compound formulas constructed from atomic formulas. These are the simplest kind of compound formula and, in fact, such compound formulas can themselves be combined in turn, using the connectives, to produce more complex compound formulas. When constructing complex compound formulas we must be careful to avoid ambiguity and to say clearly exactly what we mean. Ambiguity can easily be avoided by following some simple rules; two for the binary connectives and another two for our lone unary connective. For example, although the following two sentences are made up of the same basic sentences and use the same connectives they clearly have very different meanings:

1. Either I'll stay in bed and read my logic book or I'll have a shower.
2. I'll stay in bed and either I'll read my logic book or I'll have a shower.

Let's formalise Sentences 1 and 2. When attempting to formalise any natural language argument it is crucial first to establish exactly how many basic sentences are involved so as to determine how many sentence-letters are required. To that end, it can help to draw up a **key** (especially for those beginning formal logic). In the case of sentences 1 and 2, there are three basic sentences involved and so we require three sentence-letters. The key should look like this:

Key

(i)	I'll stay in bed.	P
(ii)	I'll read my logic book.	Q
(iii)	I'll have a shower.	R

Thus, let P, Q and R stand for Sentences (i), (ii) and (iii) respectively. Next, carefully identify the connectives involved. The only connectives used in 1 and 2 are 'and' and 'or'. So, we will only need the logical connectives '&' and 'v'.

Now consider the original sentences carefully. Sentence 1 is clearly a disjunction: 'Either . . . or --- '. However, the first gap, the first **disjunct**, is itself a compound formula. Look closely at the sentence which plugs the first gap before the 'or'. That sentence is the conjunction: 'I'll stay in bed *and* read my logic book', i.e. P & Q. As noted, when we form a complex PL formula we should enclose that formula in brackets, i.e. (P & Q). Shortly, we will see why this is an important practice. But for now let's complete the formalisation at hand. The rest of Sentence 1 is simply our third basic sentence: 'I'll have a shower', i.e. R. So, Sentence 1 is a disjunction, whose first disjunct is the conjunction (P & Q) and whose second disjunct is just R. Formally:

1′ (P & Q) v R

But look closely at Sentence 2. It is not a disjunction. Rather, it is a conjunction, i.e. . . . & --- . The first gap, the first **conjunct**, is simply the basic sentence: 'I'll stay in bed', i.e. P. However, the second conjunct is a compound. In fact, it is the disjunction: 'either I'll read my logic book or I'll have a shower', i.e. (Q v R). Formally then, we have a very different compound formula:

2′ P & (Q v R)

This simple example clearly illustrates the rules for binary connectives:

1. Always make sure that each binary connective connects two formulas.
2. If any one of those formulas is compound put it in parentheses first.

In fact, we are all already perfectly familiar with these grammatical rules of thumb. Why? Just because we follow exactly the same rules in arithmetic. $(10 + 5) \times 3$ is a very different operation and gives a different result from $10 + (5 \times 3)$ and we mark that difference precisely by using brackets to specify which operation we mean. Now look closely at the formal sentences 1′ and 2′. In both cases we have two connectives but in each case one is inside brackets while the other is not. In effect, the connective inside the brackets only connects the formulas within those brackets. But the connective outside the brackets connects the whole bracketed expression with everything else. Clearly, there is a difference here. The difference lies in what the logician calls the **scope** of the connective:

The scope of a connective consists of the connective itself together with what it actually connects.

So, the scope of '&' in 1′ is just (P & Q). But the scope of 'v' in 1′ is the whole sentence (P & Q) v R (work out the scope of each connective in 2′ for yourself).

Given the notion of scope we can easily define another important notion, that of a **main connective**:

> For any formula of any complexity, the main connective is just the one whose scope is the entire formula.

In other words, the main connective will not occur within the scope of any other connective in the formula. This notion is extremely useful because the main connective tells us just what type of formula we are dealing with, i.e. if the main connective is '&' the formula is a conjunction, and so on (identify the main connectives in 1′ and 2′ and work out which type of formula is involved in each case).

So much for the binary connectives. What of our lone unary connective? Again, we should be careful to say what we mean clearly so as to avoid any ambiguity. In practice, this means clearly identifying the scope of the negation operator in any given case, i.e. working out exactly which formula the negation sign belongs to. There are two kinds of case to consider here. First, the negation sign may simply be applied to a single sentence-letter, an atomic formula, such as P, to produce the negation of P, i.e. ~P. Alternatively, the negation sign may negate a compound formula such as (P & Q) to produce its negation, i.e. ~(P & Q). Note the difference in scope in the two different cases. In the first case, the scope of the negation operator is ~P. In the second, the scope of the negation operator is ~(P & Q).

Finally, note carefully that the negation sign is always taken to apply to the smallest following formula. So, for example, ~P & Q is properly read as: (~P) & Q rather than as: ~(P & Q). Strictly speaking, then, (~P) & Q is the correct form here. In order to minimise the proliferation of brackets I will generally omit the brackets in this and similar cases and simply write ~P & Q. However, to assert ~(P & Q) brackets must first be placed around P & Q.

To sum up:

1. To negate an atomic formula simply connect the negation sign with the sentence-letter.
2. To negate a compound formula be sure to place that formula in brackets before connecting the negation sign with it.

Hence, for both binary and unary connectives, we can always use brackets to make clear exactly which type of formula is involved. Fortunately, brackets form part of the language of PL. So, we can always avoid ambiguity when formalising in PL. As Wittgenstein once said:

> Everything that can be thought at all can be thought clearly. Everything that can be put into words can be put clearly.[1]

Before we turn to the final topic of this section, let's reflect briefly on the formulas available to us in PL. The atomic formulas of PL are simply sentence-letters. Every complex formula formed from atomic formulas using the connectives, in terms of the grammatical rules given above, is a compound formula of PL. All such formulas belong to PL. In Chapter 4 we will consider a new type of variable which ranges over the well-formed formulas of PL and, in the final section of that chapter, we will go on formally to define the class of well-formed formulas exactly. But already we have reached the very heart of PL and the essence of propositional logic. For the plain truth of the matter is that propositional logic is the logic of sentences formed from sentence-letters, by the appropriate rules, using what logicians call an **adequate** set of connectives. We will consider adequacy in more detail later. For the moment, rest assured that our set of five connectives {&, v, →, ↔, ~} is certainly an adequate set. Indeed, we might even call it a *generous* set. As we shall see, we could have used three connectives, two connectives or even just one binary connective. Consider the contents of Box 2.1 carefully.

Finally, let's consider another way of analysing complex compound formulas into their constituent parts. Every complex formula can be

BOX 2.1

- The basic elements of PL are the *sentence-letters* P, Q, R, etc., which may be combined into *compound formulas* using the *logical connectives*: &, v, →, ↔, ~.
- When forming a compound formula always make sure that (i) each *binary* connective connects two formulas and (ii) that if any of those formulas is itself compound it is put in parentheses first.
- The *scope* of a connective consists of the connective itself together with the formula(s) it connects.
- The *main connective* in any formula is the one whose scope is the entire formula.

represented as a sort of tree, known as a **syntactical tree**, which clearly displays the overall syntactic structure of a formula in diagrammatic form. To set up the trees we need some new symbols for some old ideas. First, we have considered two kinds of formula: compound formulas and atomic formulas. For convenience as regards tree-construction, we will simply designate any formula, compound or atomic, by '**F**' and adopt the symbol '**A**' specifically for any atomic formula.

Equally, we have talked about two kinds of connective: unary and binary connectives. For present purposes, we will simply abbreviate these as '**U**' and '**B**' respectively. This is all we need to get our trees off the ground (pardon the pun). In fact, our trees are not really normal trees. Rather, they are upside-down trees. This is because we will start off from a compound formula, as large as you like, which we will just represent as '**F**'. Below that symbol we will construct branches with a symbol at the end of each branch in order to spell out what the compound formula is made up of, i.e. smaller compound formulas '**F**'s, atomic formulas '**A**'s, connectives, and so on.

In turn, those branches will themselves branch until we finally make the actual symbol explicit. The following **development rules** (for PL formulas)[2] make clear how to rewrite any given symbol. In the dullest case, our formula '**F**' will itself be an atomic formula. So, we have a rule which allows us to rewrite '**F**' as '**A**':

```
1. F
   |
   A
```

More interestingly, the original formula might be a compound formula. Any compound formula will either be the negation of some formula or it will consist of two formulas connected by one of the other connectives. Hence, we can go on to add two further rules:

```
2.    F            3.    F
    / | \              /   \
   F  B  F            U     F
```

The only unary connective in PL is '~'. So,

```
4. U
   |
   ~
```

However, PL has four binary connectives, so:

```
5. B     6. B     7. B     8. B
   |        |        |        |
   &        v        →        ↔
```

Finally, we need to be able to spell out exactly which atomic formula is involved in any actual case. For present purposes, the following rules will suffice:

```
9. A    10. A    11. A    12. A
   |        |        |        |
   P        Q        R        S
```

We are now in a position to consider some examples. Here is the (upside-down) tree for the compound formula P & Q:

1. P & Q

```
      F
    / | \
   F  B  F
  /   |  |
 A    &  A
 |       |
 P       Q
```

Next, consider the tree for ~(P → Q):

2. ~(P → Q)

```
     F
   /   \
  U     F
  |   / | \
  ~  F  B  F
     |  |  |
     A  →  A
     |     |
     P     Q
```

Question 4 of Exercise 2.1 contains some examples which you can try for yourself.

EXERCISE 2.1

1. Identify the main connective in each of the following formulas. State which kind of formula you consider each to be, e.g. conjunction, disjunction, etc.

(i)	P & Q	(vi)	(P & Q) v (Q & R)
(ii)	~(P & Q)	(vii)	P → Q
(iii)	~P & ~Q	(viii)	(P → Q) & (Q → R)
(iv)	~(P & Q) & (P & ~Q)	(ix)	(P & Q) ↔ (Q & P)
(v)	P v Q	(x)	~((P & Q) ↔ (Q & P))

2. Specify the scope of each and every occurrence of the negation operator in formulas (ii), (iii), (iv) and (x).

3. Translate the following natural language sentences into PL. In each case make your key explicit:

   (i) Blind Lemon Jefferson is not the only bluesman.

   (ii) Dr Strangely Strange is not a bluesman and neither is Mr Oddly Normal.

   (iii) If Blind Lemon Jefferson is a milkman then he certainly isn't a bluesman.

   (iv) It's just not true that if Blind Lemon recorded it then it's a blues album.

   (v) Blind Lemon was either a bluesman or a milkman.

   (vi) Blind Lemon was neither a bluesman nor a milkman.

4. Construct syntactical trees for formulas (i) to (x) of Question 1 above using the development rules given at the end of the last section.

III
Arguments and Sequents

The formal language PL employs a slightly different terminology for describing arguments. Premises are still premises and a conclusion remains a conclusion but arguments, pieces of reasoning in PL, are represented by *sequents*. A **sequent** consists of a finite (possibly empty) set of well-formed formulas (the premises) together with a single well-formed formula (the conclusion). So, just like an argument, a sequent consists of a finite (possibly empty) set of premises together with a conclusion which may or may not follow logically from those premises. We can think of a sequent of the formal language as capturing a putative or purported proof whose validity is not yet decided. Therefore, sequents really just represent comprehensible pieces of reasoning in PL, valid or not.

Let's formalise the Blind Lemon Jefferson argument and represent it as a sequent in PL:

1. If it's a Blind Lemon Jefferson album then it's a blues album.
2. It's a Blind Lemon Jefferson album.

Therefore,

3. It's a blues album.

Let the sentence-letter P stand for the sentence 'It's a Blind Lemon Jefferson album' and the sentence-letter Q stand for the sentence 'It's a blues album.' Hence, we can rewrite this particular argument as follows:

1. If P then Q
2. P

Therefore,

3. Q

Given our earlier discussion of the logical connectives, PL allows us to formalise a little further. The first premise is clearly a conditional, i.e. a combination of P and Q using 'If . . . then --- '. So, we can replace 1 with: 1′ $P \rightarrow Q$. If we stretch the whole argument out on a single line we may write:

$P \rightarrow Q$, P Therefore, Q

We can eliminate the last natural language expression from the argument simply by choosing a symbol to mark the distinction between the premises and the conclusion. I will use the **colon** ':' to mark just that distinction. So, we replace 'therefore' with ':' and, at last, we can represent the Blind Lemon Jefferson argument in purely formal terms as a sequent in PL:

$$P \rightarrow Q, P : Q$$

Quite generally, anything consisting of a bunch of formulas of PL (the premises), a colon symbol and a single formula of PL (the conclusion) is a sequent in PL. Hence, sequents simply represent comprehensible pieces of reasoning in PL. But this raises an obvious question: how are we to distinguish good, valid reasoning from bad, invalid reasoning in the language PL? Well, if we just do not know whether a given conclusion really does follow logically from a given set of premises in PL then we simply separate the conclusion of that sequent from the premises using the colon symbol ':'. This reflects the fact that we are entertaining or considering that sequent but do not as yet know whether the conclusion of the sequent really does follow from that set of premises in PL. (You might think of the colon almost as a question mark, i.e. does that conclusion really follow from those premises? Or even as a challenge, i.e. can you show that the conclusion really does follow from just those premises?) When a particular conclusion really does follow logically from a given set of premises in PL then we may replace the colon symbol with another symbol which indicates precisely that the conclusion really does follow logically from those premises. For obvious reasons this symbol is known as the **turnstile**: '⊢'. Remember: only if the conclusion really does follow logically from those premises in PL can we replace the colon with the turnstile symbol '⊢'. Further, note that the turnstile says something about PL. We could subscript the turnstile with 'PL' to make clear that it is exactly that formal language we're talking about. I'll say more about that shortly but for the moment, the point is simply that anything consisting of a finite set of formulas of PL (premises), a turnstile symbol and a single formula of PL (the conclusion) is explicitly asserted to be a sequent whose conclusion really does follow logically from those premises in the formal language PL.

The distinction marked by colon and turnstile is one which is glossed over in many introductory logic texts[3] but it is an important distinction none the less and we will make it explicit. The presence of the turnstile in a sequent represents a sort of guarantee that the conclusion really does follow from those premises in the formal language PL (or at least represents the assertion that this is so). The question is how to show that the conclusion really does follow logically from the premises. When does a conclusion follow logically from a set of premises in the formal language PL? If the conclusion does follow logically from just that set of premises in PL then a **proof** of that conclusion can be constructed from those premises in PL without the need for any additional premises. A more precise definition will be given later, but for the moment we can take a proof in PL to be a step-by-step way of getting from the premises to the conclusion, each step being justified by a rule.

In PL, proofs prove sequents. If a sequent can be proved in PL then that sequent is said to be *provable* or *derivable* in PL (though some logicians distinguish between the two, we will use these terms interchangeably). Hence, the presence of the turnstile in a sequent guarantees the existence of a proof or **derivation** of that conclusion from just those premises in PL, according to the rules of the game, as it were. The turnstile therefore asserts *provability* or *derivability* in PL. Moreover, for deep reasons which we will consider later, we know that if a sequent is provable in PL then it is indeed a valid sequent of PL. Hence, when the turnstile applies the conclusion really does follow from those premises in PL, i.e. the turnstile characterises the relation of logical consequence in PL. Therefore, provability or derivability in PL provides one answer to the fundamental question in logic, at least in the context of PL, namely, when does a sentence follow logically from some other sentences? In PL, we can answer that fundamental question precisely in terms of provability, i.e. a given formula follows logically from a set of formulas if the former is derivable from the latter. In other words:

> A conclusion follows logically from a set of premises in PL if there is a proof of that conclusion from just those premises in PL.

More precisely, the notion of derivability, of provability, represents logical consequence in PL syntactically. *Syntactic* just means 'concerning the symbols of the formal language PL considered purely formally and not in terms of the meanings of those symbols'; as formal logicians put the point, 'concerning the symbols of PL as wholly uninterpreted'. So, odd as it may sound, **syntax** is not really concerned with questions about truth or meaning (though, as we shall see, syntactic rules and principles may themselves embody meaningful intuitions). First and foremost, syntax is concerned with the form, literally, the shape of formulas. Indeed, we can construct formal languages with vocabularies of squares and circles, triangles and rectangles and so on.[4] The notion of syntax will become clearer as we go. For the moment concentrate on the idea that it is the shape of the formulas which will be important to us in the rest of this chapter and in the next.

Consider the contents of Box 2.2 carefully before you attempt Exercise 2.2.

EXERCISE 2.2

Represent the following arguments as sequents of PL. In each case, construct a key specifying precisely which sentence-letter of PL stands for which natural language sentence. Replace 'so', 'therefore', etc., with '∴' in each of your answers. (At the end of this chapter you might consider whether there is any case in which we can replace '∴' with '⊢'.)

1 Big Bill Broonzy is either a Delta bluesman or a Chicago bluesman. If he was not born in Mississippi then he is definitely a Chicago bluesman. He wasn't born in Mississippi. Therefore, Big Bill Broonzy was not a Delta bluesman.

BOX 2.2

- Arguments or inferences in PL are represented by *sequents*.
- A sequent consists of a finite (possibly empty) set of well-formed formulas (premises), the *colon* symbol ':' and a single well-formed formula (the conclusion).
- Every sequent of PL either is or is not valid.
- A valid sequent of PL is a sequent whose conclusion is a logical consequence of its premises in PL.
- A sequent of PL can be shown to be valid by constructing a *proof* of the conclusion of the sequent from all and only the premises of that sequent.
- The presence of the *turnstile* symbol '$\vdash$' in a sequent indicates the provability of that sequent.
- A valid sequent consists of a finite (possibly empty) set of well-formed formulas (premises), the turnstile symbol and a single well-formed formula (the conclusion).
- The turnstile symbol represents the notion of logical consequence in PL *syntactically*.

2 It is plainly not true both that Etta James was an angel and that Robert Johnson sold his soul to the devil. But Etta James was an angel. So, Robert Johnson surely sold his soul to the devil.

3 If there's no light on in the Venue, the band aren't on stage yet. It's not true both that there is a light on in the Venue and that it's not going to be a great night. It really is going to be a great night. So, the band are definitely on stage.

4 Either there's a punk rock band playing at the Venue tonight or the music is strictly classical. If the punk rock band are playing then there will be no blues at the Venue tonight. But if the music is strictly classical there won't be any blues at the Venue tonight. So, there will be no blues at the Venue tonight.

5 If there's a band on stage and the music is groovy then it can't be the Nasal Flute Orchestra. So, if it's the Nasal Flute Orchestra then, if there really is a band on stage, then the music certainly is not groovy.

IV
Proof and the Rules of Natural Deduction

Because proofs prove sequents, each proof will begin with the relevant set of premises and will end with the conclusion proved from those premises (at this early stage, it may help to think of literally *making* the conclusion out of the raw material of the premises). Proofs are constructed in PL by manipulating the premises we have been given by applying certain rules for such manipulations. These rules are known as **rules of inference** in PL, just because the rules determine which inferences can legitimately be drawn in PL. That part or, better, *fragment* of propositional logic which is concerned with actually constructing proofs in terms of the rules of inference is known, rather solemnly, as the **propositional calculus**. The version of propositional calculus which we will consider here involves twelve simple rules of inference.

For each of the binary logical connectives in PL, there is both an **introduction-rule**, which brings that connective into a line of proof, and an **elimination-rule**, which takes the connective back out of a line of proof. In addition, a small number of important rules govern the behaviour of the negation symbol, '~', in proofs. Such rules are syntactic, i.e. via the connectives, each rule exploits the form and shape of formulas in PL. However, many logicians feel that these principles of inference, to a great extent, fix the meaning of the logical connectives just by spelling out exactly what we can and can't infer from them. On this view, the introduction- and elimination-rules for each connective are thought of as capturing our core understanding of each connective in a very immediate way. This is a subtle point which requires a little reflection, but each of the introduction- and elimination-rules is perfectly simple and intuitive. In the process of proof-construction, these simple rules can be combined in complex strategies just as each of the pieces in chess has a simple move and yet they combine intricately. In this way, important traditional laws of logic and some very complex sequents become provable.

Every proof consists of a number of **lines of proof**. Each line of proof consists of four parts. The first is a **line number**. Lines are simply numbered consecutively: 1, 2, 3 . . . with each new number written on the line below the last until the final line, the line of the conclusion, is reached, i.e.:

1.

2.

3.

.

.

The second part of any line of proof is a **formula** on that line. In the simplest case, this would be a single sentence-letter, i.e. an atomic formula. For example:

1. P

Each and every step in a proof, from one line to the next, must be made in terms of a rule of inference and the actual rule used must be cited. Citing the *name* of the rule, the **rule annotation**, is the third part of the line of proof. The first and most basic rule of inference is the rule of **Premise-Introduction**. It simply allows us to enter the relevant premises of any sequent onto lines of proof. According to the rule of premise-introduction any well-formed formula may be entered as a premise on any line of a proof. This rule is simply cited as 'Premise'. To introduce P as a premise on line 1, for example, we write:

1. P Premise

So far then, we know which formula is on which specific line of proof and that we entered the formula on that line by a particular rule. But the line of proof is still incomplete, for we do not yet know whether that formula itself depends upon any other formulas. In practice, the formula on the line of proof frequently does depend upon other formulas and it is crucial to be able to express that fact and to specify exactly which other formulas are being leaned on. To that end, we set aside a space to the left of the line number and there we simply list the line numbers of any other formulas which that formula depends upon. These numbers are the formula's **dependency-numbers**. Because a formula often has more than one dependency-number, and so that we do not confuse dependency-numbers with line numbers, we place the set of dependency-numbers inside curly brackets '{ }'.

In practice, this part of a line of proof is always very easy to get hold of just because every rule of inference contains a recipe for making the set of dependency-numbers. For example, here is the rule of premise-introduction in full:

> Premise-Introduction: Any well-formed formula may be introduced as a premise on any line of proof. The dependency-number of that line is identical with the line number of that line of proof.

So, the dependency-number of any formula introduced as a premise is always the same as the line number of the line where it is introduced. In an important sense, every premise is self-sufficient. It is not derived from anything else and its presence on a line of proof depends upon nothing other than the fact that we have asserted that premise. Hence, the formula on any line annotated 'premise' depends only upon itself. Now, dependency-numbers are always entered in brackets behind the line number. In the example above, where P was introduced as a premise on line number one, its dependency-number is the same as its line number, so the set of dependencies is simply {1}. The complete line of proof looks like this:

{1}	1.	P	Premise

Suppose I now introduce Q as a second premise on line number two. The proof now looks like this:

{1}	1.	P	Premise
{2}	2.	Q	Premise

(Work out how the proof would look if I next introduced R as a premise on line 3.)

So far, the proof consists of two premises: P on line 1 and Q on line two. But remember: PL is a language and just as I may conjoin two sentences in English using 'and' so two formulas in PL can be conjoined using '&'. In both English and PL, we create a *conjunction*. Every conjunction consists of two conjuncts. The rule of inference which permits this move is called **&-introduction** (and-introduction). To cite it, write the line number of each line used, followed by '&I'. Here is the rule in full:

> &I: Any two well-formed formulas may be conjoined. The relevant line numbers and '&I' must be cited. The dependency-numbers of the new line consist of all the dependency-numbers of both lines used.

For example, to conjoin P from line 1 with Q from line 2, on a new line, say 3, I simply cite lines 1 and 2, write '&I', and pool the dependency-numbers of both to get the dependency-numbers of the new line 3, i.e.:

{1,2}	3.	P & Q	1,2 &I

To recap:

{1}	1.	P	Premise
{2}	2.	Q	Premise
{1,2}	3.	P & Q	1,2 &I

I have now constructed a proof of the conclusion P & Q from premises P and Q. Therefore, I may write P, Q ⊢ P & Q. This simple proof illustrates a number of important points about proof-construction: first, proofs proceed from premises to conclusion. The premises are just those formulas listed on the left-hand side of the turnstile. The conclusion is the formula on the right of the turnstile. Distinct premises are separated on the left by commas. Each distinct premise should be asserted on its own line of proof. The natural way to do this is just to enter each premise on a separate line of proof in the order in which they are given, i.e. just as originally listed on the left-hand side of the turnstile or colon symbol.

Note that the proof proceeds via the introduction of premises P and Q on lines 1 and 2 and ends with the conclusion P & Q on the last line of proof. Note also that, in this case, the conclusion is literally constructed from the premises using the rules of inference. Finally, note that all of this information is encoded in the last line. Recall the four parts of any line of proof. In this instance, what these tell us is exactly that the conclusion P & Q was constructed on line 3 by the rule of &-introduction and that the set of its dependencies is {1,2}. These dependency-numbers just refer back to earlier line numbers. If we check back, we can see that the line numbers cited are line numbers 1 and 2. These lines contain the premises P and Q respectively. Therefore, the conclusion has been derived from just those premises and does not depend upon any other formula. Further, note the way in which the introduction rule for '&' works. As noted earlier the introduction rule brings the connective in, in a line of proof. But recall that each connective also has an elimination-rule. The elimination-rule for '&', **&E**, is as follows:

> &E: One conjunct may be removed from a conjunction by one application of &E. The line number of the conjunction must be cited together with '&E'. The dependency-numbers of the new line are identical with those of the original line containing the conjunction.

Suppose I introduce P & Q on line 1:

{1}	1.	P & Q	Premise

&E allows me to infer:

{1}	2.	P	1 &E

Hence, I have proved that:

P & Q ⊢ P

Equally, introducing P & Q on line 1:

{1}	1.	P & Q	Premise

&E allows me to infer:,

{1}	2.	Q	1 &E

This time I have proved that:

P & Q ⊢ Q

&E allows me to take either conjunct out of the conjunction and, in fact, I can take both conjuncts out. The only constraint is that I cannot take both out with a single application. Each conjunct requires a separate application of the rule. Note how &E as an elimination-rule effectively removes the connective from the line of proof which we use it to construct: line 2 does not contain '&'. In both cases, the application of the elimination-rule removes the connective from the new line. As we shall see, this holds quite generally for the rules governing each connective. Now consider the following proof which combines both rules for &:

(Q & R), P ⊢ (P & Q) & R

{1}	1.	(Q & R)	Premise
{2}	2.	P	Premise
{1}	3.	Q	1 &E
{1}	4.	R	1 &E
{1,2}	5.	P & Q	2,3 &I
{1,2}	6.	(P & Q) & R	4,5 &I

So, &I can both be used to conjoin two atomic formulas, as in line 5, and to conjoin an atomic formula with a compound formula. Equally, &I can also be used to conjoin two compound formulas. Note also the pattern of the proof: first, the premises are set out. Next, compound formulas, like the formula on line 1, are broken down into their simple constituents. Finally, those simple constituents are put back together slightly differently so as to construct the desired conclusion depending only on its premises. In effect, we construct a formula with a certain specific shape (the conclusion) from the premises. This pattern is common to many proofs in propositional calculus and reflects the attempt to derive the conclusion from the premises.

V
Defining: 'Proof-in-PL'

We are now in a position to begin to construct a rigorous and precise definition of what exactly counts as a proof in the formal language PL. Because we have not yet considered every rule of inference the definition will not be

BOX 2.3

♦ A *proof* consists of a set of consecutively numbered lines of proof.

♦ Each *line of proof* is composed of four parts: (i) a line number; (ii) a formula on that line; (iii) the name of the rule in virtue of which the formula was entered on that line; (iv) the formula's dependency-numbers, i.e. a set of line numbers each of which refers back to a formula upon which the current formula depends.

♦ Each step in a proof must be made in terms of a rule of inference. So far, we have outlined the following rules:

♦ *Premise-Introduction*: Any well-formed formula may be introduced as a premise on any line of proof. The dependency-number of that line is identical with the line number of that line of proof.

♦ &I: Any two well-formed formulas may be conjoined. The relevant line-numbers and '&I' must be cited. The dependency-numbers of the new line consist of all the dependency-numbers of both lines used.

♦ &E: A conjunct may be removed from a conjunction by an application of '&E'. The line number of the conjunction must be cited together with '&E'. The dependency-numbers of the new line are identical with those of the original line.

as specific as possible. But we can arrive at a useful, general definition: a **proof-in-PL** is a finite sequence of consecutively numbered lines, each consisting of a well-formed formula of PL, together with a set of numbers known as the *dependency-numbers* of that line, the entire sequence being constructed using the rules of inference of PL.

In order to make the definition fully specific we replace the last four words '... of inference of PL' with a list of the names of the rules. At this stage we might begin as follows 'Premise-Introduction, &I, &E ... ', but we cannot fill in the dots until we have considered the remaining rules of inference. In the meantime, study the contents of Box 2.3 carefully before attempting Exercise 2.3.

EXERCISE 2.3

1 Name and explain each of the four parts of a line of proof.

2 Prove that the following sequents are valid sequents of PL. If you succeed in constructing a proof you may replace ':' with '⊢' in the sequent. (The numbers in brackets next to each sequent indicate the number of lines in my proof of that sequent. If your proof turns out to be longer or shorter than that check each line of proof carefully.)

1.	P, Q : P & Q	(3)
2.	P, Q, R : (P & Q) & R	(5)
3.	P, Q, R, S : (P & Q) & (R &S)	(7)
4.	P & Q : P	(2)
5.	P & Q : Q	(2)
6.	(P & Q) & R : P	(3)
7.	(P & Q) & (R & S) : P	(3)
8.	(Q & R), P : (P & Q) & R	(6)

VI
Conditionals 1: MP

The next rules of inference to consider govern the connective marked by the arrow '→' which is read as 'If ... then --- '. In PL, sentences of this type are known as *conditionals*. The arrow is a binary connective which connects two formulas. In the simplest case, for example, $P \rightarrow Q$. In any conditional, the formula occurring before the arrow symbol is known as the **antecedent** of the conditional. The formula occurring after the arrow is the **consequent** of the conditional.

Like every other binary connective in PL there is an introduction-rule for arrow and an elimination-rule. In the case of this connective, however, these rules are honoured with traditional names. We could simply replace these traditional names with **arrow-introduction** and **arrow-elimination**. But most logicians still use the traditional terms and it is more difficult to impress your friends with rather mundane names such as 'arrow-elimination'. The traditional name for arrow-elimination is *modus ponendo ponens*, which is usually shortened to **modus ponens**, or just **MP**. This particular Latinism is most readily understood as 'way of putting', i.e. way of putting the consequent (on a line), having already put the antecedent (on an earlier one). This rule of inference is already perfectly familiar: the conclusion of the Blind Lemon Jefferson argument is inferred by MP:

> MP: Given a conditional formula on one line of proof and its antecedent on another line, its consequent may be inferred. The line numbers of each must be cited together with MP. The dependency-numbers of the new line consist of the dependency-numbers of both cited lines.

Here is a simple example of MP which should look familiar:

$$P \rightarrow Q, P \vdash Q$$

{1}	1.	$P \rightarrow Q$	Premise
{2}	2.	P	Premise
{1,2}	3.	Q	1,2 MP

This particular sequent of PL is of exactly the same logical form as the original Blind Lemon Jefferson argument. Further, note that MP genuinely is the elimination-rule for arrow: it eliminates arrow from the line of proof which it is used to construct. Many logicians refer to MP as the rule of *detachment*. This is very appropriate, i.e. it is almost as if, given P on line 2, we use MP literally to detach the consequent Q from the conditional $P \rightarrow Q$ on line 1, to end up on its own on line 3.

Again, the simplest conditionals will contain only atomic formulas, e.g. $P \rightarrow Q$. But note that the antecedent of any conditional may be a compound formula, e.g. $(P \& Q) \rightarrow R$. Equally, the consequent may be compound, e.g. $P \rightarrow (Q \& R)$. Indeed, both may be compound. MP applies none the less. Be it simple or complex, provided that the antecedent of the relevant conditional is also on a line of its own then the consequent of that conditional may be inferred by MP.

Now try Exercise 2.4.

EXERCISE 2.4

1 Prove that the following sequents are valid in PL. If you succeed in constructing a proof you may replace ':' with '⊢' in the sequent. (Again, the numbers in brackets next to each sequent indicate the number of lines in my proof of the sequent. If your proof is longer or shorter check each line of proof carefully.)

1. $P \to Q, P : Q$ (3)
2. $P \to (P \to Q), P : Q$ (4)
3. $P \to (P \,\&\, Q), P : Q$ (4)
4. $P \to (Q \to R), P \to Q, P : R$ (6)

2 Consider the following arguments carefully. Note that although I have identified the premises and the conclusion in the first two, you must identify the premises and the conclusion in the remaining cases. Represent each argument as a sequent of PL and, in each case, prove that the sequent in question is valid in PL:

(i) Premise 1: If Blind Lemon Jefferson was a bluesman then I'm a Dutchman and I will eat my hat.

Premise 2: Blind Lemon Jefferson was a bluesman and I am a Dutchman.

Therefore: Blind Lemon Jefferson was a bluesman and I will eat my hat.

(ii) Premise 1: If it's really true that if Blind Lemon Jefferson was a bluesman then he lived in utter poverty, then if he wanted to have an income then he undoubtedly worked as a part-time postman in Mississippi.

Premise 2: If Blind Lemon Jefferson was a bluesman then he did live in utter poverty.

Premise 3: Blind Lemon Jefferson was a bluesman and he wanted to have an income.

Therefore: Blind Lemon Jefferson undoubtedly worked as a part-time postman in Mississippi.

(iii) Miss Scarlet was in the kitchen. If Miss Scarlet was in the kitchen then Reverend Green was in the conservatory. Therefore, Miss Scarlet was in the kitchen and Reverend Green was in the conservatory.

(iv) Reverend Green was in the conservatory. If Reverend Green was in the conservatory then the murder weapon was the knife. If Reverend Green was in the conservatory then if the murder weapon was the knife then Colonel Mustard was the murderer. Therefore, Reverend Green was in the conservatory and Colonel Mustard was the murderer.

(v) The murder weapon was the knife. If the murder weapon was the knife then Professor Plum is plainly innocent. If the murder weapon was the knife then if Professor Plum is plainly innocent then Colonel Mustard was the murderer. If Colonel Mustard was the murderer then Miss Scarlet was in the kitchen. Therefore, the murder weapon was the knife and Professor Plum is plainly innocent. And Colonel Mustard was the murderer and Miss Scarlet was in the kitchen.

VII
Conditionals 2: CP

The introduction-rule for arrow is known as **conditional proof** or **CP**. For both theoretical and practical reasons, this is by far the most important rule of inference we have considered to date. Before examining the formal rule of inference itself it will help if we briefly explore, informally, the way in which we reason with conditional sentences in general. This is a strategy which we will often exploit: appealing to ordinary practices and our intuitions first, and then moving on to formal considerations. Justifying a particular conditional sentence is not the same thing as justifying the entire rule of conditional proof. None the less, some important aspects of ordinary reasoning with conditionals are involved in the use of CP as a formal rule of inference.

For example, recall the conditional sentence which we considered as the first premise of the very first example of valid argument:

If it's a Blind Lemon Jefferson album then it's a blues album.

Under what conditions might this conditional sentence be justified? Note carefully just what kind of claim a conditional sentence makes: the claim is not that the antecedent *is actually* true but just that *if it is* then the consequent should also be true, i.e. if it is a Blind Lemon album then it is a blues album. The consequent, we might say, is conditional upon the antecedent in the sense that if the antecedent is true then we do expect the consequent to be true also. We may not be able to establish the truth of the consequent just on the basis of the antecedent alone. Some other premises may be required. For example, we might also need to know that Blind Lemon was a truly miserable bluesman who only ever played the blues. Be that as it may, if the conditional is justified then we would expect that when the antecedent is true the consequent is also true. Then, surely, we are justified in asserting the conditional sentence itself. But how are we to determine whether or not this is the case? The natural strategy, surely, is first to assume that the antecedent is true and then consider whether or not the consequent is also true, i.e. assume that the album really is by Blind Lemon and then ask: is it a blues album? Given what we already know in this case, we know that the consequent certainly is true, that it is indeed a blues album, and so this particular conditional really is justified.

At this point it is worth warning against a very popular mistake. Perhaps you are tempted to reason *erroneously* as follows: 'Well, if P entails Q then obviously Q follows from P. So, the conditional $P \rightarrow Q$ just means that Q is a logical consequence of P, i.e. that $P \vdash Q$.' Making this particular mistake could certainly result in failing a logic exam. Why is it a hanging matter? Just because it confuses and conflates two very different relations between formulas. Think hard for a moment about the meaning of the turnstile symbol. We know that it replaces the word 'Therefore' but note that it also has a quite different function: the presence of the turnstile in a sequent represents the assertion that the conclusion of that sequent really does follow logically from its premises in PL. So, we might say, the turnstile means something like: 'In the formal language PL this conclusion is a logical consequence of these premises.'

Note how this assertion begins: 'In the formal language PL . . . '. Clearly, the turnstile is saying something *about* PL. It is not itself an assertion *in* PL. In fact, the turnstile belongs to a language beyond PL which we use to talk about PL. The language to which the turnstile belongs is called the **metalanguage**. The prefix 'meta' derives from Ancient Greek and means 'after'. Therefore, the metalanguage is a language coming after PL which we use to talk about PL. For our purposes, we can simply consider English supplemented with '$\vdash$' and the other formal symbols we have introduced to date as forming the metalanguage, i.e. in Haskell B. Curry's terms, we can understand the metalanguage as 'the totality of linguistic conventions which, at the moment, we understand'.[5] From the viewpoint of the metalanguage, PL is known as the **object language**. So, why shouldn't we confuse turnstile and arrow? For starters, the turnstile is a symbol belonging to the metalanguage while '$\rightarrow$' is a logical connective belonging to the object language. So, these two don't even belong to the same language. Moreover, the turnstile represents the relation of provability, derivability and therefore consequence in PL while the arrow represents no such thing. The turnstile is one way of representing logical consequence between formulas of PL. Arrow is not. The arrow is simply a logical connective which represents a much weaker, grammatical, relation known as **material implication**. If P materially implies Q, that just means that Q is not false when P is true – and that's all. And this is exactly what we would expect. Take the Blind Lemon Jefferson case: the conditional is justified, we said, as long as the consequent is true when the antecedent is true. So, the conditional itself will not be true if the consequent is false when the antecedent is true. And that is just what is meant by material implication, i.e. the relation which the arrow represents.

To recap: in general, a conditional will have been justified if, having assumed the antecedent, the consequent can also be shown to hold. Equally, if we assume the antecedent and the conclusion does not hold then, surely, the conditional is not justified. Now, it is exactly this kind of reasoning which is involved in any application of conditional proof: to prove a conditional, first assume its antecedent, then try to derive its consequent from the antecedent alone or together with some other premises. Only if you succeed in deriving the consequent can you take the conditional itself as having been proved.

The practical significance of this particular rule is substantial: conditional proof isn't simply a rule, it is a *strategy*, and a highly economical one at that. Why? Given *any sequent* with a conditional as its conclusion we know exactly how to prove it: assume the antecedent and then attempt to derive the consequent using the antecedent and whatever else is available. Most importantly, if, having assumed the antecedent, you succeed in deriving the consequent, conditional proof allows you to enter that conditional on a line of proof *without it depending on that assumption*. Why? Remember that the conditional does not involve the claim that its antecedent is actually true. So, we do not need to assert that it is true either.

Rather, we simply *assume* the antecedent and then try to show that under those circumstances the consequent also holds. So our reasoning, we might say, is only **hypothetical**. Hence, though we may seem to have an extra *premise* in the form of the antecedent, in reality we do not. We have only an extra *assumption* and CP allows us to eliminate or *discharge* that assumption. For precisely that reason logicians refer to CP as a **discharge rule**: the assumed antecedent is simply discharged once the conditional itself is proved. The distinction between premises and assumptions is again one which is often glossed over in introductory logic texts but it is a good distinction none the less: premises are explicitly asserted to be true and conclusions are proved from them. Such reasoning is **categorical reasoning**. But assumptions are merely assumed. Reasoning with assumptions is *hypothetical*. So, any assumption that we might make for CP will certainly be involved in a proof, but not as a premise. For that reason we require a new rule which allows us to make clear that a formula has been assumed rather than asserted as a premise. Unsurprisingly, the new rule is known as the **rule of assumptions** and its annotation is just 'assumption' or, even more briefly, '**A**'. Here is the rule in full:

> A: Any formula may be assumed on any line of proof. The line must be annotated 'A' for 'assumption'. The dependency-number of the assumed formula is identical with the line number of the line on which it is assumed.

Obviously, this rule-statement is similar in some respects to the one given for premise-introduction and you might be tempted to think of assumptions as being rather like premises, perhaps as *temporary* premises. For example, you might think of an assumption as a premise for any formula which is derived from it. That thought is not entirely false but it is potentially misleading in so far as it obscures a useful distinction. Any formula on a line annotated 'A' is only assumed to be true. Hence, it is an assumption. It is not explicitly asserted to be true. So, it is not a premise. Most importantly, always remember that, at the last line of proof, the conclusion should never include among its dependencies any number which refers back to an assumption. Why not? Because all assumptions should have been discharged by that stage, leaving the conclusion to rest only on its premises. As we have seen, the distinction between premises and assumptions is ultimately well grounded in ordinary reasoning in natural language and I hope that you find it intuitive. We will return to the same distinction again in the next chapter, where we will find that we can sharpen that distinction a little further, again, just in terms of the role and use of premises and assumptions in the **proof-theory** of PL.[6]

We are now in a position to consider the rule of conditional proof formally, but let's recap on the strategy. To prove a conditional:

1 First, assume the antecedent. When you assume the antecedent annotate that line 'assumption' *and* write 'for CP' alongside. Then enter the line number as the dependency-number (this practice makes clear why the assumption has been made and allows you to keep track of the number you will later discharge).

2 Next, use the antecedent, and anything else available, to derive the consequent.

3 Finally, take the antecedent from its line, the consequent from its line and introduce the arrow between them. Annotate the new line with those line numbers and 'CP'. *But always remember*: CP is a discharge

rule. Once CP is applied the dependency-number of the antecedent should always be removed from the set of dependencies of the line annotated CP.

Every proof with a conditional conclusion can be proved using CP. Here is a statement of the rule in full. It is well worth learning by rote:

> CP: Assume the antecedent. Derive the consequent. Enter the conditional on a line along with 'CP'. The line numbers are both those where the antecedent is assumed and those where the consequent is derived. Discharge the dependency-number of the antecedent. Pool the remaining dependency-numbers to complete the line.

Note: In any application of CP, the dependency-number of the antecedent must always appear in both lines used in the CP. Otherwise, the consequent has not been proved to follow from the antecedent, and CP cannot be used. This is an important constraint but one that is easy to meet in practice. Consider some simple examples:

1. $P \rightarrow Q, R \rightarrow P \vdash R \rightarrow Q$

{1}	1.	$P \rightarrow Q$	Premise
{2}	2.	$R \rightarrow P$	Premise
{**3**}	3.	R	Assumption for CP [antecedent of $R \rightarrow Q$]
{2,**3**}	4.	P	2,3 MP
{1,2,**3**}	5.	Q	1,4 MP
{1,2}	6.	$R \rightarrow Q$	**3**,5 CP

Note how MP detaches the consequent of the conditional to which it is applied and eliminates the arrow from the line it is used to construct (lines 4 and 5). Note also that the application of CP introduces the arrow to the line it is used to construct. So, we can think of CP as taking two formulas, one from line 3, the other from line 5, and introducing the arrow between them. And note very carefully that there is indeed a number in common between the two lines used for CP (3, in bold type). Moreover, that common number is precisely the dependency-number of the assumed antecedent.

Consider another example. This time the conditional we want to prove has a complex consequent. But the other rules make it easy to derive that consequent:

2. $P \rightarrow Q, P \rightarrow R \vdash P \rightarrow (Q \& R)$

{1}	1.	$P \rightarrow Q$	Premise
{2}	2.	$P \rightarrow R$	Premise
{**3**}	3.	P	Assumption for CP
{1,**3**}	4.	Q	1,3 MP [detaching Q]
{2,**3**}	5.	R	2,3 MP [detaching R]
{1,2,**3**}	6.	$Q \& R$	4,5 &I [deriving the consequent]
{1,2}	7.	$P \rightarrow (Q \& R)$	3,6 CP

Note again how MP detaches the two formulas needed to derive the consequent. This time they must be conjoined using &I (line 6). Again, the common number is 3. And again that is just the dependency-number of the antecedent, which is precisely what CP allows us to discharge. The conclusion $P \rightarrow (Q \& R)$ on line 7 now has as dependency-numbers only 1 and 2. It is easy to check back up the proof to see that lines 1 and 2 are just where the original premises lie. Therefore, $P \rightarrow (Q \& R)$ has been proved from $P \rightarrow Q$ and $P \rightarrow R$ alone.

Consider a final example of CP, $(P \& Q) \rightarrow R \vdash P \rightarrow (Q \rightarrow R)$. This time the formula on the right-hand side of the turnstile is a conditional. Its antecedent is just P but its consequent is itself a conditional, $Q \rightarrow R$. Conditionals of this level of complexity and greater levels of complexity are known as **nested conditionals**. CP makes it very easy to handle nested conditionals of this kind in proofs. Recall CP as a strategy: *assume the antecedent then derive the consequent*. To handle nested conditionals we simply keep applying the strategy; we *iterate* CP. So, first assume the antecedent of the whole conditional to the right of the turnstile, i.e. P. But note that the consequent we want to derive is also a conditional: $Q \rightarrow R$. So, we simply assume its antecedent, i.e. Q, and then derive its consequent, R. The rest is perfectly straightforward. Here is the complete proof:

{1}	1.	$(P \& Q) \rightarrow R$	Premise
{**2**}	2.	P	Assumption for CP [antecedent of conclusion]
{**3**}	3.	Q	Assumption for CP [antecedent of $Q \rightarrow R$]
{**2**,**3**}	4.	$P \& Q$	2,3 &I [for MP]
{1,**2**,**3**}	5.	R	1,4 MP [deriving the consequent of $Q \rightarrow R$]
{1,**2**}	6.	$Q \rightarrow R$	3,5 CP [discharging 3]
{1}	7.	$P \rightarrow (Q \rightarrow R)$	2,6 CP [discharging 2]

Note that I simply assume each antecedent in turn, derive the consequent, R, by &I and MP, and then discharge each of the 'extra' dependency-numbers in turn, leaving the conclusion to depend only on its premise, dependency-number 1. You should now study each of the proofs given in this section very carefully, noting down what the sequent is and making a mental note of how each is proved (you should learn CP as a strategy by rote). Next, turn the book over and try to prove each sequent on your own, without cheating! Once you've mastered these proofs you can recap with Box 2.4 and then hone your skills on the sequents in Exercise 2.5.

EXERCISE 2.5

1 Consider the following ten sequents carefully. In each case the formula to the right of the turnstile is a conditional. However, in sequents 4–10, that conditional is either nested or composed of other compound formulas. In constructing a proof of a nested conditional it is important to

BOX 2.4

♦ Each and every sequent with a *conditional conclusion* can be proved using CP.

♦ The *strategy* for CP is simply this: assume the antecedent then derive the consequent.

♦ Reasoning from *assumptions* is *hypothetical*. Reasoning from *premises* is *categorical*.

♦ Assumptions may be *discharged* and CP allows us to discharge assumptions at the point at which we apply that rule, i.e. at the line annotated 'CP'.

Learn the following rule-statements by heart:

♦ *A*: Any formula may be assumed on any line of any proof. The line must be annotated 'A'. The dependency-number of the assumed formula is identical with the line number of the line on which it was assumed.

♦ *CP*: Assume the antecedent. Derive the consequent. Enter the conditional on a line along with 'CP'. The line numbers are both those where the antecedent is assumed and those where the consequent is derived. Discharge the dependency-number of the antecedent. Pool the remaining dependency-numbers and complete the line.

assume each antecedent in turn. In constructing a proof of a complex conditional composed of other compound formulas remember to assume the whole antecedent each time. Prove that each of these sequents is valid in PL. (Again, the numbers in brackets indicate the number of lines in my proof of the sequent.)

1. $P \rightarrow (Q \& R) : P \rightarrow Q$ (5)
2. $(P \& Q) \rightarrow R, P : Q \rightarrow R$ (6)
3. $(P \& Q), (P \& R) \rightarrow S : R \rightarrow S$ (7)
4. $(P \& Q) \rightarrow R : P \rightarrow (Q \rightarrow R)$ (7)
5. $P \rightarrow Q : (P \& R) \rightarrow (R \& Q)$ (7)
6. $P \rightarrow Q : (Q \rightarrow R) \rightarrow (P \rightarrow R)$ (7)
7. $R \rightarrow P, Q \rightarrow S : (P \rightarrow Q) \rightarrow (R \rightarrow S)$ (9)
8. $P \rightarrow Q : (P \rightarrow R) \rightarrow (P \rightarrow (Q \& R))$ (8)
9. $P \rightarrow (Q \rightarrow R) : (S \rightarrow Q) \rightarrow (P \rightarrow (S \rightarrow R))$ (10)
10. $P \rightarrow Q : ((R \& Q) \rightarrow S) \rightarrow ((R \& P) \rightarrow S)$ (10)

VIII
Augmentation: Conditional Proof for Exam Purposes

Earlier I noted the following crucial constraint on the use of CP: in any application of CP, the dependency-number of the antecedent must always be included among the dependencies of *both* lines used for CP. Otherwise, the consequent has not been derived from the antecedent. The proofs we have considered so far have posed no problems in this respect. Looking back, the common dependency-number (of the antecedent) is obvious. But that may not always be the case. For example, suppose I want to prove that: $Q \vdash P \rightarrow Q$. Given that the conclusion to be proved is a conditional, the strategy to crack the proof is CP. So, assume the antecedent and derive the consequent. But note carefully what results:

{1}	1.	Q	Premise
{2}	2.	P	Assumption for CP
	3.	$P \rightarrow Q$	???

First, we introduce the premise, Q. Next, we assume the antecedent, P, for CP. We now want to apply CP. But we cannot: there is no number common to the sets of dependency-numbers for lines 1 and 2. So, the formulas on lines 1 and 2 have nothing whatever to do with one another. To apply CP, the formulas must work together such that we *derive* the consequent from the antecedent. Only in that way will we ever get the desired common dependency-number that will allow us to apply CP. In practice, this stumbling block is easily overcome. We can always get the formulas to work together by &I-ing them together and then &E-ing them apart again, i.e. we exploit the rules for '&' to get the dependency-numbers just as we want them, like so:

$Q \vdash P \rightarrow Q$

{1}	1.	Q	Premise
{**2**}	2.	P	Assumption for CP
{1,**2**}	3.	P & Q	1,2 &I
{1,**2**}	4.	Q	3 &E
{1}	5.	$P \rightarrow Q$	2,4 CP

Note that the formulas on lines 1 and 4 are the same, namely Q. But note also that each occurrence of the sentence-letter Q has different dependency-numbers. The set of dependencies belonging to line 4 is larger than that of line 1 because the formula on line 4 is the result of conjoining Q with P and then taking Q back out of that conjunction. In virtue of those moves, the rules *do* now permit us to apply CP to lines 2 and 4 just because there is a common number, namely, 2. That number is precisely the dependency-number of P and P is precisely the antecedent. It is tempting to describe this useful strategy as 'cheating with CP for exam purposes'. However, according to certain logicians, known as 'relevance' logicians, if one cheats in order to get the right answer by illicit means then this is worse than cheating, for in their view it is just to get the wrong answer.

That is not to say that relevance logicians object to the rules of &I and &E. These rules are not considered problematic in themselves. Rather, the objection is that exploiting the rules in that way legitimates CP on the result. Again, however, this is controversial territory and, for our purposes, this strategy for CP is a perfectly legitimate one. Indeed, it has the solemn and impressive title of **augmentation**, i.e. Q may be *augmented* by P using the & rules. Moreover, the legitimacy of augmentation as a strategy is underpinned by the definition of validity which we stated in Chapter 1. Remember: the truth of the premises is sufficient to guarantee the truth of the conclusion. But if the conclusion follows from *some* true premises and we simply add a few *more* true premises the conclusion must still follow. More generally:

> If a given formula follows from a particular set of formulas then the same formula follows from any augmentation of that set of formulas.

This property of PL is known as **monotonicity**: if a given formula follows from some set of formulas then any addition of further formulas to that set results in a set from which the original formula will still follow. Fortunately for those facing exams in PL, PL is a monotonic logic.

IX
Theorems

CP is the first rule of inference we have encountered which allows us to discharge assumptions. But it will not be the last. Discharge rules are especially useful for proving sequents which represent the traditional laws of logic or logical truths in PL. Why should this be so? Remember that CP is a discharge rule which allows us to discharge assumptions, and that applications of that rule may be iterated. Suppose we only have assumptions to worry about in a given case. Further suppose that we can and do keep iterating CP until there are no assumptions left for the conclusion to depend upon. What are we to say of a conclusion derived in this way? Does it follow from nothing? After all, sequents representing such formulas will have nothing on the left-hand side of the turnstile, i.e. no premises will be involved. So, they will consist just of a turnstile followed by a formula of PL. Certainly, these formulas require no premises, but that does not mean that they follow from nothing. Rather, the mere form of such a formula is always sufficient to guarantee its truth.

Those sequents of PL whose set of premises is the empty set, i.e. which have no premises, are known as **theorems** of PL. The theorems of PL include the traditional laws of logic. And this is unsurprising. Theorems are *logical truths*. Recall the definition of validity: an argument is valid if and only if it is impossible for the conclusion to be *false* when the premises are true. But logical truths are *never* false. No matter how we interpret the sentence-letters involved in any logical truth it still comes out true.

For example, consider the **law of identity**: $\vdash P \rightarrow P$. There is no possible interpretation under which the formula $P \rightarrow P$ is false. If it's raining then it's raining. If the sun is shining then the sun is shining. And so on, ad nauseam. The very *form* of this formula ensures that no matter how we translate or interpret its constituent sentence-letters it will always be true. Hence, the law of identity is a logical truth. It is also a theorem of PL and because the formula is a conditional we prove it by CP. In the following proof of the law of identity note how a well-formed formula in effect takes *itself* as assumption. This is a perfectly respectable application of CP and one which is eminently useful at times:

$\vdash P \rightarrow P$

{1}	1.	P	Assumption for CP
----	2.	$P \rightarrow P$	1,1 CP

Keep this strategy in mind when you attempt Exercise 2.6 (particularly when you try to prove the first theorem).

EXERCISE 2.6

1. Prove that the following sequents are theorems of PL:

1.	$: ((P \rightarrow P) \rightarrow Q) \rightarrow Q$	(5)
2.	$: (P \rightarrow Q) \rightarrow ((Q \rightarrow R) \rightarrow (P \rightarrow R))$	(8)
3.	$: (Q \rightarrow R) \rightarrow ((P \rightarrow Q) \rightarrow (P \rightarrow R))$	(8)
4.	$: P \rightarrow (Q \rightarrow (P \,\&\, Q))$	(5)

X
The Biconditional

While the nature of material implication is still fresh in your mind and the strategy for conditional proof is still echoing round your head and tripping off your tongue, the time is just right to consider another connective: the biconditional.

The very name of this connective, the 'biconditional', at once suggests a connection with conditionals and that is entirely appropriate. However, when it comes to asserting a biconditional, rather than merely asserting 'if P then Q' I am asserting 'P *if* and *only if* Q'. Let's unpack this a little using an example. What exactly do I imply when I assert that 'Logic students pass their logic exams if and only if logic students do their homework'? Well, surely I imply two things, i.e. I imply *both* that logic students pass their logic exams *if* they do their homework *and* that logic students pass logic exams *only if* they do their homework. The first of these implications, 'Logic students pass logic exams if they do their homework', is just equivalent to the conditional: 'If logic students do their homework then logic students pass their exams.' And it's easy to see that the second implication simply reverses that conditional, i.e. it is equivalent to: 'If Logic students pass their exams then they have done their homework!'

This point is even easier to see formally. Consider the assertion: 'P if and only if Q'. This implies both 'P if Q' and 'P only if Q'. 'P if Q' is equivalent to 'If Q then P' while 'P only if Q' is equivalent to 'If P then Q'. It follows logically that to assert 'P if and only if Q' is to assert *both* 'If Q then P' *and* 'If P then Q' or, equivalently, 'If P then Q' and 'If Q then P'. In virtue of its equivalence to the *conjunction* of these two conditionals, we refer to '↔' as the *bi*conditional.

Moreover, because 'P if and only if Q' is equivalent to 'if P then Q and if Q then P', i.e. in formal terms to $(P \rightarrow Q) \& (Q \rightarrow P)$, we can always rewrite the former as the latter whenever it occurs and be sure that the truth-value is preserved. For exactly that reason, a number of systems of propositional logic make the equivalence explicit and *define* the biconditional in terms of that equivalence. Such a definition is considered to make the meaning of '↔' explicit in terms of the equivalence. Therefore, whenever a biconditional features as a premise (or anywhere else for that matter) in a proof, such formal logicians simply rewrite the biconditional formula as the conjunction of the relevant two conditionals on the next line and annotate the new line with the original line number and 'Df. ↔', i.e., 'Df' for 'Definition'. The dependency numbers of the new line formed in this way are exactly those of the old line. Thereafter, the conjunction of the two conditionals is just handled using the original rules as, for example, in the following proof:

1. $P \leftrightarrow Q \vdash Q \leftrightarrow P$

{1}	1.	$P \leftrightarrow Q$	Premise
{1}	2.	$(P \rightarrow Q) \& (Q \rightarrow P)$	1 Df. ↔
{1}	3.	$P \rightarrow Q$	2 &E
{1}	4.	$Q \rightarrow P$	2 &E
{1}	5.	$(Q \rightarrow P) \& (P \rightarrow Q)$	4,3 &I
{1}	6.	$Q \leftrightarrow P$	5 Df. ↔

Given $P \rightarrow Q$ on a line (such as line 3) and $Q \rightarrow P$ on another (such as line 4) it is perfectly valid to infer that $P \leftrightarrow Q$. After all, the latter is equivalent to the former pair of conditionals.

However, certain systems of propositional logic do make explicit an introduction-rule for '↔',**↔I**. The rule-statement is just as we'd expect:

> ↔I: Given a pair of conditionals on two lines such that the antecedent of the first is the consequent of the second and the consequent of the first is the antecedent of the second you may write '↔' between the two formulas, antecedent and consequent, on a new line. The new line is annotated with the line numbers of both lines used and '↔I'. The dependency-numbers of the new line are all of those of both lines used.

On this approach, we must also go on to make explicit an elimination-rule for '↔', **↔E**. Unsurprisingly, we eliminate the biconditional by rewriting that formula in terms of the conjunction of the pair of conditionals to which the formula is equivalent. So:

> ↔E: Given a biconditional on a line we may rewrite that formula as the conjunction of the relevant pair of conditionals on the next line of proof. The new line is annotated with the line number of the old line and takes as dependency-numbers all and only those of the old line.

This describes just the kind of inference we made on line 2 of the previous proof, but given that, what's the point in adopting introduction and elimination-rules for '↔'? First, the fact that such rules can be specified keeps faith with my earlier claim that there is an introduction-rule and an elimination-rule for each connective. Moreover, the introduction-rule will always save us one line of proof against the definition-based approach, just because it does not require us to make explicit the conjunction of the pair of conditionals as the definition-based approach does. So, we can abbreviate our six-line proof to a mere five-line proof. As you will see, there is much fun to be had in trying to find the shortest proof; not least when you succeed (as you undoubtedly will) in finding a shorter proof than the logic teacher – myself included!

1.′ $P \leftrightarrow Q \vdash Q \leftrightarrow P$

{1}	1.	$P \leftrightarrow Q$	Premise
{1}	2.	$(P \rightarrow Q)\ \&\ (Q \rightarrow P)$	1 ↔ E
{1}	3.	$P \rightarrow Q$	2 &E
{1}	4.	$Q \rightarrow P$	2 &E
{1}	5.	$Q \leftrightarrow P$	4,3 ↔I

Exercise 2.7 gives you the opportunity to construct some proofs for yourself. Before attempting the exercise you should first establish whether the system of formal logic you are using involves a definition-based approach or explicit introduction/elimination rules.

Note: Please note carefully that those who are only concerned to master the mechanics of proof-construction can safely ignore the final section of this chapter and jump ahead to the beginning of the next; once they have completed Exercise 2.7!

EXERCISE 2.7

1 Prove that the following are valid sequents of PL:

1.	$P, P \leftrightarrow Q : Q$	(5)
2.	$P \& (P \leftrightarrow Q) : P \& Q$	(7)
3.	$(P \& Q) \leftrightarrow P : P \rightarrow Q$	(7)
4.	$P \rightarrow Q : (Q \rightarrow P) \rightarrow (P \leftrightarrow Q)$	(4)
5.	$P \rightarrow (Q \leftrightarrow R) : (P \& Q) \rightarrow R$	(9)
6.	$P \leftrightarrow Q, Q \leftrightarrow R : P \leftrightarrow R$	(17)

XI
Entailment and Material Implication

In Section VII we took some time to distinguish entailment carefully from material implication in PL. As we noted, the turnstile which represents entailment and the arrow which represents material implication belong to different languages, i.e. the turnstile belongs to the metalanguage while the arrow belongs to the object language. That distinction is an important one, but students of first logic are often intrigued by these two notions and many will wonder if the two are not unrelated none the less. In fact, the two are related and, having spent so much time considering conditionals, biconditionals and conditional proof, it may be insightful to come clean at this point about the nature of the connection.

The relation between the two can be seen vividly in what is known as the **deduction theorem for PL**. This particular theorem is not a theorem *of* PL but is rather a theorem *about* PL, i.e. it is a *meta*theorem. Intuitively, the theorem, which was first proved by the Polish logician Alfred Tarski in 1921, asserts that if a set of PL formulas together with a particular PL formula entails another formula then the original set of formulas entails a conditional with the first particular formula as antecedent and the second as consequent. This is undoubtedly clearer in formal terms. But now we need a way of representing both a set of PL formulas and an arbitrary particular PL formula. For the first purpose, let 'X' be any set of PL formulas. For the second purpose, we introduce a new set of variables: 'A', 'B', 'C' and so on, each of which simply stands for any PL formula. These variables are names of PL formulas and so are not themselves formulas of PL. In other words, these variables belong to the metalanguage and are as such **metalinguistic variables**. With this much formal vocabulary at hand, the deduction theorem for PL can simply be stated as follows:

If X, A $\vdash$ B then X $\vdash$ A $\rightarrow$ B

A little reflection should now reveal the nature of the connection between the two notions we are concerned with here. Technically, however, the theorem is not properly stated here. Again, the problem is just the one we noted in Section VII. The symbols involved belong to different languages, i.e. the variables and the turnstile belong to the metalanguage while the arrow belongs to the object language. Logicians usually overcome this kind of problem by placing such hybrid sentences in *corner* or *quasi* quotes thus:

$\ulcorner$ If X, A $\vdash$ B then X $\vdash$ A $\rightarrow$ B $\urcorner$

Such parentheses can perturb students and may distract from the content of the sentences the quotes enclose. So, in what follows I will not make such quotes explicit, i.e. I will 'continue to speak with the vulgar'.[7] Where necessary, the reader might supply the relevant quotes. Finally, note carefully that the deduction theorem for PL also holds, as I might put it, in the other direction, i.e.:

If X $\vdash$ A $\rightarrow$ B then X, A $\vdash$ B

Therefore, we can clarify the relation between the turnstile and the arrow in terms of the following equivalence:

A $\vdash$ B if and only if $\vdash$ A $\rightarrow$ B

Notes

1 Wittgenstein, Ludwig, [1961], *Tractatus Logico-Philosophicus*, London, Routledge & Kegan Paul, Remark 4.116.
2 The following set of rules is based on a set designed by John Slaney.
3 On this point see Read, Stephen, [1988], *Relevant Logic*, Oxford, Blackwell, pp. 54–5. The distinction was first pointed out to me by John Slaney.

4 See, for example, Hunter, Geoffrey, [1971], *Metalogic: An Introduction to the Metatheory of Standard First-Order Logic*, London and Basingstoke, Macmillan, Part One, Section One.
5 Curry, Haskell B., [1976], *Foundations of Mathematical Logic*, New York, Dover Publications, p. 28. To many readers this account of the metalanguage will seem rather vague. However, as Curry also points out: 'This may seem vague, but in that vagueness we are no worse off than in any other field of study. Every investigation in any subject whatsoever, must presuppose that same datum' (p. 28).
6 See below, Chapter 3, Section VI.
7 Boolos, George S., and Jeffrey, Richard C., [1996], *Computability and Logic*, third edition, Cambridge, Cambridge University Press, p. 98.

3
How to Prove that You Can Argue Logically #2

3
How to Prove that You Can Argue Logically #2

I
Conditionals Again

To connect P with Q using the arrow '$\rightarrow$' in PL is to form the conditional $P \rightarrow Q$. In any conditional, the formula occurring before the arrow is known as the *antecedent* of the conditional, and the formula occurring after the arrow is the *consequent*. So, in the case in question, P is the antecedent and Q the consequent. Here, I want to consider four different inferences which people may make using conditionals. All four inferences are commonplace in ordinary discourse – despite the fact that two of them are obviously invalid! To understand fully the four inferences in question we first need to understand the nature of negation. It is intuitive to think of negation as working by simply reversing truth-values. In other words, if the sentence 'I am now reading this book' is true then the negation of that sentence 'I am not now reading this book' must be false. Similarly, if the sentence 'It's raining in Old Aberdeen today' is true then the negation of that sentence, namely 'It's not raining in Old Aberdeen today', will, again, be false. But this is only half of the story. If the negation of a given sentence is true then the original sentence must be false, i.e. if the sentence 'I am not now reading this book' is true then the sentence 'I am now reading this book' is false. Equally, if the sentence 'It's not raining in Old Aberdeen today' is true, then the sentence 'It's raining in Old Aberdeen today' is false.

These intuitions about negation are central to the account of negation given in PL. The kind of negation involved in PL is called **classical negation**. Classical negation is *denial* and the net effect of negating a sentence is precisely to reverse the truth-value of that sentence. To deny that P is true is to assert that ~P is true. To deny that ~P is true is to assert that P is true. We can easily tabulate this state of affairs for an arbitrarily chosen formula, say P, and represent the reversal of truth-value which negation effects as follows:

P	~P
True	False
False	True

With these points about negation in mind, we can now consider four basic inferences which people may make. Note carefully the way in which these differ just in terms of whether the antecedent or the consequent is affirmed or denied in the course of the inference. Crucially, note also how this affects the validity of each inference.

First, recall the original Blind Lemon Jefferson argument and its formal counterpart in PL. It is itself a prime example of one of the four inferences, i.e. recall the natural language argument:

1. If it's a Blind Lemon Jefferson album then it's a blues album.
2. It's a Blind Lemon Jefferson album.

Therefore,

3. It's a blues album.

As you know, this argument can be formally represented in PL as $P \rightarrow Q, P : Q$. In this case, given $P \rightarrow Q$, we assert the antecedent, P, and then infer the consequent, Q. In more traditional logical terms, we *affirmed* the antecedent P and then concluded that Q. Does Q logically follow? It certainly does. Indeed, just this kind of reasoning is represented by one of our rules of inference, modus ponens. So **affirming the antecedent** is a perfectly valid form of inference.

But suppose that, rather than affirm the antecedent, we affirm the consequent instead, i.e. that, given $P \rightarrow Q$, we affirm Q and infer P. Is this inference a valid inference? It certainly is not! To affirm the consequent is to commit a traditional fallacy, i.e. to reason in a perfectly invalid way. The following counterexample demonstrates that this so:

1. If all cats are black then Tiffin is black.	$P \rightarrow Q$
2. Tiffin is black.	Q
Therefore,	———
3. All cats are black.	P

Hence, **affirming the consequent** is invalid. (You might remember that this particular example featured earlier in Exercise 1.1. In the course of that exercise you should have proved yourself that the form of argument involved is indeed an invalid one.)

The two remaining inferences both involve negation and hence denial rather than assertion or affirmation. The first case involves **denying the antecedent**. For example, consider the following sequent: P → Q, ~P : ~Q. Is this a valid sequent of PL? Again, it certainly is not. Here is a counterexample:

1. If all cats are black then Zebedee is black. P → Q
2. It's *not* the case that all cats are black. ~P

Therefore, ———

3. It's not the case that Zebedee is black. ~Q

Again, we met this example in Exercise 1.1 and so you should already have proved there that this particular argument is indeed an instance of an invalid logical form. Hence, denying the antecedent is fallacious.

One last possibility remains, namely, that of **denying the consequent**. Is the following sequent valid in PL, P → Q, ~Q : ~P? This sequent is indeed valid in PL. And again, we frequently make inferences of this type in natural language. For example:

1. If it's a Blind Lemon Jefferson album then it's a blues album. P → Q
2. It's *not* a blues album. ~Q

Therefore, ———

3. It's *not* a Blind Lemon Jefferson album. ~P

This particular argument is certainly a valid argument and it is also an instance of a valid logical form of argument. Moreover, like the first type of inference we considered in this section, inferences of this last type also have a traditional name, not modus ponens but modus *tollens*. Further, we can easily incorporate this mode of inference into our proof-theory for PL simply by adding a new rule. Logically enough, the new rule is called **modus tollens** and its annotation is just '**MT**'. Here is the rule-statement in full:

> MT: Given a conditional on one line and the negation of its consequent on another, infer the negation of the antecedent. Annotate the new line with the line numbers of both lines used and 'MT'. The dependency-numbers of the new line are all those of both lines used.

Consider the simplest possible example of the rule MT in action in the proof of the sequent P → Q, ~Q ⊢ ~P:

$P \rightarrow Q, \sim Q \vdash \sim P$

{1}	1.	$P \rightarrow Q$	Premise
{2}	2.	$\sim Q$	Premise
{1,2}	3.	$\sim P$	1,2 MT

MT works well with our existing stock of rules. For example, consider the proof of the following sequent $P \rightarrow Q \vdash \sim Q \rightarrow \sim P$. Because the main connective in the conclusion is a conditional the strategy for proof is CP. So, assume the antecedent, ~Q, and try to derive the consequent, i.e. ~P. Given MT, ~P is easy to derive:

{1}	1.	$P \rightarrow Q$	Premise
{**2**}	2.	$\sim Q$	Assumption for CP
{1,**2**}	3.	$\sim P$	1,2 MT
{1}	4.	$\sim Q \rightarrow \sim P$	2,3 CP

In the next case (Exercise 3.1), MT works together with MP to give us the desired conclusion. Try this one yourself.

EXERCISE 3.1

1 Construct a proof of the following sequent (as ever, the number in brackets indicates the number of lines in my proof):

$P, P \rightarrow (Q \rightarrow R), \sim R : \sim Q$ (5)

II
Conditionals, Negation and Double Negation

MT is a useful and intuitive rule which works well with our existing stock of rules. Moreover, it is the first rule of inference we have considered which involves negation. Indeed, because any conclusion inferred by MT is always and only a negation you might think that MT is something of an introduction-rule for negation. But this is not the case. Certainly, MT yields a negated conclusion but only if we already have a negated formula, i.e. MT transfers negation from consequent to antecedent. So, although MT allows us to infer a negated conclusion for the first time, it is not the introduction-rule for negation. The actual introduction-rule for negation has the rather solemn Latin title of 'reductio ad absurdum'. We will consider this rule in detail a little later, in Section VII of the present chapter.

While it is not appropriate to consider the introduction-rule for negation just yet, we can usefully consider the elimination-rule for negation at this stage (recall that every connective has both an introduction-rule and an elimination-rule). So, what of negation-elimination? In order to appreciate fully the force of this particular rule we must explore the nature of classical negation a little further. In the process, we will also make clear that MT is even more useful than it has appeared to date.

The question about negation which is of particular interest to us now is just: what happens when we negate a negation? Logically enough, when we negate a negation we produce what's called a **double negation**. Quite simply, this is achieved by putting another negation sign in front of the original negation sign. Classical negation is therefore desperately easy to use. To negate P we simply connect the negation sign to it, i.e. ~P. To negate ~P, again, just connect the negation sign to that negated formula, i.e. ~~P.

> To negate any formula in PL simply place a negation sign in front of it. If the formula is already a negation, say, ~P, negating it produces a double negative, in this case ~~P.

But suppose that I wanted to negate ~~P. Could I just prefix a third negation sign to it? We could indeed adopt that practice if we liked but, in an important sense, anything beyond a double negative is really totally redundant. Why? Remember the point we noted earlier: classical negation reverses truth-value. So, consider P. If P is true then ~P is false. And, if ~P is true, then P is false. Now consider the negation of ~P, i.e. ~~P. If it is true then ~P is false. But if ~P is false then P must be true. So, adding a single negation to a formula reverses the truth-value of that formula. And negating that negation simply reverses the truth-value back to 'true'. It follows that when ~~P is true, so is P. So, although we could continue to prefix negation signs to double negatives, because we recognise only two truth-values, any and every string of negation signs beyond two must reduce back to one of those two truth-values. Even numbers of negation signs would just reduce to 'true' while odd numbers would just reduce to 'false'.

Moreover, the point is not simply that if ~~P is true then so is P. In fact, the opposite is also true and the two are logically equivalent. This is intuitive and is easily shown. If I asked: 'Is it raining?' and you, having just read your logic book, replied: 'Well, it's certainly *not* the case that it *isn't* raining', I might feel you were being a bit awkward or pedantic but I could easily work out from what you said that it *was* raining. Here, we might say, two negatives make a positive. The elimination-rule for negation embodies just this intuition. To keep things crystal clear, the rule is known as **double negation-elimination (DNE)**. This rule allows us to collapse any double-negative back into the original positive formula. Here is the rule in full:

> DNE: Given the double negation of a formula on any line of proof you may write the original un-negated formula on a new line. Annotate the new line 'DNE' together with the line number of the line containing the double negative. The dependency-numbers of the new line are identical with those of the old line.

Again, this is only half of the story. Remember, if P is true then ~P is false. But if ~P is false then *both* P *and* ~~P must be true. For example, if 'It's raining' is true then 'It is not raining' must be false. So, the negation of that negation must be true, i.e. it must also be true that it's not the case that it isn't raining! In short, any formula is logically equivalent to its double negation. So, given any un-negated formula, we can always validly infer the double negation of that formula. This is just the *converse* of the point we made above about double negation-elimination. There we noted that given a double negative we could validly infer the original un-negated formula. Here, we note that, given an un-negated formula, we can always validly infer or *introduce* its double negative. Hence, we can adopt another rule as regards double negatives: **double negation-introduction (DNI)**. Here is the rule in full:

> DNI: Given an un-negated formula on any line of proof you may write the double negative of that formula on a new line. Annotate the new line 'DNI' together with the line number of the line containing the original formula. The dependency-numbers of the new line are identical with those of the old.

Let's have a quick look at these rules in practice. DNE will allow us to prove immediately that ~~P ⊢ P as follows:

{1} 1. ~~P Premise

{1} 2. P 1 DNE

DNI allows us to make precisely the same move in the opposite direction, i.e. to prove that P ⊢ ~~P:

{1} 1. P Premise

{1} 2. ~~P 1 DNI

You will quickly find that MT works particularly well with both DNE and DNI in the proof-theory of PL. For example, DNI can often be used to provide a negation when we need one in order to apply MT. The usefulness of DNI for enabling the application of MT is clearly illustrated in the following proof:

P → ~Q, Q ⊢ ~P

{1}	1.	P → ~Q	Premise
{2}	2.	Q	Premise
{2}	3.	~~Q	2 DNI
{1,2}	4.	~P	1,3 MT

Note carefully that we cannot simply infer the conclusion ~P from the formulas on lines 1 and 2 just as they stand. Why not? Just because Q on line 2 is not a negated formula. Therefore, it certainly is not the negation of the consequent of the conditional on line 1. However, Q is logically equivalent to a negated formula, namely, its own double negation ~~Q. That formula is the negation of the consequent of the conditional on line 1. (Remember: the negation of a given formula always has one more negation sign than that given formula.)

Therefore, we can now infer the negation of the antecedent of the conditional on line 1 by MT and that gives us the desired conclusion, ~P. Those facing exams in PL should note this enabling application of DNI very carefully. Many logicians delight in deducting marks from students who fail to make use of DNI to enable the proper application of MT. Although this is a relatively minor logical sin it is a logical sin none the less. Further, also note carefully that MT works equally well with DNE. This time, it's not that DNE enables the application of MT. Rather, DNE allows us to derive a positive conclusion when MT has supplied us with a double negative. Again, we must make each step explicit; particularly the DNE step. Consider the following proof:

~P → Q, ~Q ⊢ P

{1}	1.	~P → Q	Premise
{2}	2.	~Q	Premise
{1,2}	3.	~~P	1,2 MT
{1,2}	4.	P	3 DNE

Here, the important point is that we cannot infer line 4 directly from lines 1 and 2 using MT. The conclusion we want to prove is P. But MT does not give us P on line 3. It only gives us ~~P. So we must go on to use DNE to collapse that double negative back into the positive formula P.

Use Box 3.1 to recap the chapter so far.

BOX 3.1

- *Negation* reverses the truth-value of any sentence to which it is applied.
- To negate a formula in PL simply connect the negation sign '~' to the formula.
- The negation of a formula always has *one more* negation sign than the original formula.

So far, three rules of inference allow us to manipulate negated formulas:

- *MT*: Given a conditional on a line and the negation of its consequent on another, infer the negation of the antecedent. Annotate the new line with the line numbers of both lines used and 'MT'. The dependency-numbers of the new line are all those of both lines used.
- *DNE*: Given the double negation of a formula on any line of proof you may write the original un-negated formula on a new line. Annotate the new line 'DNE' together with the line number of the line featuring the double negative. The dependency-numbers of the new line are identical with those of the old line.
- *DNI*: Given an un-negated formula on any line of proof you may write the double negative of that formula on a new line. Annotate the new line 'DNI' together with the line number of the line featuring the original formula. The dependency-numbers of the new line are identical with those of the old line.

Exercise 3.2 contains ten proofs for you to try yourself. The first proof simply involves the double negation rules without MT. The following five proofs can all be proved quite straightforwardly by various combinations of MT and the rules for double negation. As ever, you will have to ensure that in addition to the relevant conditional you also have the *negation* of the consequent, if you are to apply MT properly.

The remaining four proofs are considerably more difficult. Each of these proofs requires that you first *derive* the relevant conditional and *then* apply MT. In each case, you will have to derive the conditional by augmenting the premises. In these cases it is crucial to identify clearly the formula you ultimately want to derive, i.e. the conclusion of the particular sequent you are trying to prove. If that formula is a sentence-letter try assuming the *negation* of that sentence-letter. If the formula you want is *already* a negated sentence-letter try assuming the original sentence letter *un-negated*. Finally, make sure that the conditional you derive has as its *antecedent* the formula which you ultimately want to negate by MT.

Before attempting this particular exercise it is crucial to appreciate that, at this stage, mastery of the first six proofs alone is perfectly adequate. The remaining four proofs really are very difficult. But they are challenging and the attempt to crack them will sharpen your proof-theoretical abilities. Later in this chapter we will go on to consider another rule of inference which enables those same sequents to be proved much more easily. So, it is extremely unlikely that you will ever be asked to attempt these proofs without that new rule. If you succeed in constructing the proofs at this stage reward yourself enormously. But the point is not to be put off if you do not succeed. Consider the hints about strategy in the preceding paragraph carefully before you begin, and good luck!

EXERCISE 3.2

1 Prove that the following sequents are valid in PL:

1. $\sim\sim(P \,\&\, Q) : \sim\sim(Q \,\&\, P)$ (6)
2. $\sim P \rightarrow \sim Q : Q \rightarrow P$ (6)
3. $: (P \rightarrow Q) \rightarrow (\sim Q \rightarrow \sim P)$ The principle of transposition (5)
4. $Q \rightarrow R : (\sim Q \rightarrow \sim P) \rightarrow (P \rightarrow R)$ (9)
5. $(P \,\&\, Q) \rightarrow \sim R : R \rightarrow (P \rightarrow \sim Q)$ (11)
6. $P : [(\sim(Q \rightarrow R) \rightarrow \sim P)] \rightarrow [(\sim R \rightarrow \sim Q)]$ (9)
7. $P, \sim Q : \sim(P \rightarrow Q)$ (6)
8. $P, \sim P : Q$ (8)
9. $: \sim P \rightarrow (P \rightarrow Q)$ The law of Duns Scotus (10)
10. $P \rightarrow \sim P : \sim P$ (11)

III Introducing Disjunction

We are now in a position to consider the rules which govern 'v', the formal counterpart of 'or', in one sense of that term anyway, in natural language. As ever, there is an introduction-rule for this connective, which brings 'v' into a line of proof, and an elimination-rule, which takes the connective out of a line of proof. Let's consider the introduction-rule, **vIntroduction** or **vI**, first.

In essence, given any formula on any line of proof, vIntroduction allows us to infer immediately the disjunction of that formula with any other well-formed formula we care to choose. In other words, vIntroduction allows us to take any formula from any line of proof, to write that formula on a new line together with 'v' *and* to complete the disjunction with absolutely any other well-formed formula we might like the look of.

You may feel that this is remarkably generous on the part of the logician, for we seem to be getting something for nothing here. After all, it's entirely up to us which formula we pick to complete the disjunction. In fact, the generosity is only apparent. Logically speaking, the disjunctive conclusion is *weaker* than the original premise. We can think of **logical strength** as a relation between formulas measured in terms of the logical consequences of those formulas. If each entails the other, then the two have the same degree of logical strength. If the first entails the second but the second does not entail the first then the first is logically stronger than the second. So, consider the formulas P and P v Q. P entails P v Q and we can always use vIntroduction to infer P v Q from P. But P v Q does not entail that P. Clearly, P v Q might well be true even if P is *not* true, i.e. just in the case where Q is in fact true. So, P v Q is logically weaker than P and the logician's generosity is, indeed, only apparent.

It follows logically that all that is required for a disjunction to be true is just that *at least one* of its disjuncts is true, e.g. even if P is false P v Q can still be true, as long as Q is true. Of course, both P and Q might be true and the sense in which we use 'v' in PL certainly *includes* the possibility that both disjuncts are true. For that reason, the formal logician refers to our use of 'v' as the **inclusive** sense of disjunction. Logically enough, inclusive disjunction contrasts with **exclusive** disjunction, which excludes the possibility that a disjunction is true when both its disjuncts are true. Either way, a disjunction is true if even *one* of its disjuncts is true.

It is precisely this fact that underwrites the validity of the type of inference we make using vIntroduction. For example, it's raining in Old Aberdeen today so 'It's raining in Old Aberdeen today' is true. But suppose I now infer that: '*Either* it's raining in Old Aberdeen today *or* the moon is made of green cheese.' Now, the moon is not made of green cheese, but because the first disjunct is true the whole disjunction is true anyway. So it really doesn't matter which sentence we choose to complete the disjunction. If the original premise is true then any conclusion derived from it by vIntroduction must also be true. Therefore, inferences made using vIntroduction are indeed truth-preserving, valid inferences.

Finally, note carefully that when I formed the disjunction 'Either it's raining in Old Aberdeen today *or* the moon is made of green cheese' from the premise 'It's raining in Old Aberdeen today' I did so by writing the 'or' *after* that sentence followed by the second disjunct. In fact, I might equally have written the 'or' *before* the premise and in effect reversed the order of the disjuncts like so: 'Either the moon is made of green cheese *or* it's raining in Old Aberdeen today.' The fact that the order of the disjuncts is reversed makes no difference to the truth-value of the conclusion. Hence, it should be no surprise to find that in PL we can introduce 'v' *to the right of* a given formula, i.e. *after* the original formula, and that we can equally well introduce 'v' *to the left of* a given formula, i.e. *before* the original formula. For obvious reasons, I will refer to these as **right-handed vIntroduction** and **left-handed vIntroduction** respectively.

Here is the rule in full:

> vI: Given a formula on a line of proof you may infer the disjunction of that formula with any other well-formed formula on a new line of proof. Annotate the new line with the line number of the old line and 'vI'. The dependency-numbers of the new line are identical with those of the old line.

Consider two simple proofs which illustrate the distinction between right-handed and left-handed vI. First, P ⊢ P v Q:

P ⊢ P v Q

{1}	1.	P	Premise
{1}	2.	P v Q	1 vI

Imagine taking the formula P from line 1, putting it on a new line, introducing 'v' to the right of that formula and then choosing Q to complete the disjunction on the new line. This is right-handed vI. Compare the following proof of P ⊢ Q v P:

P ⊢ Q v P

{1}	1.	P	Premise
{1}	2.	Q v P	1 vI

This time imagine taking the formula P from line 1 and putting it on a new line, introducing 'v' to the left of that formula and choosing Q to plug the gap *before* the 'v' to complete the disjunction. This is left-handed vI.

These simple proofs illustrate the bare minimum we can infer using vI. But we are by no means confined to introducing atomic formulas. Remember: we can disjoin the original formula with *any* other formula, no matter how complex. The two proofs also show that vI can be useful when we want to derive a disjunction as conclusion, i.e. when the formula on the right-hand side of the turnstile has 'v' as its main connective. You should not be surprised to learn that logic exams rarely contain such easy proofs and, as we shall see, arriving at a disjunction as conclusion usually requires a bit more work.

Again, vI works well with the existing set of rules allowing us to infer disjunctions as required. Look closely at the following sequent (P v Q) → R : (P → R) & (Q → R). In this case, the conclusion we want to derive is a conjunction of which each conjunct is a conditional. So we must derive each conditional separately and then use '&I' to conjoin them. As ever, we derive each conditional using CP. Hence, in each case, assume the antecedent and try to derive the consequent. But note that the consequent of *both* conditionals is R. How are we to derive R? As you will see from the first few steps of the following proof (Exercise 3.3), this is precisely where we can enlist the help of 'vI'. However, *all* you will see of that proof is the first few steps. I have deliberately left the proof incomplete. My version of the proof has ten lines. Fill in the gaps in the 'gappy proof' to complete my version. Where vI is involved specify whether the application is right- or left-handed.

EXERCISE 3.3

1 Complete the following proof:

(P v Q) → R ⊢ (P → R) & (Q → R)

{1}	1.	(P v Q) → R	Premise
{2}	2.	P	Assumption for CP
{2}	3.	P v Q	2 vI (right-hand)
{1,2}	4.		
	5.		2,4 CP
	6.		Assumption for CP
{6}	7.	P v Q	
	8.		1,7 MP
{1}	9.		6,8 CP
	10.	(P → R) & (Q → R)	

IV
vElimination

In the previous section we considered the introduction-rule for 'v' in PL. There we saw how to use vI to arrive at a disjunction as conclusion and noted that vI can be useful within proofs in conjunction with the other rules. In each of these cases we were only really concerned with reasoning *to* a disjunction either as conclusion or as a helpful step in a proof of something else. What of reasoning *from* a disjunction? When can we legitimately infer a conclusion from a disjunction? This is a question about disjunctions not as conclusions but as *premises*. We already know that under certain circumstances we can infer a disjunction as conclusion using vI; but under which circumstances can we infer a conclusion from a disjunction as a premise? Well, we can already draw a number of valid inferences from disjunctive formulas using the existing rules of inference in PL. For example, we can use the rule we've just considered, vI, to prove in a mere two lines that P v Q ⊢ (P v Q) v (R v S):

P v Q ⊢ (P v Q) v (R v S)

{1}	1.	P v Q	Premise
{1}	2.	(P v Q) v (R v S)	1 vI (right-hand)

However, inferences of this kind (inferences from a well-formed formula to the disjunction of that formula with another well-formed formula) can be made from *any* kind of well-formed formula in PL, i.e. such inferences do not depend upon the *disjunctive* logical form of the premise. Moreover, the existing rules do not permit us to represent every kind of inference which can be validly drawn from a disjunction and so certain obviously valid sequents cannot yet be proved, e.g. P v Q ⊢ Q v P. How are we to demonstrate the validity of such sequents? The answer to this question lies in the elimination-rule for disjunction: **vElimination, vE** for short. If the formal logician seemed very generous in the previous section, apparently giving us something for nothing in applications of vI, the same logician may now seem rather mean-spirited, penurious and pedantic. Why? Imagine asking the formal logician: when, in general, can I legitimately infer a conclusion from a disjunction in virtue of its having a disjunctive logical form? The reply will be as follows:

> 'You may infer a conclusion from a disjunction if you can prove to me *first* that, assuming the first disjunct, the conclusion follows from that disjunct *and second* that, assuming the second disjunct, the same conclusion also follows from that disjunct.'

If one and the same conclusion follows from *both* disjuncts then we can certainly say that it follows from *either* and, therefore, that the same conclusion follows from the disjunction itself. This is certainly perfectly intuitive but a bit laborious in practice, for in any such derivation from a disjunction we must derive the conclusion from each disjunct in the course of the proof *before* we can derive that same conclusion from the original disjunction. This time then, we certainly do not get something for nothing. As we shall see, everything must be spelled out step by step.

Recall the formal logician's claim: we can infer a conclusion from a disjunction as premise in virtue of its disjunctive logical form only if we can show *both* that (1) having assumed the first disjunct, the conclusion can be shown to follow from it *and* that (2) having assumed the second disjunct, the conclusion can also be shown to follow from it. This is precisely how vElimination works in practice as a strategy for proof.

First, we begin with a *disjunction as premise* on a line of proof. Next, we *assume the first disjunct*. Third, we use any and every available rule and formula to derive the *conclusion from the first disjunct*. Fourth, we *assume the second disjunct*. Fifth, we derive the *conclusion from the second disjunct*. Finally, we simply *repeat* the conclusion on the next line of proof. That final line is the line which we annotate 'vE'.

We must also take care to keep track of each of the lines we use in the process of vElimination. So, in addition to 'vE', we annotate the final line with *five* numbers. What are these? Just look at the first five italicised phrases in the paragraph above. These are the five numbers. Just to spell things out:

> *The first number is*: the line number of the original disjunction as premise.
> *The second number is*: the dependency-number of the first disjunct assumed.
> *The third number is*: the line-number of the conclusion derived from the first disjunct.
> *The fourth number is*: the dependency-number of the second disjunct assumed.
> *The fifth number is*: the line number of the conclusion derived from the second disjunct.

Finally, the beauty of vE is that it is a *discharge* rule. Remember that we entered each disjunct into the proof as an *assumption*. So, our reasoning here was ultimately hypothetical. Hence, just as CP allows us to discharge any assumption we make for it, so vE allows us to discharge the dependency-numbers of both assumed disjuncts. Having done so, we are free to infer the conclusion from the original disjunctive premise and its dependency-numbers; together with anything else we might have used other than the two assumed disjuncts. And this is quite natural. Remember, if we can derive the conclusion from the assumption of the first *and* second disjuncts then we can infer that conclusion from the original disjunctive premise.

Let's have a look at vE in action in the proof of the sequent which frustrated us earlier, i.e. P v Q ⊢ Q v P. Before we begin, note carefully that the premise is the disjunction P v Q. The conclusion we want to derive is also a disjunction, i.e. Q v P. But if we want to derive that conclusion from the premise P v Q we will first have to assume each of the disjuncts P and Q and derive that same conclusion from each of them. So:

{1}	1.	P v Q	Premise
			– enter the disjunctive premise
{**2**}	2.	P	Assumption
			– assume the first disjunct
{2}	3.	Q v P	2 vI
			– derive the conclusion from the first disjunct
{**4**}	4.	Q	Assumption
			– assume the second disjunct
{4}	5.	Q v P	4 vI
			– derive the conclusion from the second disjunct
{1}	6.	Q v P	**1, 2, 3, 4, 5 vE**
			– now apply vE and discharge

Note the pattern of the proof: enter the disjunction as a premise as usual. Assume the first disjunct. Try to derive the conclusion from it. (Here we used left-hand vI to derive the conclusion from the first disjunct straight away on line 3.) Next, assume the second disjunct and derive the conclusion from it. (Here we used right-hand vI to derive the same conclusion on line 5.) Now repeat the conclusion on a new line, annotate that line 'vE' and check back through the proof for the five numbers. Remember to discharge the dependency-numbers of both assumed disjuncts and to check carefully to see whether any formulas other than the disjuncts were used (here none was). Finally, we can infer the conclusion from the original disjunction and its dependencies. You might well find it useful to annotate each line as appropriate to the right of the rule-annotation, as I did, when you construct your own proofs. This practice allows you (and examiners!) to keep track of exactly what you are up to.

The strategy for any vE is always the same and, like CP, it is one which is well worth learning by rote: assume the first disjunct, derive the conclusion. Assume the second disjunct, derive the conclusion. Repeat the conclusion on a new line. Annotate the new line vE. Select the five numbers. Next, discharge the dependency-numbers of the assumed disjuncts and replace them with the dependency-numbers of the original disjunctive premise. Finally, *remember to include the dependency-numbers of any other formula you have used, if any, in the course of deriving the conclusion from each disjunct.* Exercise 3.4 gives you an opportunity to try some simple vEliminations for yourself. First, however, study Box 3.2 carefully.

BOX 3.2

To infer a disjunction use vI:

♦ vI: Given a formula on a line of proof you may infer the disjunction of that formula with any other well-formed formula on a new line of proof. Annotate the new line with the line number of the old line and 'vI'. The dependency-numbers of the new line are identical with those of the old line.

♦ Note carefully that you may introduce the new formula and the disjunction sign 'v' on either the *right* or the *left* of the original formula and that you can only introduce *one* disjunction sign per application of vI.

To draw an inference from a disjunction use vE:

♦ vE: To draw an inference from a disjunction as such you must derive the desired formula from each disjunct first, i.e. assume each disjunct in turn and derive the desired formula from each. Having done so, you may repeat the conclusion on a new line of proof. Annotate the new line with five numbers followed by 'vE'. The five numbers are: (i) the line number of the disjunction; (ii) the dependency-number of the first disjunct assumed; (iii) the line number of the conclusion derived from the first disjunct; (iv) the dependency-number of the second disjunct assumed; (v) the line number of the conclusion derived from the second disjunct.

♦ Note carefully that vE is a *discharge* rule. Hence, at the line annotated 'vE' you may discharge the dependency-numbers of each disjunct and replace them with the dependency-number of the original disjunction together with the dependency-number of any other formula you used to derive the conclusion.

EXERCISE 3.4

1 Each line of proof annotated 'vE' is also annotated with five numbers. What exactly do these numbers refer to?

2 Prove that the following sequents are valid in PL:

1.	P v Q : (P v R) v (Q v R)	(8)
2	(P & Q) v (P & R) : P & (Q v R)	(12)
3	P v (P & Q) : P	(5)
4	P v P : P	(3)

V
More on vElimination

To date we have only considered vEliminations of the simplest kind. In every case, our check for *extra* dependency-numbers beyond those of the original disjunction has been fruitless. This reflects the fact that no premise other than the disjunctive premise was involved. It should come as no surprise, however, that other premises can be involved and that vE works well with our existing stock of rules. Let's now consider cases in which the check for extra dependencies will bear fruit. For example, consider the sequent P v R, P → S ⊢ R v S. This time, in addition to a disjunctive premise, we also have a conditional as premise, namely P → S. Because we still have a disjunction as premise we will still require vE. But note how and when the second premise helps in applying vE. Let's consider the proof carefully:

P v R, P → S ⊢ R v S

{1}	1.	P v R	Premise
{**2**}	2.	P → S	Premise
{3}	3.	P	Assumption [first disjunct]
{**2**,3}	**4.**	**S**	**2,3 MP**
{**2**,3}	5.	R v S	4 vI [conclusion from first]
{6}	6.	R	Assumption [second disjunct]

{6}	7.	R v S	6 vI [conclusion from second]
{1,**2**}	8.	R v S	1, 3, 5, 6, 7 vE

The second premise comes into the proof at line 4. There, using the first assumed disjunct from line 3 and MP, we derive the first disjunct of the conclusion R v S. This means that the dependency-number of the second premise will feature in the set of dependencies of the conclusion derived from the first disjunct on line 5. Now, although vE allows us to discharge the dependency-numbers of each of the *assumed disjuncts*, that is all that vE allows us to discharge. It follows that we *cannot* discharge dependency-number 2. After all, that number refers back to the formula P → S, which certainly is not one of the assumed disjuncts. And so we must include dependency-number 2 among the set of dependencies belonging to the final line.

Just as the last proof illustrates how MP can help in the process of an vE, the next proof (Exercise 3.5) illustrates how MT can also assist in applying vE. In fact, the next proof brings together *all* of the rules we have considered so far in this chapter (*and* MP). This time, there are two premises other than the disjunction. So, there should be two dependency-numbers other than that of the disjunctive premise on the final line of proof. Try this one for yourself.

EXERCISE 3.5

1 Prove that the following sequent is a valid sequent of PL:

1. R v S, ~Q → ~R, S → Q : Q v P (12)

As we have seen, vE works well with the existing rules. These allow us to manipulate any other premises we might have when deriving the desired conclusion. So far, we have only considered examples involving conditionals as premises working together with a disjunction as premise in a proof. But there is nothing to prevent a disjunction as premise working together with another disjunction as premise in the course of a proof. If we want to derive a conclusion from a disjunction we must apply vE. So, if we want to derive a conclusion from two disjunctions we must apply vE in both cases, i.e. we must make *two* uses of vE. In the course of such a proof it is crucial to keep track of the dependencies involved at each stage and to be clear about which dependency-numbers are discharged in each application of vE. However, provided we are rigorous in applying the strategy for vE such proofs can be perfectly straightforward. The proof of the following sequent constitutes a very clear example of two disjunctions working together as premises:

P v Q, P v R ⊢ P v (Q & R)

{1}	1.	P v Q	Premise
{2}	2.	P v R	Premise
{3}	3.	P	A [note that this is the first disjunct of *both* premises]
{3}	4.	P v (Q & R)	3 vI [conclusion derived from the first disjunct of *both*]
{5}	5.	Q	A [the second disjunct of the first premise assumed]
{6}	6.	R	A [the second disjunct of the second premise assumed]
{5,6}	7.	Q & R	5,6 &I
{5,6}	8.	P v (Q & R)	7 vI [conclusion derived from the second disjunct of *both*]
{**2**,5}	9.	P v (Q & R)	**2**,3,4,6,8 vE [first elimination on premise 2]
{**1**,2}	10.	P v (Q & R)	**1**,3,4,5,9 vE [second elimination on premise 1]

This proof has a number of interesting features. Because P is the first disjunct of *both* disjunctions I need only derive the conclusion from P once. This came straight away on line 4 by vI. At this stage, half the job is already completed. To complete the task I must now derive the conclusion from the *second* disjunct of both disjunctions. Hence, I assume the second disjunct of each disjunction on separate lines of proof. But I can now employ another labour-saving device: I can use &I to conjoin both disjuncts and now, with a single application of vI, I can derive the desired conclusion from the conjunction of *both* disjuncts at line 8. Moreover, because both disjuncts are involved, there is no need to repeat the process. The task has been completed very rapidly here (the conclusion only actually features twice in the proof!) but it is complete none the less. It only remains to make the appropriate applications of vE. Hence, at line 9, I carry out my first application of vE, repeat the conclusion, and infer the conclusion directly from the second premise. Thus, the dependency number of the second premise, 2, enters into the set of dependencies on line 9. I can now discharge the dependency-numbers of the assumptions relevant to that premise. But note that I also made essential use of the formula on line 5. So, its dependency-number must also be included among the set of dependencies belonging to line 9. Further, note that I cannot discharge dependency-number 5 at this point just because that number refers back to line 5 and to the assumption of a formula which is no part of the second premise. Check back for yourself. You will see that the formula sitting on line 5 is just Q. But that formula is the second disjunct of the *first* disjunctive premise and I have not yet applied vE to that premise.

So, at line 10, I move to apply vE for a second time and infer the conclusion directly from the first premise. When I do, I can legitimately discharge 5 just because it is the dependency-number of the second disjunct of the first premise. However, 2 must remain among the set of dependencies belonging to line 9. That number clearly features among the dependencies of the line which was cited as the conclusion derived from the second disjunct of the first disjunction. Further, it is not the dependency number of any assumption we made for the elimination on the first disjunction. So, it cannot be discharged. Hence, line 10 depends upon lines 1 and 2, which are exactly the dependency-numbers of the original premises. Therefore, the proof is complete.

Finally, it's worth noting the order in which applications of vE are made when more than one application of the rule is involved in a proof. Note that I worked from the inside out, as it were, dealing with the second premise first, i.e. the first application of vE to 2, the second to 1. You will find that the order of application can make a difference in such cases. Work out for yourself what the implications would be for the set of dependency-numbers belonging to the final line if I had reversed the order of application of vE. In general, you will find that it pays to work from the inside out in such cases.

The best way to become adept at using vE is to practise using the rule in proofs. Equally, practise with double vEs is the only way to end up feeling really at home with inferences of that complexity. In the latter case even more than the former, it is crucial to annotate your proof briefly so as to keep track of where things are. To that end, try the proofs in Exercise 3.6 for yourself.

Note: the first proof is only a slight modification of the one we have just considered. So, study that proof carefully. Write down the new sequents. Turn the book over and try these for yourself first.

EXERCISE 3.6

1 Prove that the following are valid sequents of PL:

1. (P v Q) & (P v R) : P v (Q & R) (11)
2. P v (Q v R) : Q v (P v R) (12)

VI
Arguing Logically for Exam Purposes: How to Construct Formal Proofs

(i) Initial Advice

We have now considered most of the rules of inference for PL. In fact, only one further rule remains to be added. In the course of this chapter and the previous one we have also considered quite a number of different proofs and noted how well the simple rules of inference work together to produce proofs of quite complex sequents. Now, you may say, that is all very well. But, seated in the examination hall facing a number of daunting proofs some Monday morning, how should I go about trying to crack those proofs? How do I know which rules to apply on any given occasion? What *strategy* should I use? These (crucially important!) questions are not about how to apply a given rule. Rather, they are questions about knowing which rules to try to apply and when: how, in any given case, do I identify the correct strategy for proof? Fear not. We will now consider questions of strategy from the very beginning.

As you will no doubt have noticed, certain kinds of proof are very straightforward. For example, recall the proofs involving only &I and &E in Exercise 2.3. These are all perfectly easy to prove. There, we simply used the relevant elimination-rule to dismantle the premises and then applied the relevant introduction rule, one step at a time, to derive the desired conclusion. In general then, when we want to arrive at a conjunction as conclusion, the strategy will just be to derive each conjunct on a separate line of proof before conjoining the formulas on those lines to arrive at the desired conclusion.

Equally, proofs which only involve MP, or MP together with the & rules, such as those in Exercise 2.4, are also perfectly straightforward. Further, proofs which involve a disjunctive conclusion can often be proved very quickly. In such cases, the strategy will just be to derive either disjunct and then to arrive at the whole disjunction by vI. However, you should not be surprised to learn that such simple proofs are not always at the top of the examiner's agenda! What if your exam contains more difficult proofs?

1 The first point to make about strategy is simply this: *a conclusion must be derived from its premises*. In any given case in which there are premises, i.e. when we are not trying to prove a theorem, we always know exactly what those premises are. We also know that each premise must be entered on a line of proof in the order given. So, no matter how complex

or difficult a proof might turn out to be, we always know how to start the proof.

Moreover, we also know exactly which conclusion should be derived from those premises. Hence, we always know which formula should appear on the last line of proof. Further, even though we do not know in advance what the actual line number of that final line will be, we do know the dependency-numbers of that line. Remember: the conclusion must depend only on its premises. Premises must be entered at the outset of a proof, on separate lines, using premise-introduction. So, if we have one premise, it will be entered on line 1 and, by premise-introduction, its dependency-number is just 1. Hence, the dependency-number of the conclusion inferred from that premise will also be 1. If we have two premises, these will be entered on lines 1 and 2; hence, the dependency-numbers of the conclusion will be both 1 and 2; and similarly if we have three, four or more premises. In all those cases where the set of premises is non-empty, then, we know both how to start the proof and, to a large extent, what the last line of proof should look like.

So, begin the proof by entering the premises. Leave a gap of a few lines under the premises, take a new line, imagine that it's the last line, and number it 'n'. Enter the conclusion as the formula on that line. Pool the dependency-numbers of the premises and write those in as the dependency-numbers of the final line 'n'. Now the trick is to bridge the remaining gap.

2 But how do we build the bridge? In two ways. By working in two directions: both from the top down and from the bottom up. By exploiting all available resources we try to make the proof grow in both directions, rather like stalactites and stalagmites!

First, try to develop the proof from the top down like a stalactite. Check whether or not there are any obvious routes from the premises to the conclusion. Ask yourself: are there any obvious inferences I can make? If so, make them and look again for a route or some further step. If you come to a halt try the stalagmite approach: look again at the formula on the last line. Ask yourself whether there is an obvious rule which you might have applied to arrive at that conclusion. If so, what else would you need to apply that rule? Remember: you can always enter any formula on any line of proof by rule of assumptions. In practice, you should be able to pick up some points for your efforts here even if, in the end, you are still unable to build the bridge and fully complete the proof.

But what if, having followed all these instructions, you do come to a grinding halt and fail to bridge the gap? And what if you are faced with an empty set of premises? How should you proceed?

3 Fear not. In just those situations we can appeal to a very useful rule of thumb. The rule is not absolutely guaranteed to provide the right strategy for every proof straight off, hence, 'rule of thumb'. But those faced with ticklish proofs in exams should find it invaluable. So, I will call this rule the **golden rule**. The rule addresses the question of the strategy for proof quite generally and can as readily be used at the very outset of proof-construction as at any subsequent stage. (We should always be ready to exploit all available resources at *every* stage of proof-construction. After all, in constructing a proof we are always trying to bridge the gap.) In fact, you should quickly find that merely by scrutinising the sequent and reflecting on the rule you can crack the question of strategy before you even put pen to paper.

However, there are certain proofs whose construction is eased by following the instructions above *first* and then applying the rule. So, if you experience a little difficulty in getting started on a proof under exam conditions follow the original instructions first, just to get going, and then apply the rule if you fail to build that bridge between premises and conclusion. Rest assured: when following the initial instructions above has still not delivered the goods, the golden rule will always suggest a strategy for deriving the conclusion from the premises.

(ii) The Golden Rule

The golden rule consists of three parts: two questions and a recommendation. The first part and the first question to ask yourself is just this:

1. Is the main connective in the conclusion a conditional?

If the answer to that question is 'yes' then the overall strategy is conditional proof. This is familiar territory. As we noted in Chapter 2, the strategy for CP is just to assume the antecedent of the conditional and then try to derive the consequent. If we succeed in deriving the consequent then we can simply discharge the 'extra' dependency-number of the assumption for CP, leaving the conclusion to rest only on its premises. This part of the golden rule clearly gives the right overall strategy for all the proofs we considered in Exercise 2.5 and for proofs 2–7 in Exercise 3.2. But note that it will also give the right strategy for the theorems we proved in Exercise 2.6 just because question 1 only concerns the form of the conclusion. So, it really doesn't matter whether the set of premises is empty or not. But perhaps we knew this much already from our understanding of CP. What if the conclusion is not a conditional? What if our answer to the first question is 'no'? In that case move to part two of the golden rule.

The second part of the golden rule and the second question to ask yourself is just:

2. Is the main connective in any or all of the premises a disjunction?

If the answer to that question is 'yes' then the overall strategy may well involve vE, and this is familiar territory. The strategy for vE, we noted, is just: assume the first disjunct and try to derive the conclusion. Assume the second disjunct and try to derive the conclusion. If you succeed, you can always discharge the 'extra' dependency-numbers of those assumptions when you apply vE. Again, the conclusion will depend only on its premises. This part of the golden rule immediately yields the correct strategy for every vE proof we have considered so far in this chapter, excepting only the first proof in Exercise 3.6 (for more on that one see Exercise 3.8 below).

However, note carefully that simply identifying the conclusion as a disjunction is *not* sufficient to establish vE as the appropriate overall strategy. Remember that we can always infer a disjunction by vI. Therefore, the real clue to identifying vE as the appropriate strategy is having a disjunction as a premise rather than as a conclusion.

It should be clear even from the first two parts of the golden rule that even if the rule is not absolutely foolproof or exceptionless (and it isn't!) it does cover an enormous number of cases and is extremely useful none the less. In fact, it is even more useful than it might first appear, and before considering the last part of the rule it's worth making a little more of its utility clear.

(iii) Digging Deeper: Proof and Sub-Proof

Imagine for a moment that you find yourself facing a proof in a logic exam. Suppose that you have followed the initial instructions above but that you have failed to bridge the gap and must have recourse to the golden rule. Further suppose that the answer to the first question is 'Yes, the main connective in the conclusion is a conditional.' You decide to employ CP and assume the antecedent. How is the consequent to be derived? Again, remember that the conclusion must be derived from its premises and that all available resources should be exploited, in both directions, at each and every stage of proof-construction. In certain proofs we may only need to apply MP in order to derive the consequent from the assumed antecedent, as, for example, in the proof of $P \rightarrow Q, R \rightarrow P \vdash R \rightarrow Q$ (check for yourself that this is the case). In others, we may only need MP together with rules for '&', e.g. $P \rightarrow Q, P \rightarrow R \vdash P \rightarrow (Q \& R)$ (again, check for yourself that this is so). Similarly, certain CP proofs will require

MT or MT with DNE or DNI or both, e.g. as in the proof of $\sim P \rightarrow \sim Q \vdash Q \rightarrow P$.

More complex CPs may require a little more in the way of inference to get us from the assumed antecedent to the derived consequent. Just how to proceed once we've assumed the antecedent may not always be clear immediately. Here's where the golden rule can help. This time we're asking questions not about the *overall* strategy for proof but about the appropriate strategy for deriving one formula from another within a proof. Let's call this idea of a 'proof within a proof' a **sub-proof**. Again, whenever the moves are not simple or obvious we can apply the golden rule. So, first ask: is the formula we want to infer in the sub-proof a conditional? If it is, assume *its* antecedent and apply CP *again* (note that this strategy covers all those cases in Chapter 2 where we *iterated* CP). But suppose that the consequent is not a conditional. What then? Simply move to question two: is the formula we're inferring *from* a disjunction? If so, apply vE. Consider the following sequent, for example, $P \rightarrow (Q \vee R), Q \rightarrow R : P \rightarrow R$.

Let's follow the initial instructions: set out the premises *and* the conclusion and try to bridge the gap:

{1}	1.	$P \rightarrow (Q \vee R)$	Premise
{2}	2.	$Q \rightarrow R$	Premise
	.		
	.		
	.		
	.		
{1,2}	n.	$P \rightarrow R$	??

In this case, there doesn't appear to be any obvious move to make straightaway. So, apply the golden rule. First, ask yourself: is the conclusion a conditional? Here, the conclusion is indeed a conditional. So the *overall* strategy for proof is CP. Hence, assume the antecedent P and try to derive the consequent R:

{1}	1.	$P \rightarrow (Q \vee R)$	Premise
{2}	2.	$Q \rightarrow R$	Premise
{3}	3.	P	Assumption for CP

Remember: we must always be prepared to exploit all available resources to bridge the gap. So, again, check for obvious moves. This time, there is an obvious move which we can make straight away, namely MP with 1 and 3, so let's make that move:

{1}	1.	$P \to (Q \vee R)$	Premise
{2}	2.	$Q \to R$	Premise
{3}	3.	P	Assumption for CP
{1,3}	4.	$Q \vee R$	1,3 MP

We are now faced with deriving R from Q v R. But how should we go about doing so? If the correct strategy is not immediately obvious, simply apply the golden rule a second time: is the formula we want to derive a conditional? No. Is the formula we are deriving from a disjunction? This time the answer is 'yes'. It follows that the correct strategy for *sub-proof* in this case is vE. So, we proceed to derive R from Q v R by vE:

{5}	5.	Q	Assumption [first disjunct]
{2,5}	6.	R	2,5 MP
{7}	7.	R	Assumption [second disjunct]
{1,2,3}	8.	R	4,5,6,7,7 vE
{1,2}	9.	$P \to R$	3,8 CP

To sum up, the overall strategy for proof is indeed CP. But there is an important sub-proof of the consequent R from Q v R by vE which runs from line 4 to line 8. Note that Premise 2 becomes involved during the vE in the derivation of the conclusion from the first disjunct. Note also that because R is both the second disjunct and identical with the formula we want to derive from that disjunct, there is no need to do anything other than assume it and cite that line number twice in the annotation for vE.

Finally, a little reflection on the completed proof highlights an important point. Note that the strategies for proof and sub-proof are CP and vE respectively. The reasoning involved in applying such strategies is, of course, hypothetical. Hence, we make some assumptions and proceed with a sub-proof towards the final overall proof. But note that although assumptions are involved within such strategies, those assumptions are *only* used within the context of those strategies and are not *exported* out of those strategic contexts. When all hypothetical reasoning is complete all assumptions have been discharged. In sharp contrast, premises (in this case Premise 2, for example) are often *imported* into hypothetical contexts *and* may also be used outwith those contexts. When all the hypothetical reasoning is complete the dependency-numbers of the original premises will still feature among the dependencies of the final line. And this is quite intuitive. After all, a conclusion must follow from its premises. To sum up, we shall refer to this principle as the **import–export law** (of logic!).[1] Quite simply:

> While premises may be freely imported into contexts involving hypothetical reasoning, assumptions may not be exported outside contexts involving hypothetical reasoning.

The following proofs (Exercise 3.7) involve similar sub-proofs to the one we have just considered. Try these for yourself.

EXERCISE 3.7

1 Prove that the following are valid sequents of PL:

1. $P \rightarrow (Q \vee R), R \rightarrow S : P \rightarrow (Q \vee S)$ (11)
2. $Q \rightarrow R : (P \vee Q) \rightarrow (P \vee R)$ (9)

By now you should be beginning to get something of a feel for strategic thinking. Invariably, what we are looking for is the right strategy for deriving one formula from another (or others), be it as an overall strategy for proof or as a strategy for sub-proof. At the moment, there are really only two important strategies to note: CP and vE. So, at each stage of proof-construction, we are invariably looking to see whether the formula we are trying to derive is a conditional or whether a disjunction is lurking among the set of formulas we are trying to derive that formula from. These are the only *kinds* of formula we are looking for at this stage.

Further, it's worth making a final, simple, point about strategic thinking in this context: *always scrutinise each of the formulas you are concerned with very carefully*. Remember the kinds of formula we are concerned with at the moment: conditional conclusions and disjunctive premises. Always scrutinise both premises and conclusions very carefully – look *inside* the formulas. Are any of the *component* formulas conditionals or disjunctions? This awareness will develop as you practise constructing proofs, and it should help you to see the applicability of the golden rule even in cases which are technically exceptions to the Rule. For example, consider the sequent P & (Q v R) : (P & Q) v (P & R).

The conclusion is not a conditional. The premise is not a disjunction. It is a conjunction. But the second conjunct of that conjunction is a disjunction! Following the initial instructions about bridge-building at each and every stage of proof-construction should lead you to carry out the required &E and make the disjunction explicit in any case; but always study the formulas you are given very carefully and try to see where the work will lie before you actually begin. This should at least take care of that class of exceptions to the rule which results from simply conjoining premises (note that our earlier exception, Proof 1 of Exercise 3.6, is a case in point here). Further, careful scrutiny of formulas should also help with those exceptions which result from conjoining *conclusions*. With that advice in mind consider each of the sequents in Exercise 3.8 carefully, scrutinise the formulas involved, try to work out your strategy for proof first, and then attempt to construct the proof.

EXERCISE 3.8

1 Prove that the following are valid sequents of PL:

1. P & (Q v R) : (P & Q) v (P & R) (10)
2. (P v Q) → R : (P → R) & (Q → R) (10)

With even this much strategic thinking under your belt you will be well placed to tackle many of the proofs which occur in first-level logic exams in propositional logic. But there remains one possibility which we have not yet considered. Remember: no matter how complex it might turn out to be, we know both what the beginning of the proof should look like and what the final line of proof should look like. Always be prepared to exploit all available resources to bridge the remaining gap and take any obvious steps you can in both directions. Look closely at the formulas which feature as premises and conclusion. Scrutinise their components. Ask: (1) Is the conclusion a conditional? If not, ask: (2) Are all or any of the premises disjunctions? But what if the answer to that question is 'no'? If the answer to that question is 'no' the final part of the golden rule makes the following recommendation: try **reductio ad absurdum**.

VII Reductio Ad Absurdum

The Latin name for this particular rule of inference simply means 'reduction to absurdity'; a phrase which accurately and succinctly describes the strategy for enabling the application of that rule in a proof. More specifically, if assuming a particular formula leads us to a contradiction then, by reductio, we may infer that the formula in question is false.

In logical terms, a **contradiction** is just the conjunction of any formula with the negation of that same formula. In many logic courses, you will find the formula P & ~P given as an example of a contradiction. But note that this is just the simplest possible case. (P & Q) & ~(P & Q), for example, is just as much a case in point, even though compound formulas are involved.

Remember: a contradiction consists of the conjunction of a formula with its negation; there is no stipulation on the complexity of the formula involved. Now, if assuming some formula enables us to derive a contradiction within a proof then reductio allows us to conclude that that formula must be false. The reasoning involved is surely highly intuitive. After all, no contradiction can possibly be true. Aristotle makes the point strikingly around one possible type of contradiction in his *Metaphysics* [1005b19–20] when he identifies the 'First Principle of all First Principles':

> the same attribute cannot at the same time belong and not belong to the same subject and in the same respect.

Contradictions are absurd or *logically false* sentences. Any sentence which entails a contradiction or logical falsehood must itself be false; and it is exactly this fact which is exploited in any application of reductio: if a contradiction can be shown to be derivable from a formula then, on that basis alone, we may validly assert the negation of that formula. Hence, reductio is often referred to as 'proof by contradiction'.

Certain logic texts,[2] and certain courses in formal logic, introduce a special symbol at this stage to stand for *any* contradiction. Typical notational devices include: '#', 'Λ' and '⊥'. The first two symbols are somewhat arbitrary but we can think of the third symbol as an upside-down 'T'. If 'T' stands for 'true' then we can think of '⊥' as the opposite, i.e. as 'the false'. Either way, all these symbols represent contradiction. Here, I will not introduce a special symbol. There is an argument against doing so simply from linguistic economy, and adopting one particular symbol might mask the possibility of different kinds of contradiction, i.e. among atoms, among conjunctions and so on.

The foregoing considerations about the status of contradictions as logical falsehoods quickly lead to an obvious strategy for proof. Suppose you want to show that a given formula is false. If you could demonstrate that a contradiction followed from that formula then you could certainly validly infer that the formula in question is false. So, first assume the formula whose falsity you want to prove. Next, try to derive a contradiction from that assumption. If you succeed, reductio will enable you to infer the negation of that assumption.

But note two points carefully. First, reductio always and only allows us to infer the negation of a formula. Hence, reductio is **negation-**

introduction and a number of systems of formal logic refer to it explicitly as the introduction-rule for negation. (We will denote it **RAA**.) Second, reasoning with reductio is *not* categorical. Rather, it is hypothetical: we *assume* a particular formula and look to see whether or not a contradiction *would follow* from that assumption. If so, reductio allows us to conclude that the assumption is false.

Further, and crucially, not only does reductio allow us to derive the negation of a given assumption, if we successfully derive a contradiction from it, it also allows us to *discharge the dependency-number* of that assumption. So, the final addition to our propositional proof-theory is a third discharge rule. As we shall see, the addition of reductio to the system also adds a last, enormously helpful, strategy for proof which we will frequently utilise. Moreover, that strategy is both simple and intuitive: to demonstrate the negation of a formula, assume the formula and try to derive a contradiction from it. Reductio will then allow you to derive the desired negation. For example, consider the sequent $P \rightarrow Q$, $P \rightarrow \sim Q : \sim P$. The conclusion we want to derive is a negation. We know that we can prove the negation of a formula if that formula can be shown to entail a contradiction. The formula whose negation we are after is P. So, assume P and attempt to derive a contradiction. In this case, deriving the contradiction is perfectly straightforward:

$P \rightarrow Q, P \rightarrow \sim Q \vdash \sim P$

{1}	1.	$P \rightarrow Q$	Premise
{2}	2.	$P \rightarrow \sim Q$	Premise
{**3**}	3.	P	Assumption for RAA
{1,**3**}	4.	Q	1,3 MP
{2,**3**}	5.	$\sim Q$	2,3 MP
{1,2,**3**}	6.	$Q \mathbin{\&} \sim Q$	4,5 &I
{1,2}	7.	$\sim P$	3,6 RAA

The 'extra' dependency number, 3, of the assumption for RAA is indeed discharged when we apply RAA on line 7, leaving the conclusion resting only on the premises on lines 1 and 2. Note also that RAA does indeed introduce the negation sign (to the formula on line 3) and that applications of RAA are usually preceded by &I; just because we must always explicitly form the contradiction before RAA can be applied. Further, note that the line annotation for RAA involves two line numbers: the line number of a contradiction and the line number of the formula from which that logical falsehood was derived.

Note the way in which the set of dependency-numbers grows in the course of the proof up to line 6. The two uses of modus ponens bring the dependency-number of the assumption for RAA into the sets of dependencies belonging to each of Q and ~Q. Those formulas are then conjoined on line 6 to form a contradiction. The presence of the dependency-number of the assumption for RAA, 3 in bold type, among the dependencies of line 6 reflects the fact that that assumption is integral to the derivation of the contradiction. And that is crucial. The contradiction must depend upon the assumption if we are to apply RAA legitimately to that assumption. Finally, the set of dependencies shrinks on line 7 when we deduce the negation of that assumption by RAA and discharge the dependency-number of the assumption. Here is the rule-statement in full:

> RAA: If a contradiction has been shown to be derivable from an assumption we may write the negation of that assumption on a new line of proof. Annotate the new line with the line number of the line of the contradiction, the line number of the assumption and 'RAA'. The dependency-numbers of the new line will consist of all those of the old lines except that of the assumption from which the contradiction has been derived.

You will quickly discover that reductio is an extremely useful strategy for proof. In fact, any proof which can be constructed using MT coupled with DNI and/or DNE can be constructed using RAA coupled with DNI and/or DNE instead. What's more, RAA will often provide a more obvious strategy for proof and may also shorten the proof of that sequent significantly. You will appreciate the extent to which that is the case if, at the end of this section, you consider again the four very difficult MT proofs included in Exercise 3.2!

One last aspect of the utility of RAA remains to be made clear. What if the conclusion we want to infer is not a negated formula?

RAA may still be applicable none the less. All we need do is assume the *negation* of the formula we are interested in and then try to derive a contradiction from that negated formula. If we succeed, RAA will allow us to negate that negation, i.e. to infer the *double negation*. We can then exploit DNE to convert that double negative back into the positive formula we were after in the first place. The proof of the following sequent is a case in point:

$\sim P \rightarrow P \vdash P$

{1}	1.	$\sim P \rightarrow P$	Premise
{2}	2.	$\sim P$	Assumption for RAA
{1,2}	3.	P	1,2 MP
{1,2}	4.	$P \;\&\; \sim P$	2,3 &I

{1}	5.	~~P	2,4 RAA
{1}	6.	P	5 DNE

Even here, reductio remains negation-introducing. The formula we infer by RAA on line 5 is the double *negative* ~~P. What then allows us to infer that P is not RAA itself but DNE, which we apply on the next line. Because RAA is always going to give us a negation it is wise to assume the opposite of what you want before attempting to apply RAA. So, if you want to derive the negation of a formula, assume the formula un-negated. If you want to derive an un-negated formula, assume its negation. There are a number of clues to look out for when you are trying to determine when to apply RAA. If the desired conclusion is a negated formula, RAA may well be worth a try. Moreover, when the conclusion is an atomic formula, a negated atomic formula or a conjunction of such formulas and the proof is not of a very simple kind, RAA will be your best bet. This is when RAA really comes into its own. And that should be no surprise. Remember the golden rule: if it's not a terribly obvious proof, ask: is it a CP? If not, ask: is it a vE? If not, try RAA!

Finally, note that, as ever, applications of RAA can be iterated, RAA works well with the existing stock of rules, and RAA is often a useful strategy for a sub-proof even if it itself is not the appropriate overall strategy for the whole proof. With all these points in mind, Exercise 3.9 gives twenty proofs involving RAA in roughly ascending order of difficulty for you to try yourself. Again, the numbers in brackets next to each sequent indicate the number of lines in my proof of that sequent. (*Note:* this is not a guarantee that there is no shorter proof of the sequent in question!) Try Exercise 3.9 after studying Box 3.3.

BOX 3.3

♦ *RAA* rests on the principle that if a **contradiction** is derivable from a formula that formula must be false. Hence, to infer a *negation* as conclusion use RAA:

♦ *RAA*: If a contradiction is shown to be derivable from a formula you may write the negation of that formula on a new line of proof. Annotate the new line with the line number of the line of contradiction, the line number of the relevant formula and 'RAA'. The dependency-numbers of the new line consist of all those of the old lines except that of the formula from which the contradiction was derived.

♦ Note carefully that we can use DNE to convert double negatives into un-negated formulas. Hence, to derive an *un-negated formula,* assume *its* negation, apply RAA and, finally, apply DNE to the result.

EXERCISE 3.9

1 Prove that the following are valid sequents of PL:

1.	$: \sim(P \,\&\, \sim P)$		(2)
2.	$P \rightarrow \sim P : \sim P$		(5)
3.	$P \rightarrow Q, Q \rightarrow \sim P : \sim P$		(7)
4.	$P \rightarrow Q, \sim P \rightarrow Q : Q$		(8)
5.	$\sim(P \vee Q) : \sim P$		(5)
6.	$\sim(P \vee Q), R \rightarrow P : \sim R$		(7)
7.	$(P \,\&\, Q) \rightarrow \sim R : R \rightarrow (P \rightarrow \sim Q)$		(10)
8.	$P \rightarrow (Q \rightarrow (R \,\&\, \sim R)) : P \rightarrow \sim Q$		(7)
9.	$\sim(P \,\&\, \sim Q) : P \rightarrow Q$		(8)
10.	$P \rightarrow Q : \sim(P \,\&\, \sim Q)$		(7)
11.	$\sim(P \rightarrow Q) : P \,\&\, \sim Q$		(12)
12.	$P \rightarrow Q : (Q \rightarrow \sim P) \rightarrow \sim P$		(8)
13.	$P \rightarrow R, Q \rightarrow \sim R : \sim(P \,\&\, Q)$		(9)
14.	$\sim P : P \rightarrow Q$		(9)
15.	$P, \sim P : Q$		(8)
16.	$: P \vee \sim P$	law of excluded middle	(9)
17.	$P \vee Q : \sim(\sim P \,\&\, \sim Q)$		(11)
18.	$\sim(P \vee Q) : \sim P \,\&\, \sim Q$		(10)
19.	$\sim(\sim P \,\&\, \sim Q) : P \vee Q$		(14)
20.	$: ((P \rightarrow Q) \vee (Q \rightarrow R))$		(20)

VIII The Golden Rule Completed

We can now update our earlier discussion of strategy for proof-construction. In the first instance, try to apply the golden rule straight away before you begin the proof. On most occasions you will find that you can crack the question of strategy before you even put pen to paper. If not, begin the proof by entering the premises, write out the last line as fully as possible and use all available resources to try to bridge the remaining gap. Remember to scrutinise the formulas involved carefully and to make any obvious moves in either direction. Finally, if you still cannot complete the proof, try again to apply the Golden Rule, i.e. ask yourself:

1. Is the main connective in the conclusion a conditional? If so, apply the strategy for CP, i.e. assume the antecedent and try to derive the consequent. If not, ask:
2. Is the main connective of any member of the set of premises a disjunction? If so, apply the strategy for vE, i.e. assume the first disjunct and try to derive the conclusion, assume the second disjunct and try to derive the conclusion. Finally, draw that same conclusion from the original disjunctive premise by vE. If not:
3. Try RAA. Remember: the trick here is to assume the *opposite* of what you want and *then* try to derive a contradiction from that assumption together with any other formula or formulas already available in the proof. The double negation rules will allow you to finish things off to suit your purposes.

Finally, remember that *some or even all* of these strategies can work together within a single proof, one strategy providing a sub-proof of something useful for another. Remember also that the golden rule is not absolutely failsafe and that in some cases we just have to keep bashing away, trying first one line of attack then another.

In the last analysis, the true 'golden rule' of proof-construction is simply that practice makes perfect. To that end, I present four sets of revision exercises below. Within each exercise the level of difficulty is generally on the increase from one proof to the next, and each exercise involves proofs of a higher level of difficulty than its predecessor. In conjunction, Exercises I–III recap fairly comprehensively on the proof-theory of the last two chapters, but Exercise IV consists of sequents we have not considered so far. In all honesty, cases 5–10 of Exercise IV are very difficult. None the less, practice at proof-construction in Exercises I–III should help to develop your intuitions about strategies for proof and sub-proof and your skill in actual proof-construction so as to enable you to attempt Exercise IV. Once you have mastered every proof you have nothing whatsoever to fear from the proof-theory of propositional logic.

Before you try Revision Exercises I–IV, study Box 3.4 carefully.

BOX 3.4

Strategies for proof-construction:

♦ Always scrutinise the *component formulas* of any sequent carefully *before* attempting a proof.

♦ In *un*obvious cases, set out the premises, leave a gap then set out the conclusion and apply the *golden rule*, i.e. ask: (i) *is the conclusion a conditional?* If it is, apply CP. If not, ask: (ii) *are any or all of the premises disjunctions?* If so, apply vE. If not, assume the negation of the desired conclusion and try the RAA strategy.

♦ Never lose sight of the fact that each and all of the above strategies can work *together* in a single proof, i.e. the pursuit of an overall strategy may necessitate a sub-proof which itself requires a distinct strategy. Hence, apply the golden rule at the outset to identify overall strategy and then reapply as necessary throughout the process of proof-construction.

REVISION EXERCISE I

1 Prove that the following are valid sequents of PL:

1. $P \rightarrow Q : ((R \,\&\, Q) \rightarrow S) \rightarrow ((R \,\&\, P) \rightarrow S)$ (10)
2. $(P \,\&\, Q) \rightarrow {\sim}R : R \rightarrow (P \rightarrow {\sim}Q)$ (10)
3. $: (P \rightarrow Q) \rightarrow ({\sim}Q \rightarrow {\sim}P)$ (5)
4. $P \text{ v } Q : (P \text{ v } R) \text{ v } (Q \text{ v } R)$ (8)
5. $P \rightarrow R, Q \rightarrow S : (P \text{ v } Q) \rightarrow (R \text{ v } S)$ (11)
6. $P \rightarrow (Q \text{ v } R), Q \rightarrow R : P \rightarrow R$ (9)
7. $(P \text{ v } Q) \rightarrow R : (P \rightarrow R) \,\&\, (Q \rightarrow R)$ (10)
8. ${\sim}(P \,\&\, {\sim}Q) : P \rightarrow Q$ (8)
9. $P \rightarrow (Q \leftrightarrow R) : (P \,\&\, Q) \rightarrow R$ (9)
10. $: {\sim}P \rightarrow (P \rightarrow Q)$ (10)

REVISION EXERCISE II

1 Prove that the following are valid sequents of PL:

1. P v P : P (3)
2. P : (P → Q) → Q (4)
3. P : (~(Q → R) → ~P) → ((~R → ~Q)) (9)
4. P → (Q v R), R → S : P → (Q v S) (11)
5. ~Q → ~R, R v S, S → Q : Q v P (12)
6. ~(P v Q) : ~P & ~Q (10)
7. ~~(P v ~Q) : (P → ~Q) v (~Q → P) (12)
8. (P v Q) ↔ P : Q → P (7)
9. (P & Q) v (P & R) : P & (Q v R) (12)
10. : P v ~P (9)

REVISION EXERCISE III

1 Prove that the following are valid sequents of PL:

1. : ((P → P) → Q) → Q (5)
2. ~(P → Q) : P & ~Q (12)
3. (P v Q) & (R v S) :
   ((P & R) v (P & S)) v ((Q & R) v (Q & S)) (15)
4. P v Q, ~Q : P (11)
5. P v Q, ~P : Q (11)
6. : ((~P → R) & (~Q → R)) → (~(P & Q) → R) (15)
7. P v Q, P v R : P v (Q & R) (10)
8. P ↔ Q, Q ↔ R : P ↔ R (17)
9. : P v (P → Q) (16)
10. : ((P → Q) v (Q → R)) (20)

REVISION EXERCISE IV

1 Prove that the following are valid sequents of PL:

1.	$: (P \vee Q) \rightarrow (Q \vee P)$	(4)
2.	$: \sim(P \vee Q) \rightarrow \sim P$	(6)
3.	$\sim(P \,\&\, Q), P : \sim Q$	(6)
4.	$\sim(P \,\&\, Q) : \sim P \vee \sim Q$	(14)
5.	$P \vee (Q \vee R) : (P \vee R) \vee Q$	(12)
6.	$\sim P, \sim Q : \sim(P \vee Q)$	(14)
7.	$P \rightarrow (Q \vee R) : (P \rightarrow Q) \vee (P \rightarrow R)$	(14)
8.	$P \rightarrow \sim Q, P \rightarrow \sim R, Q \vee R : P \rightarrow (\sim Q \vee R)$	(16)
9.	$P \vee \sim Q, P \vee \sim R, Q \vee R : P$	(23)
10.	$(P \,\&\, Q) \rightarrow R : (P \rightarrow R) \vee (Q \rightarrow R)$	(24)

IX
A Final Note on Rules of Inference for PL

We have now completed the set of rules of inference for PL and by now you should be familiar both with each individual rule and with the ways in which those simple rules can be combined in proof-construction. However, so far in our discussion, we have omitted any mention of an important distinction which may be drawn between two different kinds of rule of inference for PL. It is precisely that distinction which I consider in the present section.

The distinction in question is that between **primitive rules** and **derived rules**. In the present context, we can simply consider our set of rules of inference for PL as a set of primitive rules. To say that a rule is primitive is just to say that the addition of that rule to the set of rules of inference for a formal system allows new sequents to become provable in that system. In contrast, a derived rule is a rule whose addition to that set of rules will not allow anything new to be proved in the formal system. In all honesty, drawing the distinction in this way turns out to imply that not all of our existing stock of rules are primitive, i.e. certain of those rules can, in fact, be derived within the system, given the remaining rules. This should not be too surprising. As I admitted earlier, any proof involving RAA can, with a little hard work, also be proved using MT and DNE (in the not too distant future you will have the opportunity to prove that this is so).

For present purposes, however, focus on the idea of a derived rule of inference as a rule whose addition to the system does not allow anything new to be proved in that system. As such, derived rules may seem rather dull: aren't such rules, in a sense, redundant? Indeed. But they offer important advantages none the less. As we shall see, derived rules can be used to abbreviate long, tedious proofs. As such, they constitute a logical (proof-theoretic) economy. The two particular derived rules which I consider here are designed precisely in order to maximise the possibility of exploiting that proof-theoretic economy.

The first such rule is known as **theorem-introduction**, or **TI**. Quite simply, the rule allows us to enter any theorem which we have already proved on any line of any proof with an empty set of dependency-numbers and annotated 'TI'. The point about the dependency-numbers here reflects the familiar fact that a theorem is a logical truth, as we discussed in Chapter 2. So, at any point in the process of proof-construction, we may enlist the help of any potentially useful theorem free of charge. Something of the usefulness of TI should be obvious already but, in fact, this rule is even more useful than might first appear. For not only can we introduce a particular proved theorem using TI but we can also introduce any formula of that form. To illustrate, not only can we introduce on an arbitrary line – call it line 'n' – the theorem $\vdash P \rightarrow P$, as follows:

______	n	$P \rightarrow P$	TI

but we could also introduce:

______	n	$(P \rightarrow Q) \rightarrow (P \rightarrow Q)$	TI

Equally, we could introduce:

______	n	$((P \rightarrow Q) \rightarrow (R \rightarrow S)) \rightarrow ((P \rightarrow Q) \rightarrow (R \rightarrow S))$	TI

In fact, provided the formula in question can be constructed from the original theorem simply by uniformly replacing the same original constituent formula with the same new well-formed formula, then that resulting formula may be introduced by TI. This practice of careful replacement, careful swapping, of old constituent formulas with new constituent formulas is known as **uniform substitution**. Simple as it may seem, uniform substitution is of crucial importance in formal logic generally and we will have cause to appeal to that notion again later.

The second derived rule we will consider here extends still further the possibilities for introducing formulas to lines of proof during the process of proof-construction. This rule is not theorem-introduction but **sequent-introduction, SI**. In short, the rule of sequent-introduction allows us to exploit our existing stock of proved sequents analogously to the way in which theorem-introduction allowed us to exploit our existing stock of proved theorems. Thus, if in the process of proof-construction, you were to come across the formulas $P \rightarrow Q$ and $P \rightarrow \sim Q$ on separate lines of proof, SI would enable you to write ~P on the next line of proof without having to go through the obvious RAA which would usually be required. Instead, you could simply enter ~P, annotate the new line with the line numbers of the two formulas involved and 'SI', and complete the line by pooling the dependency-numbers of the old lines to create the set of dependency-numbers of the new line. Again, something of the usefulness of SI is apparent and, again, via the notion of uniform substitution we can amplify the utility of SI by allowing it to apply not only to proved sequents but also to any substitution-instance of a proved sequent.

This particular pair of derived rules maximises the possibilities for making proof-theoretic economies in the process of proof-construction, and both are extremely useful in allowing us to see how we might prove new, complex sequents which we have not already proved. However, a few words of warning are apt here. First, many logicians forbid the use of both rules for exam purposes, i.e. 'TI and SI may not be used' is a commonplace of formal logic exam papers. Alternatively, a numbered list of legitimate theorems and/or sequents may be made explicit in the exam paper and applications of TI and/or SI may be restricted to that list (there is also no guarantee that such a list will be available prior to the exam).

These particular rules can also be the source of some cruel (if amusing) trickery on the part of logicians. In my own first logic course, for example, the logic lecturer (who shall remain nameless) pointed out that we should feel free to use TI as we pleased during the logic exam. Further, he also helpfully pointed out that our textbook contained a numbered list of theorems which we could use and, indeed, encouraged us to learn that list by rote (the course textbook was in fact *Elementary Logic* by Benson Mates [1972] which, in a number of respects, is a truly excellent text). When I and fellow students consulted the text we found that Mates began his list over pp. 98–9 where theorems 1–7 were stated and continued over pages 100–1. All in all, ten theorems were stated over these pages. But we quickly found that the list continued up to p. 106 until no fewer than *100* theorems had been listed. Finally, look out for the equally amusing strategy of allowing the use of TI in an exam provided you also include a proof of the theorem in your exam paper!

In the light of these facts, I will not give any more emphasis to TI and SI in particular but will instead look very briefly at derived rules in general and the question of how to show that a rule is derivable in particular. To exhibit the *derivability* involved in derived rules, i.e. to show a rule to be derivable, is really just to show a way of getting from the inputs for the rule to the outputs for the rule without actually using the rule itself. Hence, we show a rule to be derivable within a system by deriving the output of the rule from the input for the rule using only the other rules of that system, i.e. by exploiting *only* the remaining stock of primitive rules. In the present context, for any derived rule, we will separate the input from the output by a line thus:

$$\frac{\text{input}}{\text{output}}$$

Looked at in this way, showing a rule to be derivable comes to showing how to derive the output from the input. For that purpose, we need not adapt or extend our notation any further, i.e. both inputs and outputs will be given in the usual notation. Moreover, to show that a rule is derivable in PL we can simply construct a proof of the given output from the given input using only the other rules of the system. Exercise 3.10 contains some examples of derived rules. Consider each rule carefully. Many of these rules exemplify principles of inference which are honoured with traditional names and, as appropriate, I have stated the name next to the rule. In each case, construct a proof of the stated output for the rule from the stated input in the usual manner. Of course, TI and SI may *not* be used!

EXERCISE 3.10

1 Show the following rules to be derivable in PL:

1. $\dfrac{P \vee Q, \sim P}{Q}$ disjunctive syllogism

2. $\dfrac{P \rightarrow Q, Q \rightarrow R}{P \rightarrow R}$ hypothetical syllogism

3. $\dfrac{P \rightarrow Q, \sim P \rightarrow Q}{Q}$ constructive dilemma

4. $\dfrac{P \rightarrow Q, P \rightarrow \sim Q}{\sim P}$ destructive dilemma

5. $\sim P \rightarrow P$

$\overline{\quad P \quad}$ consequentia mirabilis

6. $P \rightarrow \sim P$

$\overline{\quad \sim P \quad}$

X
Defining 'Formula of PL': Syntax, Structure and Recursive Definition

We turn now to the problem of defining accurately just what we mean by a *well-formed formula* of PL. You will already have a good understanding of just what it is to be a well-formed formula of PL in virtue of your understanding of the proof-theory of PL. And, in fact, the notion of a formula is completely defined by the set of rewrite rules that allowed us to construct syntactical trees in Chapter 2. So, in an important sense, we have already completed this task. Hence, you might think that we could simply avoid this issue here. However, our earlier approach was not the standard one to supplying the definition we require, and the standard approach exemplifies a special kind of logical definition which it would be a great pity to overlook.

Indeed, the particular logical construction involved in this definition may represent the most important and insightful of all the formal devices we have considered to date. Formal definitions of this type may have a fundamentally important role to play in explaining how it is that human beings learn natural languages. Detailed consideration of these issues lies beyond the present text and what follows is not even a thumbnail sketch. In all honesty, it is no more than a hint at a fascinating and potentially momentous area of study which is aptly termed *mathematical linguistics*. None the less, it would be misleading, I think, to omit any mention of the promise which lies in this area and I will conclude with some pointers to useful texts for interested parties.

Throughout our study of formal propositional logic we have taken very seriously the idea that PL is a formal language: formal certainly, but a language none the less. I have also argued that we might understand a formal language like PL as being analogous to certain natural languages. For example, we might consider PL as an analogue of English. PL is certainly not identical with English, but according to many formal logicians, linguists and philosophers of language there is an important sense in which formal languages like PL are analogous to natural languages. According to such authors, PL-type deductive structures share some genuinely analogous structural features with natural languages at the deepest level.

Most famously, the great American linguist Noam Chomsky [1928–], Institute Professor of Linguistics and Philosophy at the Massachusetts Institute of Technology, has pointed out that in *language-acquisition* a child is only ever exposed to a *finite* amount of linguistic input before it becomes able to produce a potentially *infinite* linguistic output. Moreover, once it has achieved language-mastery, the child is able both to recognise and to construct meaningful sentences which it has never before confronted. How is this possible? This problem has come to be known as the **creativity problem**, i.e. how are we to explain the extraordinary creative ability involved in language-acquisition?

If we closely examine the syntactical structure of a formal language such as PL, we can see one possible way in which an infinite linguistic output might be achieved on the basis of a finite amount of input, given some simple linguistic (grammatical) rules whose application can be iterated. Such a set of simple rules or basic grammar for a language is provided in terms of a special kind of definition called a **recursive definition**. As regards English, for example, John Lyons, one of Chomsky's commentators, notes:

> We may begin by defining the *language* that is described by a particular grammar as the set of all the sentences it generates . . . We will also assume that the number of distinct operations that are involved in the generation of English sentences is finite in number. There is no reason to assume that this is an implausible assumption; and if they were not, this would mean that the sentences of English could not be generated by means of a specifiable set of rules.
>
> If the grammar is to consist of a finite set of rules operating upon a finite vocabulary and is to be capable of generating an infinite set of sentences, it follows that at least some of the rules must be applicable more than once in the generation of the same sentence. Such rules and the structures they generate are called *recursive*. Once again, there is nothing implausible in the suggestion that the grammar of English should include a certain number of recursive rules.[3]

What is being asserted here is not just that English itself may ultimately be a recursive structure but also that it is with reference to that property that we can begin to construct a solution to the creativity problem and, thus, an explanation of language-acquisition itself. Moreover, recursion is exactly the kind of definition we will exploit in order to determine what is to count as a well-formed formula of PL.

To get our definition off the ground here we need to be able to represent formally a new level of generality, for which purpose our present notation is inadequate. Hence, we introduce a new set of variables: 'A', 'B', 'C'. . . . These variables simply range over all the well-formed formulas of PL, simple and complex. So, 'A' stands for any well-formed formula of PL and 'B' stands for any well-formed formula and so on. Because these variables represent formulas of PL they do not themselves belong to PL. Rather, the variables 'A', 'B', 'C', etc., belong to the metalanguage – call it 'ML' in which we talk about PL as *object* language. Logically enough, the new variables 'A', 'B', 'C' . . . are known as *metalinguistic variables*. As such, the values of those variables are simply the well-formed formulas of PL. Armed with our new class of metalinguistic variables we can proceed with the recursive definition as follows.

First, we define a *bracket* as one of the two marks '(', ')', left-hand and right-hand respectively. Next, we define the following familiar marks, purely ostensively (i.e. directly, just as those marks), as *logical connectives*: '&', 'v', '~', '→', '↔'. We now define P, Q, R . . . as *sentence-letters*, leaving the list open-ended to emphasise that there is no theoretical upper limit to it. In fact, it is easy to extend the list beyond the alphabet simply by using a stroke to mark a distinct sentence-letter: P, P′, P″. . . .

A *symbol* is then defined as any mark already defined and a *sequence* of symbols as an ordered list of symbols. Finally, we can define a formula as a sequence of symbols and pick out the class of well-formed formulas in terms of a few simple rules. To do so, we exploit the metalinguistic variables 'A' and 'B'. The rules we require are as follows:

1 Any sentence-letter is an atomic well-formed formula.

2 If A is a well-formed formula then ~A is a well-formed formula.

3 If A and B are well-formed formulas then (A & B) is a well-formed formula.

4 If A and B are well-formed formulas then (A v B) is a well-formed formula.

5 If A and B are well-formed formulas then (A → B) is a well-formed formula.

6 If A and B are well-formed formulas then (A ↔ B) is a well-formed formula.

7 Nothing else is a well-formed formula.

We have now defined the grammar of PL, i.e. we have explicitly defined *all and only* those constructions in the language PL which are properly or grammatically well-formed formulas. Note also that the type of syntactical structure which I have defined here actually allows me to generate an indefinitely large number of well-formed formulas, i.e. a potentially *infinite* number of well-formed formulas, from a *finite* base. The problem of explaining how, in the process of language-acquisition, a child who is only ever exposed to a finite amount of linguistic input becomes able to produce a potentially infinite linguistic output is at the very heart of the creativity problem. Hence, definitions of this kind may well give us some insight into the nature of language-acquisition. After all, it was just the possibility of there being a similar grammar at the heart of the English language which Lyons hinted at above. Therefore, definitions of this kind, recursive definitions, might provide some basis at least for a syntactical answer to Chomsky's question: creative abilities in a particular language might well be accounted for in terms of the speakers' possession of a grammar in which the set of grammatically well-formed sentences is defined recursively, i.e. by the repeated application of a finite set of rules to a finite vocabulary in ways which make possible the generation of an infinite set of sentences.

To some extent the possibility of the kind of explanation I am alluding to here remains controversial. What is uncontroversial is that in the process of outlining the recursive definition given above, we have arrived at a formal definition of the class of well-formed formulas for PL. Unfortunately, that definition also exposes what may seem to be something of a lack of rigour as regards constructing well-formed formulas of PL in this text so far. For note carefully that Rules 3–6, which cover the binary connectives, actually generate formulas which are enclosed in pairs of outer brackets. Strictly speaking, then, many formulas in this chapter and earlier chapters of the text are not well-formed!

In mitigation, as E.J. Lemmon notes in his logic text: 'human beings cannot stand very much proliferation of brackets'.[4] Moreover, as noted in Chapter 2, much of the usefulness of brackets consists in their potential for disambiguating formulas. But no ambiguity is introduced simply by omitting outermost pairs of brackets. Hence, we adopt that convention here and so legitimate our earlier practice. With this in mind, it is easy to convince yourself that the definition really does capture the class of well-formed formulas of PL simply by picking a few examples and working out which particular subset of {1 . . . 7} above sanctions well-formedness. For example, consider the following formula:

$$(P \rightarrow Q) \& (R \& S)$$

First, the well-formedness of the whole conjunction is sanctioned by Rule 3 above. The well-formedness of the first conjunct is sanctioned by Rule 5 while the well-formedness of the second conjunct is again sanctioned by Rule 3. Finally, the well-formedness of each individual sentence-letter as a formula of PL is sanctioned by Rule 1.

Interested parties with no background in Linguistics would do well to begin with the rigorous but none the less accessible account of Chomsky's work in this area given by V.J. Cook and Mark Newsom in *Chomsky's Universal Grammar: An Introduction* [1996]. Chomsky himself, of course, is the author of a number of seminal texts and articles, prime among them: *Syntactic Structures* [1957], *Aspects of the Theory of Syntax* [1965], *Rules and Representations* [1980], *Knowledge of Language: Its Nature, Origin and Use* [1986], and *Language and Problems of Knowledge* [1988]. Readers will find that Cook and Newsom provide an extensive bibliography for Chomsky. However, that bibliography does not include his most recent paper, 'Language as Natural Object', which is published in the journal *Mind* [1995] (and which might have been more aptly entitled 'Language as *Neurophysiological* Object', I think).

There now follows the first of four mock examination papers in formal logic which are designed to let you test what you have learned to date. However, you should not assume that every course in formal logic involves examinations of the same kind. Always check the methods and materials appropriate to your course carefully. Good luck!

Examination 1 in Formal Logic

Answer every question.

1 *Consider the following arguments carefully then (i) represent each argument as a sequent of PL and (ii) construct a proof of each of your sequents.*

1. If it rains again this afternoon then I'll eat my hat. Therefore, if it's true that if it snows and I eat my hat then I am probably madder than a hatter then if it snows and it rains again this afternoon then I probably am madder than a hatter.
2. Either it's raining today or the sun is shining gloriously. So, it can't both be the case that it's not raining and that the sun is not shining gloriously.
3. Either it's raining or it's not raining.

2 When two formulas mutually entail one another it is possible to derive the one from the other no matter which way round they are given. Such formulas are said to be logically equivalent and *interderivable*. Interderivability is represented by writing the turnstile in both directions between the formulas, i.e. '$\dashv \vdash$'. This represents the fact that each formula can be derived from the other. *Prove that the following formulas are interderivable:*

1. $P \rightarrow Q \dashv\vdash {\sim}P \vee Q$
2. $P \vee Q \dashv\vdash {\sim}P \rightarrow Q$

3 *The following proof has deliberately been left incomplete. Complete the proof:*

	1.	${\sim}P \leftrightarrow Q$	Premise
	2.		Assumption for RAA
{3}	3.	Q	
	4.		1 ↔E
	5.		2 ↔E
	6.	$Q \rightarrow P$	5 &E
	7.		4 &E
{2,3}	8.	P	
	9.	${\sim}P$	
{1,2,3}	10.		
	11.	${\sim}Q$	
{12}	12.	P	
	13.		5 &E
	14.	Q	12,13 MP
	15.		11,14 &I
	16.	${\sim}P$	
{1}	17.	${\sim}P \rightarrow Q$	
	18.		
	19.		11,18 &I
	20.		2,19 RAA

4 *Prove that:*

(i) *given the other rules of inference, MT can be considered to be a derived rule in PL.*

(ii) *given the other rules of inference, RAA can be considered to be a derived rule in PL.*

5 As logical falsehoods, contradictions cannot possibly be true. Hence, any

argument from contradictory premises cannot possibly have true premises and a false conclusion simply because the premises of such an argument could never be true. If it is impossible that the premises of an argument be true while the conclusion is false then that argument is valid. Hence, we may regard any argument from contradictory premises as implying its conclusion, i.e. we may regard inferences from contradictory premises to any conclusion whatsoever as exemplifying a valid form of argument. As a result of such reflections, many logicians maintain that *anything* follows from a contradiction just in the sense that any sentence whatsoever follows from a contradiction.[5] In PL, it is indeed the case that any formula can be derived from a contradiction as premise. This fact is exemplified in a principle of inference known as **ex falso quodlibet**, meaning literally, 'from the (logically) false anything follows'. *Represent ex falso quodlibet as a rule of inference as simply as possible and then show that rule to be derivable in PL.*

Notes

1 I believe that this expression is due to John Slaney.
2 See, for example, Tennant, Neil, [1978], *Natural Logic*, Edinburgh, Edinburgh University Press, p. 40.
3 Lyons, John, [1970], *Chomsky*, Fontana Modern Masters, Fontana/Collins, Ch. 5, pp. 48–9.
4 Lemmon, E.J, *Beginning Logic*, [1965], London, Thomas Nelson and Sons, p. 46.
5 I first became acquainted with this argument as a tutor in logic at the University of Edinburgh. However, it seems clear that the argument in the form I have it is in fact derived from Read, Stephen and Wright, Crispin, [1993], *Read and Wright: Formal Logic, An Introduction to First Order Logic*, fifth edition revised, departmental publication, St Andrews, University of St Andrews. See, for example, Chapter 5, 'Negation'.

4 Formal Logic and Formal Semantics #1

4
Formal Logic and Formal Semantics #1

I
Syntax and Semantics

So far, our discussion of formal logic has centred upon syntactical properties, i.e. properties belonging to the formulas and sequents of PL in virtue of their form or shape. As regards formulas, for example, we first tried to spell out our intuitions about which formulas were well formed, in terms of syntactical trees which were concerned precisely with the form or shape of formulas. Finally, in the last section of Chapter 3, we considered a recursive definition of the set of well-formed formulas of PL, i.e. a definition which again exploited the shape or form of formulas.

Moreover, in our earlier discussion of inference and reasoning in PL, we developed our understanding of formulas and the logical connectives in terms of which inferences were licensed by formulas with a particular logical connective as the main connective. In the light of these considerations, we were able to construct a set of rules of inference for PL. In formal logical terms, when we constructed that set of rules of inference for PL we added a **deductive apparatus** to the formal language PL and in so doing we established a **formal system**. Given the full set of rules of inference for PL, the formal definition of the notion of 'proof-in-PL' which we began in Section V of Chapter 2 can now be completed. Further, we can exploit that definition to go on to define **proof-theoretic consequence** *in PL* just in terms of what is provable using the rules of inference of PL. Proof-theoretic consequence characterises the notion of logical consequence in PL syntactically and it is that relation of syntactical consequence which the turnstile represents.

From first to last, then, our discussion has focused on the syntax of PL. In formal logical terms, syntax concerns the formulas of the formal language as *uninterpreted*, i.e. without primary regard to questions about the meaning or content of those formulas. In a sense, the rules of inference for PL fix the *meaning* of the logical connectives of PL and reflect our core understanding of the connectives in a very immediate way, i.e. by spelling out which kinds of inference are sanctioned by formulas of each shape or kind. But there is another way to consider the formulas of PL, namely, in terms of the conditions under which those formulas are true or false. If I know exactly those conditions under which a given well-formed sentence of a language is going to turn out to be true *and* I know exactly those conditions under which that same sentence will turn out to be false then, surely, I know what that sentence means. Therefore, to specify what we may call the **truth-conditions** for a sentence is just to spell out the *meaning* of that sentence, i.e. meaning and truth-conditions are one and the same. So, when we investigate the truth-conditions of formulas of PL we investigate the meaning of those formulas.[1] When we concern ourselves primarily with questions about the meaning of formulas of PL we go beyond the syntax of PL to the **semantics** of PL. Therefore, we are no longer concerned with the symbols of PL as uninterpreted shapes. Rather, we are explicitly concerned with the *interpretation* of the well-formed formulas of PL. In PL, an **interpretation** of a formula is just: *any assignment of truth-values to each of the atomic formulas which go to make up that formula*. So, an interpretation is simply an assignment of truth-values. Any interpretation of a formula which makes that whole formula true is a **model** of that formula. Therefore, a model of a formula is just any assignment of truth-values to the atomic constituents of the formula which brings that whole formula out as true.

II
The Principle of Bivalence

As we have seen, PL is a formal language into which we can translate sentences and arguments of natural language. PL is at least analogous to natural language; thus, more precisely, we might well think of PL as an *analogue* of that fragment of a natural language such as English which involves indicative sentences, i.e. statements or assertions. But how close is the analogy here? Much of what we will do in the course of the next few sections is to consider just what is involved in the claim that PL is an analogue of natural language.

Sentences in a natural language such as English may well be either true or false, and the same certainly holds for the formulas of PL. In fact, unlike natural language, we can say with absolute certainty that every well-formed formula of PL must be either true or false and can only be true or false. In philosophical terms, the formal language PL works under the strict governance of what is known as the **principle of bivalence**. The principle of bivalence is the most famous, perhaps infamous, of semantic principles. As a semantic principle we would expect it to say something about sentences and their truth-values, and it does: *every sentence is either true or false but not both and not neither*. In other words, each and every formula of PL has exactly one of two truth-values, i.e. *true* or *false*. Just because that principle holds true for all the formulas of PL, our formal language is said to be *bivalent*.

It is very important not to confuse the principle of bivalence with the **law of excluded middle** $\vdash$ P v ~P. The former is a semantic principle which generally asserts a sort of pre-established harmony between reality and claims in a given language about reality, i.e. that reality is such as to make any claim we make about it in that language either true or false, right here and now, as it were. In contrast, the latter simply asserts that, for any sentence, it is a logical truth that either it holds or its negation does. In general, it is certainly true that whenever the principle of bivalence holds the law of excluded middle holds. The former entails the latter. But note that the former is a logically stronger claim and the converse does not hold here. Consider a sentence in natural language about a future event: 'There will be a sea battle tomorrow'; the example is given by Aristotle in his *De Interpretatione* [IX]. Here and now, today, as it were, bivalence clearly does not hold for that sentence just because there is nothing about the world presently which could make the sentence either true or false. Bivalence fails here and it looks as if our sentence is not presently truth-valued at all or perhaps takes a new truth-value: '*neither true nor false*'. Be that as it may, the disjunction: 'Either there will be a sea battle tomorrow or there will not be a sea battle tomorrow' must be true none the less, i.e. the disjunction is true *now and forever*. Hence, the law of excluded middle still holds. Therefore, the law of excluded middle does not in general entail the principle of bivalence.

Moreover, although PL is bivalent it is not the case that every system of formal logic obeys the principle of bivalence. The Polish formal logician Jan Lukasiewicz [1878–1956], who was for many years Professor of Philosophy at Warsaw, famously rejected the principle precisely because the truth-value of such future contingent sentences was (presently) undetermined. Instead, Lukasiewicz proposed three truth-values: the true, the false and the *as-yet-undetermined*. Lukasiewicz's objection to the classical view was endorsed and developed slightly differently by another important formal logician, Arend Heyting, who proposed a distinct three-valued account.

Continuing and developing some important work in the philosophy of mathematics by L.E.J Brouwer, Heyting's[2] work led to the birth of an alternative school of formal logic known as **intuitionism**. Most famously, (or infamously perhaps) the law of excluded middle is not a theorem of the intuitionist formal system. That is *not* to say that intuitionists assert the negation of this sequent. Indeed, the double negation of the sequent is a theorem. But the intuitionist rejects double negation-elimination. Hence, the law of excluded middle is not provable within the system (recall the classical proof from Exercise 3.9). (To get a good picture of intuitionist PL subtract DNE from classical PL and replace it with ex falso quodlibet.) The intuitionist account of formal logic has been further developed by certain contemporary formal and philosophical logicians; particularly Michael Dummett.[3] We need not pursue intuitionist logic any further here, however, and, for present purposes, it is sufficient to note that in classical PL the law of excluded middle certainly is a theorem and, moreover, that bivalence always holds. Therefore, in PL no well-formed formula lacks a truth-value. Every well-formed formula is either true or false and so no well-formed formula is neither true nor false. There is, as it were, no third option in PL. Formulas always and only alternate between two poles: the true and the false.

III
Truth-Functionality

Given that PL is bivalent we know that each atomic formula of PL can only be true or false. However, this is not the end of the matter but only the beginning. As you know, we can always combine atomic formulas to form more complex formulas using the logical connectives. In turn, such complex formulas may themselves be combined, again, using the logical connectives. The interesting question is: what is the truth-value of a complex formula constructed from simpler formulas using the connectives? The answer to that question is that the truth-value of a complex formula depends upon two things and only upon those two things: first, the original truth-values of the constituent formulas; second, the particular connective used to form the complex formula. If we know both of these things, we can then work out the truth-value of the complex formula quite mechanically. Impressively, this property holds for each and every complex formula of PL.

In more formal terms, the overall truth-value of any compound formula of PL is said to be a **function** of the truth-values of its component parts. For our purposes, a *function* is just any operation which, when applied to a number of specified objects, generates another specific object. The objects which we put into the function are known as *arguments* for that function. The specific object which the function gives us back is the *value* of that function for those arguments. Functions are really already perfectly familiar from arithmetic. We give the square function 'X^2' numbers as arguments and it gives us back a number as value. For the argument '2', 'X^2' gives us back '4' as its value and so on, ad nauseam. In PL, of course, we are not concerned with numbers but with formulas and their truth-values. We put truth-values in and get a truth-value back. Hence, in Logicspeak, the complex formulas of PL are said to be **truth-functional**, which simply means that the truth-value of any such complex formula can always be computed on the basis of the assignments of truth-values to the component parts of that complex formula.

But this is very general. Different kinds of complex formula can be arrived at by using different connectives. And the resulting overall truth-value of the complex formula depends both upon the original values of the component formulas and upon the particular connective employed. Therefore, the resulting overall truth-value may well vary from connective to connective. So, to answer our original question about the truth-value of complex formulas more precisely we must consider each connective in turn.

IV
Truth-Functions, Truth-Tables and the Logical Connectives

First, let's consider together the simplest possible cases: the truth-conditions for the most basic truth-functional compounds of PL. A more general formal definition is given at the end of this section but, in fact, these simple cases exemplify the meanings of the logical connectives in PL. Recall our discussion of the nature of classical negation in Chapter 3. There we noted that negation is denial and that the effect of negating a formula is to reverse the truth-value of that formula. So, for example, when P is true ~P is false and when P is false ~P is true. Earlier, we tabulated that situation as follows:

P	~P
True	False
False	True

For the sake of linguistic economy, I will introduce two new symbols here (again, as autonyms) which we can use to produce a briefer notation for the tabulation, i.e. we will now simply enter T in place of 'True' and F in place of 'False' in any such tabulation. Hence, we can produce a leaner-looking table with the same content:

P	~P
T	F
F	T

The tabulation exemplifies both the reversal of truth-value which negation effects and the functional nature of that connective: we put a truth-value in and get another truth-value out. More specifically, for any formula of PL, given T as argument the classical negation function gives F as its value. Given F as argument the classical negation function gives T as its value. So, negation is indeed a function which takes a truth-value as its argument and gives a truth-value as its value. Any function whose arguments and values are both truth-values is a **truth-function**. Therefore, negation is a truth-function and, as we shall see, so is every other connective of PL. Before we go on to consider the remaining connectives look again at the tabulation for '~'. The table clearly spells out the effect of the negation function on the atomic formula P taking P's truth-values as its range of arguments. In the case of each connective, we will spell out the function involved using a similar tabulation. Because these tables tabulate truth-values they are known, logically enough, as **truth-tables**.

Given the principle of bivalence, the first column of truth-values in the table under P represents all the possibilities for P's truth-value. Moreover, each entry in that column represents an assignment of truth-values to the component formula P. So, each entry, T and F, is itself an interpretation of the formula P, and together those two interpretations exhaust all the possibilities for assigning truth-values to P. This is the simplest possible case. When we move to consider binary connectives we must consider assignments of truth-values to two formulas and, again, we must be sure to cover every possible case. But this is perfectly straightforward. Each atomic formula can only be either true or false. So, when we are concerned with two such formulas, both may be true, one may be true and the other false or, finally, both may be false. That exhausts the possibilities for truth-value assignments to two atomic formulas. You can see how all these possibilities are represented in a truth-table below. But, before you consider the table, recall the argument given in Section I to the effect that meaning and truth-conditions are intimately related. It should now be clear that we can use truth-tables to define the connectives just in terms of the conditions under which formulas having any particular connective as the main connective are true or false, i.e. the meanings of the logical connectives of PL are characterised truth-conditionally. Hence, for the connectives of PL, meaning is indeed identical with truth-conditions. Finally, any connective which can be completely defined by a truth-table and which takes truth-valued formulas as arguments to give other truth-valued formulas as values is a **truth-functional connective**. As we shall see, in PL, every connective is a truth-functional connective.

Now study Box 4.1, which recaps the chapter so far.

BOX 4.1

♦ *Formal Semantics* explicitly concerns the interpretation of formulas of PL.

♦ An *interpretation* of a PL formula is just any assignment of truth-values to its component atomic formulas.

♦ Any interpretation which results in the whole formula being true overall is a *model* of that formula.

♦ PL is *bivalent*, i.e. every PL formula has just one truth-value: T or F, and there are only those two truth-values.

♦ The truth-value of any compound PL formula is a *function* of (i) the truth-values of its atomic constituents *and* (ii) the particular connective/s used to form that formula.

♦ The meaning of each of the logical connectives is fixed by a truth-table which takes truth-valued formulas as arguments and gives truth-valued formulas as values. Any such connective is *truth-functional*. Every PL connective is a *truth-functional connective*.

Here is the truth-table for the simple conjunction P & Q in PL:

P	Q	P & Q
T	T	T
T	F	F
F	T	F
F	F	F

Note that the conjunction P & Q is only true when both conjuncts are true. Otherwise, it is false. This is surely intuitive: we would not want to say that a conjunction in natural language was true if either or both of its conjuncts were not true. Earlier I noted that we could consider PL as an analogue of natural language and, in this instance at least, we seem to have a very close approximation. So, does '&' in PL faithfully represent our use of 'and' in natural language? The truth-table certainly represents one aspect of the use of natural language 'and', namely, that unless both conjuncts are true we will not allow that the conjunction is true. However, there are other respects in which '&' does not represent standard natural language usage. In PL, the truth-value of P & Q is wholly indifferent to the *order* of the conjuncts. But this is by no means always the case in natural language. 'I got out of bed and had a shower' is by no means equivalent to 'I had a shower and got out of bed.' In the latter case, it seems to follow that I showered in bed! Natural language 'and' possesses a sense of temporal direction, 'and then . . . ', while '&' is strictly atemporal. Gordon Baker and Peter Hacker give the following, particularly pungent, illustration of the point:[4] 'He died and was buried'! More topically A.A. Luce points out that: '"He learned Logic and died" is not the same as "He died and learned Logic".'[5]

Perhaps we could employ a new formal connective which did this job, however. We might invent a formal counterpart for 'whereupon', for example. Clearly, there is more to the truth of 'I got out of bed whereupon I had a shower' than simply the truth of 'I got out of bed and I had a shower.' But that is precisely the problem for the formal logician: if anything other than the truth-values of the component formulas can make a difference to the truth-value of the overall formula then that connective cannot be a truth-functional connective. Adding in any non truth-functional connective would spoil the game. 'Whereupon' is not truth-functional and so adding it in would deprive PL of its impressive truth-functionality.

Moreover, 'and' does not always occur as a connective in natural language: 'The television in the hotel room was only a black and white one', etc. Worse still, while it is common to translate 'but' in terms of '&' in PL, 'but' and 'and' plainly are not equivalent in natural language. For example, try swapping 'but' for 'and' in the following sentence: 'The students all did very well and I was happy.'

What are we to conclude here? First, we must again recognise the existence of a controversy in formal logic and, perhaps, confront a limit to the success of the formal logician's enterprise. Ideally, '&' would faithfully represent every aspect of the use of 'and' in natural language. But it does not. In so far as it does, so far might it be held to be an analogue of 'and'. But we are not forced to press that point. Remember: truth-tables are used to define the meaning of the connective which stands on its own two feet, as it were. Moreover, just as PL is of great interest, importance and practical utility as an autonomous formal language independent of any natural language so too '&' is of interest and importance just in itself, without any reference to 'and'.

It may seem surprising, especially in semantics, to learn that bouncing your intuitions off natural language is not always the best way to deepen your understanding of the formal language. But there is some truth in that claim none the less. Further, classical formal logic is only one among many. We have noted that both relevance and intuitionist logicians propose rather different systems which revise classical logic. Perhaps these are better analogues. Perhaps not. At this stage, the important point to note is that unless you have a good grasp of classical formal logic you will not be in a good position to assess the merits of systems which revise it.

Moreover, recall our earlier analogies between formal logic and chess. When we learn how a given piece moves in that game we come to know what it means to be that piece in that game. It doesn't make a great deal of sense to ask: but why does the bishop move in diagonals of one colour only? Similarly, it can be more useful to accept that the truth-table definitions of the connectives define the nature of those connectives in the game of classical formal logic, in the formal logic *language-game*. Looked at in this way, each logical language-game is of interest just in its own right; and none more so than the classical one. With these points in mind let's consider the truth-tables for the most basic compounds formed using the remaining connectives.

Here is the table for the disjunction P v Q in PL:

P	Q	P v Q
T	T	T
T	F	T
F	T	T
F	F	F

Note that P v Q is only false when both disjuncts are false. The truth of one disjunct is sufficient for the truth of the whole disjunction. But note that the disjunction is also true when both disjuncts are true. This reflects the fact noted earlier that in PL disjunction is used in the inclusive sense, i.e. it includes the possibility that both disjuncts are true.

Again, there is something analogous to natural language 'or' in the truth-table for P v Q just in so far as we would want to recognise the sufficiency of the truth of one disjunct for the truth of the disjunction. However, we often use 'or' in the exclusive sense in natural language: 'Either Paul is revising for his logic exam or he is sleeping the sleep of the wicked.' In this case, while either disjunct might be true, both surely cannot be true at one and the same time! But this time the formal logician is rather better off. Exclusive disjunction is certainly definable in terms of inclusive disjunction, namely, as ((P v Q) & ~(P & Q)) and both senses are truth-functional.

Here is the truth-table for P → Q:

P	Q	P → Q
T	T	T
T	F	F
F	T	T
F	F	T

This particular truth-table exposes the arrow as the least analogous of the logical connectives to their natural language counterparts. Note that the conditional is only false if the consequent is false just when the antecedent is true. Perhaps that much seems intuitive. But what of the other interpretations of those formulas? Why should the conditional be considered true when both its antecedent and its consequent are false, for example? Here, the classical formal logician may appeal to the fact noted in earlier chapters that both Q and ~P independently entail P → Q, i.e. I proved in Chapter 2 that: Q ⊢ P → Q and you will yourself have proved in Chapter 3 that ~P ⊢ P → Q. So, the argument goes, as long as Q is true or P is false then P → Q must be true. But this is not an end to the matter. The relevance logician, for example, is quick to amend classical proof theory so that neither of these controversial sequents is provable. In truth, the conditional gives rise to a plethora of controversies and we will not have space or time even to mention them all! But cast your mind back to our chess analogy. Remember: the truth-table is used to define the conditional in the language-game of classical formal logic, and that really is an end to the matter.

Before we move to consider the table for the simplest biconditional formula in PL it's worth noting that the connectives we have considered to date can be used to represent another English connective, namely, 'unless'. For example, 'Q unless P' can fairly be rendered as 'if not P then Q' and thus as ~P → Q in PL. Moreover, this formula is in turn equivalent to the simple disjunction P v Q and so we could also faithfully translate *unless* using disjunction. Finally, here is the truth-table for the simple biconditional formula P ↔ Q:

P	Q	P ↔ Q
T	T	T
T	F	F
F	T	F
F	F	T

If the truth-table for the basic conditional formula was the least intuitive and least analogous to natural language, the truth-table for the basic biconditional formula is surely the most intuitive. The relation of material equivalence between formulas which is expressed by the biconditional only holds true if both formulas are true together or both are false together. In this case, then, there really does seem to be an analogy between the truth-table for the connective '↔' and natural language use of the 'if and only if' construction.

Actual use of 'if and only if' in natural language may seem rather thin on the ground but, in fact, the construction is extremely useful in both mathematics and the natural sciences, where it is frequently exploited in definitions of formal and technical terms. In ordinary discourse it is much more common to find elements such as 'if' and, indeed, 'only if' on their own, as it were, but we should be very wary about translating any such use in terms of the biconditional. After all, as we noted in Section X of Chapter 2, both 'P only if Q' and 'P if Q' are accurately translated as conditionals, i.e. 'P only if Q' should be rendered as $P \rightarrow Q$, while the use of 'if' reverses the order of the antecedent and consequent, and so 'P if Q' is accurately translated as $Q \rightarrow P$. As you know, the biconditional $P \leftrightarrow Q$ is equivalent to the conjunction of both these conditionals. Therefore, the biconditional does indeed translate 'if *and* only if'.

We can now summarise all of the truth-tables for the most basic truth-functional compounds formed using the connectives which we have considered in this section and condense them all into a single, rather longer table. Because there are four options for truth-value assignments to the binary connectives but only two options for our lone unary connective, the columns for that connective are only half as long as their binary counterparts:

P	~P	P	Q	P & Q	P v Q	$P \rightarrow Q$	$P \leftrightarrow Q$
T	F	T	T	T	T	T	T
F	T	T	F	F	T	F	F
–	–	F	T	F	T	T	F
–	–	F	F	F	F	T	T

Fortunately, we are by no means confined to these most basic of truth-functional compounds. Any well-formed formula of PL no matter how complex may be negated and the result of negating that formula will be precisely to reverse the truth-value. Any two well-formed formulas may be conjoined, disjoined, conditionalised or biconditionalised, to coin a phrase. In order to represent the level of generality involved here and to give truth-tables which really do define the connectives of PL quite generally we must again have recourse to the variables 'A', 'B', 'C', etc., which I first introduced in the final section of Chapter 2. As we noted there, these variables range over well-formed formulas of PL, i.e. 'A' stands for any well-formed formula of PL, 'B' stands for any well-formed formula and so on. Because these variables represent formulas of PL they do not themselves belong to PL. Rather, they belong to the metalanguage in which we talk about PL as *object* language, i.e. these are *metalinguistic variables*. We can utilise just those variables to express the effect of each connective on the well-formed formulas of PL. In so doing, we can at last construct truth-tables which formally define the logical connectives. The table in Box 4.2 summarises the relevant truth-tables.

BOX 4.2

A	~A	A	B	A & B	A v B	A → B	A ↔ B
T	F	T	T	T	T	T	T
F	T	T	F	F	T	F	F
–	–	F	T	F	T	T	F
–	–	F	F	F	F	T	T

V
Constructing Truth-Tables

Given the truth-table definitions of the logical connectives we can now go on to calculate quite mechanically the truth-value of any complex formula of PL for any assignment of truth-values to its constituent atomic formulas. This kind of semantic investigation reveals some interesting properties of certain formulas of PL. But, before we go on to look at some actual cases, let's first consider how truth-tables are constructed via three simple examples.

First, consider the simplest possible kind of case, a negated sentence-letter. Suppose that we are interested in the overall truth-values of ~P, for example. How should we proceed to construct the truth-table? Well, the very first thing to do is to draw a horizontal line at least one and a half times as long as the formula involved and write out the formula to the right of that line. So, for example, in the case of ~P we write:

~ P

Next, we identify *each and every* sentence-letter involved in the formula, list these in alphabetical order to the left of the formula and separate the two with a vertical line. In the present case we only have P to worry about so:

P	~ P

We must now assign truth-values to the sentence-letter listed on the left. The number of truth-value assignments required, i.e. the number of *rows* of assignments to the sentence-letter(s) on the left, is just 2^n where 'n' is the number of sentence-letters involved. So, if only *one* sentence-letter is involved, as in the present case, the number of rows of assignments is 2^1, i.e. 2.

Having done so, we construct the required number of truth-value assignments as follows: under the sentence-letter nearest the formula itself list the required number of truth-value assignments (in this case 2) beginning with T and alternating with F to the relevant number, e.g.:

	P	~ P
1.	T	
2.	F	

To complete the truth-table (i.e. to complete the blank box under the formula itself) consider each complete truth-value assignment, i.e. each row of the table (in this case rows numbered 1–2) very carefully, then, in each case, begin by simply writing the truth-value assigned to the sentence-letter by that assignment under each and every occurrence of the relevant sentence-letter in the formula itself. So, for example, because row 1 assigns T to P, we enter T under the occurrence of P on that line in the formula, and because row 2 assigns F to P we enter F under P on that assignment, like so:

	P	~	P
1.	T		**T**
2.	F		**F**

Next, we identify the *main connective* and highlight the column below it by enclosing it in a box with **'m.c.'** written underneath. Thus, we emphasise that this column is the **main column**, i.e. the column in which the *overall* truth-value of the whole formula will be recorded under each assignment.

	P	~	P
1.	T		T
2.	F		F
		m.c.	

Finally, for each assignment, consider the truth-value assigned to P itself and then simply exploit the truth-table for the relevant connective (in this case '~') to compute the overall values of the formula. According to the table for '~', of course, when a formula is true its negation is false and when a formula is false its negation is true. Hence, we complete the table by entering just those overall values under the main connective as follows:

	P	~	P
1.	T	**F**	T
2.	F	**T**	F
		m.c.	

We have considered the simplest possible case here and you should not be surprised to learn that your logic exam will contain slightly more complex cases. None the less, the procedure is very similar, if a little more articulated. For example, consider the formula (P & Q) → (Q v P). Again, we write the formula on a line, identify *every* sentence-letter involved in the formula and list these in alphabetical order on the left, separating the two with a vertical line:

P	Q	(P & Q) → (P v Q)

.
.

Next, we assign truth-values to the sentence-letters listed on the left. The number of truth-value assignments required is 2^n where 'n' is the number of sentence-letters involved. So, when two sentence-letters are involved, as in the present case, the number of rows of assignments is 2^2, i.e. 4. Again, we begin with the sentence-letter nearest the formula itself (in this case Q) and assign truth-values beginning with T and alternating with F to the requisite number (in this case 4).

	P	Q	(P & Q) → (P v Q)
1.		T	
2.		F	
3.		T	
4.		F	

Under the remaining sentence-letter (in this case P) we again list the required number of truth-value assignments but this time beginning with two Ts and alternating with two Fs to the required number:

	P	Q	(P & Q) → (P v Q)
1.	**T**	T	
2.	**T**	F	
3.	**F**	T	
4.	**F**	F	

As before, we begin to complete the table simply by writing the truth-value assigned to the sentence-letter by that assignment under each and every occurrence of the relevant sentence-letter in the formula itself. So, for example, because row 1 assigns T to P and to Q, we enter T under each and every occurrence of P and Q on that line like so:

	P	Q	(P & Q) → (P v Q)
1.	T	T	**T T T T**

Next, we must again identify the main connective and highlight the column below it by enclosing it in a box marked 'm.c.' to emphasise that this column is the main column, i.e. the column in which the overall truth-value of the whole formula will be recorded under each assignment:

	P	Q	(P & Q) → (P v Q)
1.	T	T	T T [ ] T T
2.	T	F	[ ]
3.	F	T	[ ]
4.	F	F	[ ]
			m.c.

The main connective in this formula is the arrow. So, the whole formula is a conditional. However, its antecedent is a conjunction and its consequent is a disjunction. In order to calculate the overall truth-value then we must first calculate the value of the antecedent using the table for '&', then calculate the value of the consequent using the table for 'v', and write these values under the relevant connectives. To that end, consider the first assignment carefully. Because both P and Q are true the conjunction P & Q is also true, and we record that fact by entering T under the connective '&'. Further, because both P and Q are true, the disjunction P v Q is also true, and we record that fact by writing T under the disjunction symbol. So far, then, line 1 looks like this:

	P	Q	(P & Q) → (P v Q)
1.	T	T	T **T** T [ ] T **T** T

In turn, these two results now provide our *input* truth-values for the final calculation, which we make simply by using the truth-table for '→'. Consider line 1, for example; the antecedent (P & Q) is true and so is the consequent (P v Q). When that is so the truth-table for the conditional assures us that the whole conditional is true. Therefore, the overall value of the whole formula (P & Q) → (P v Q) under the first assignment of truth-values is T. So, we may complete line 1 as follows:

	P	Q	(P & Q) → (P v Q)
1.	T	T	T T T [**T**] T T T

To complete the truth-table in its entirety, we simply follow the same procedure for the remaining assignments 2–4. The completed truth-table looks like this:

	P	Q	(P & Q) → (P v Q)
1.	T	T	T T T [**T**] T T T
2.	T	F	T F F [**T**] T T F
3.	F	T	F F T [**T**] F T T
4.	F	F	F F F [**T**] F F F
			m.c.

By now, you will have a fair idea of how to go about constructing truth-tables but, before we go on to tackle an exercise, I will sum up the procedure and give one last illustration, namely, the table for the formula P & (Q v R). So, to construct a truth-table observe the following procedure carefully:

1 Draw a horizontal line at least one and a half times as long as the formula(s) involved and enter the formula(s) to the right of that line, e.g.

P & (Q v R)
____

2 Identify *all* the sentence-letters involved in the formula(s), list these in alphabetical order to the left of the formula and separate the two with a vertical line, e.g.

P	Q	R	P & (Q v R)

.

.

3 Assign truth-values to the sentence-letters listed on the left. The number of truth-value assignments required, i.e. the number of *rows* of assignments to the sentence-letters on the left, is just 2^n where 'n' is the number of sentence-letters involved. So, if only one sentence-letter is involved, the number of rows of assignments is 2^1, i.e. 2. If two sentence-letters are involved, the number of rows of assignments is 2^2, i.e. 4. If, as in the present case, three sentence-letters are involved, the number of rows of assignments is 2^3, i.e. $(2 \times 2) \times 2$, so 8. Construct the required number of truth-value assignments as follows:

(i) Under the sentence-letter nearest the formula itself (in this case R) list the required number of truth-value assignments (in this case 8) beginning with T and alternating with F to the relevant number, e.g.:

	P	Q	R	P & (Q v R)
1.			**T**	
2.			**F**	
3.			**T**	
4.			**F**	
5.			**T**	
6.			**F**	
7.			**T**	
8.			**F**	

(ii) Under the next sentence-letter to the left (in this case Q) again list the required number of truth-value assignments but this time beginning with two Ts and alternating with two Fs to the required number, thus:

	P	Q	R	P & (Q v R)
1.		**T**	T	
2.		**T**	F	
3.		**F**	T	
4.		**F**	F	
5.		**T**	T	
6.		**T**	F	
7.		**F**	T	
8.		**F**	F	

(iii) Under the next sentence-letter to the left (in this case P) again list the required number of truth-value assignments but this time beginning with four Ts and alternating with four Fs to the required number, thus:

	P	Q	R	P & (Q v R)
1.	**T**	T	T	
2.	**T**	T	F	
3.	**T**	F	T	
4.	**T**	F	F	
5.	**F**	T	T	
6.	**F**	T	F	
7.	**F**	F	T	
8.	**F**	F	F	

4 To complete the truth-table consider each truth-value assignment, i.e. each row of the table (in this case rows numbered 1–8), then:

(i) In each case, begin by entering the truth-value assigned to each sentence-letter by that assignment under each and every occurrence of the relevant sentence-letter in the formula itself. In this case, because row 1 assigns T to P, Q and R, we can enter T under each and every occurrence of P, Q and R in the formula like so:

	P	Q	R	P & (Q v R)
1.	T	T	T	**T** **T** **T**

(ii) Identify the main connective and highlight the column below it by enclosing it in a box with 'm.c.' written underneath it thus:

	P	Q	R	P & (Q v R)
1.	T	T	T	
2.	T	T	F	
3.	T	F	T	
4.	T	F	F	
5.	F	T	T	
6.	F	T	F	
7.	F	F	T	
8.	F	F	F	
				m.c.

(iii) Identify the *scope* of every other connective involved and, for each assignment, using the truth-values already assigned to the sentence-letters involved and the truth-tables for the relevant connectives, establish the overall truth-value of each sub-formula.

In the present case the formula is a conjunction whose first conjunct is P and whose second conjunct is (Q v R). On assignment 1 every sentence-letter is assigned T and so we simply appeal to the truth-table for disjunction to establish that when both disjuncts are true the disjunction is true. Hence, the overall value of the formula (Q v R) is also T:

	P	Q	R	P & (Q v R)
1.	T	T	T	T T **T** T

(iv) Next, using the truth-values of the sub-formulas we have just worked out as inputs, we simply use the appropriate truth-table for the *main* connective to compute the *overall* value of the whole formula under each assignment of truth-values. In the case above, the conjunction has two true conjuncts, namely, P and (Q v R). A conjunction with two true conjuncts is itself true, so we complete this line by entering T in the column belonging to the main connective like so:

	P	Q	R	P & (Q v R)
1.	T	T	T	T **T** T T T

(v) Finally, repeat steps (i)–(iv) for every other assignment involved.

Before attempting Exercise 4.1 note carefully that while the main connective in the formula ~P & (Q v R) is '&' the main connective in the formula ~(P & (Q v R)) is '~'. Therefore, in that latter case, and in the case of every other negated formula, the overall truth-value of the formula must be stated under that occurrence of the connective. (Anyone who wants a refresher course on the notion of the **scope** of a connective might like to reread Section II of Chapter 2 before attempting the following exercise.)

EXERCISE 4.1

1 Construct and complete truth-tables for the following formulas of PL:

1. P → (P & P)
2. P & ~P
3. P v ~P
4. P → (Q → P)
5. (P & Q) ↔ (Q & P)
6. (P v Q) ↔ ~Q
7. (P → Q) → (~Q → ~P)
8. ~(P v Q) ↔ (~P & ~Q)
9. ~P & (Q v R)
10. ~(P & (Q v R))
11. ~ ~(P & ~P)
12. ((P → Q) → (Q → R)) → (P → R)
13. ~((P → Q) → (Q → R)) → (P → R)
14. (P → ((Q → R) v ~R)) v ~Q
15. ((P & Q) → (R v ~S)) → T

VI
Tautologous, Inconsistent and Contingent Formulas in PL

Your answers to Exercise 4.1 should have revealed three possible types of truth-table which any formula may have:

1 First, a formula may have a mixture of true and false overall values recorded in the main column. Whether such a formula comes out true or not depends upon the particular assignment of truth-values to its constituent atomic formulas. In logical terms, the overall truth-value of the formula is *contingent* upon that prior truth-value assignment and, logically enough, any such formula with a mixed bag of overall values is said to be a **contingent formula**. For the formal logician, a contingent formula is contingent in virtue of its logical form, i.e. that formula has a contingent form and any other instance of that form must also be contingent. This point will be easier to see when we have considered a rather different kind of truth-table.

2 You should already have noticed that certain formulas in Exercise 4.1 are true under every interpretation. No matter which particular assignment of truth-values we consider, these formulas are always and only true, i.e. such formulas have T for each and every entry in the main column. This special class of formulas of PL which come out true overall under every single assignment of truth-values to their component parts are known as **tautologies** in Logicspeak. Tautologies are formulas of a logical form such that, given the meanings of the logical connectives, those formulas cannot but be true. Hence, the formal logician's claim that tautologies are tautological in virtue of having a tautological form (and, indeed, the earlier claim about contingent formulas being of a contingent form). Moreover, because a tautology *must* be true under every interpretation tautologies are often described as empty or vacuous. What tautologies are 'empty' of is just any factual or, in philosophical terms, *empirical* content. Tautologies are not factual, empirical truths but logical truths. The laws of logic are definitively logical truths. So, it should be no surprise that the laws of logic are tautologies. This is intuitive and can easily be proved. For example, let's test the law of excluded middle which I mentioned earlier, ⊢ P v ~P. Is it a tautology?

	P	P	v	~	P
1.	T	T	**T**	F	T
2.	F	F	**T**	T	F
			m.c.		

This formula is true under every interpretation. Therefore, as we would expect, this formula is indeed a tautology.

3 There is one last possibility which we have still to consider. Certain formulas never have the overall value T. When each and every entry in the main column is F the formula is a logical falsehood. A logical false-

BOX 4.3

♦ Given the meanings of the logical connectives, we can use truth-tables to distinguish three kinds of formula in PL:

♦ Formulas which have a mixture of T and F as overall truth-values recorded in the main column are *contingent* formulas.

♦ Formulas which have T for each and every entry in the main column are *tautologies.*

♦ Formulas which have F for each and every entry in the main column are *inconsistent* formulas.

Note carefully that this classification is exhaustive.

hood cannot possibly be true. Any such formula is strictly self-contradictory. Formulas of this third type are said to be **inconsistent** formulas. As ever, it is the logical form of an inconsistent formula which guarantees that the formula is logically false.

Finally, note that the tripartite classification of formulas into tautologies, contingent and inconsistent formulas is exhaustive. Each and every formula of PL belongs to one of the three categories. This is summarised in Box 4.3.

EXERCISE 4.2

1 Consider again each of the fifteen formulas in Exercise 4.1. For each formula say whether it is (i) contingent, (ii) tautologous or (iii) inconsistent. In each case, give reasons for your answer.

2 In terms of the tripartite distinction between formulas given above, what is the status of (i) the negation of any tautologous formula and (ii) the negation of any contingent formula?

3 Suppose that you had been asked simply to test Formula 15 of Exercise 4.1 for tautologousness. How many lines of the truth-table for that formula would you have had to construct and why?

VII Semantic Consequence

So far, truth-tables have proved useful in providing us with an entirely mechanical procedure for determining the overall truth-value of any complex formula of PL. Given any formula whatsoever, we can always use the truth-tables to determine the overall truth-value of the formula for each and every interpretation of its constituents. But that is by no means an end to the usefulness of truth-tables. Reflect once more upon the definition of validity: a valid argument is an argument such that whenever the premises are true, the conclusion must be true. Hence, if an argument is valid it is impossible that its conclusion should be false when its premises are true.

In more formal terms, this amounts to saying that a sequent is valid if and only if there is no interpretation (i.e. no possible assignment of truth-values to the component formulas) under which all the premises of that sequent are true while the conclusion is false. To define validity in these terms is to define validity semantically. But recall that a sequent simply consists of some set of formulas as premises together with a formula as conclusion, the two being separated by the colon.

Now, it is easy to see that we can always use the truth-tables to determine whether or not a given sequent is semantically valid. All we need do is construct a truth-table for the formulas constituting that sequent, then look and see whether there is any interpretation under which the premises are true while the conclusion is false. If there is, the sequent is not semantically valid; but if not, then that sequent will indeed be semantically valid. In other words, we simply lay the whole sequent out on a line and construct a truth-table carefully calculating the overall truth-value of each premise and the overall truth-value of the conclusion for each assignment. Such a truth-table is often said to be a **comparative truth-table** just because we *compare* the truth-values of the premises with the truth-value of the conclusion to determine semantic validity.

For example, consider the sequent P → Q, Q : P. Is this sequent semantically valid? We can easily construct a comparative truth-table to show that it is not:

	P	Q	P	→	Q	Q	P
1.	T	T	T	T	T	T	T
2.	T	F	T	F	F	F	T
3.	F	T	F	**T**	T	**T**	**F**
4.	F	F	F	T	F	F	F

Under the third assignment of truth-values both premises are true while the conclusion is false. Therefore, there is a way in which the conclusion can be false while the premises are true. So, this particular sequent is semantically invalid. Moreover, we also learn from the truth-table exactly which assignment of truth-values to its constituent formulas demonstrates the invalidity of the sequent. Look closely at the third interpretation: here P is false, while Q is true. This assignment of truth-values is exactly that interpretation which invalidates the sequent. In Logicspeak, any such invalidating interpretation is an **invalidating PL interpretation**, or **IPLI** for short. Now, any IPLI tells us a great deal about what we might call the *general character* of a counterexample to that sequent, i.e. it tells us precisely which truth-values should be assigned to precisely which atomic formulas so as to generate true premises together with a false conclusion. In fact, to get from an IPLI to an actual counterexample all we need to do is to study the IPLI carefully, identify which atomic formulas are assigned T and which are assigned F, and then assign an actually true natural language sentence to each formula assigned T and an actually false natural language sentence to each formula assigned F.

For example, in the present case, we simply pick a false sentence for P, a true sentence for Q, and then mimic the logical form of the original sequent exactly. So, let P stand for the false sentence that 'All cats are black.' Let Q stand for the true sentence that 'Zebedee is black.' To generate an actual counterexample all that remains is to mimic the form of the sequent, making appropriate (uniform) substitutions thus:

1. If all cats are black then Zebedee is black	$P \rightarrow Q$
2. Zebedee is black.	Q
Therefore,	:
3. All cats are black.	P

Hence, we can always exploit the truth-table of any invalid sequent to arrive at an IPLI of that sequent, and it is a short step from there to an actual counterexample.

Further, suppose that we are interested in the following sequent: P v Q, ~Q : P. Is this sequent semantically valid? Again, we can easily use the truth-tables to get an answer:

	P	Q	P	v	Q	~	Q	P
1.	T	T	T	T	T	F	T	T
2.	T	F	T	T	F	T	F	T
3.	F	T	F	T	T	F	T	F
4.	F	F	F	F	F	T	F	F

Remember: the sequent is semantically valid if and only if there is no interpretation under which the premises are all true when the conclusion is false. Now, the conclusion, P, is only false under the last two of the four assignments. Is it the case that both premises are true under either of those assignments? No. In each case, only one premise is true. This time there is no interpretation under which all the premises are true when the conclusion is false, i.e. there is no IPLI for this sequent. Therefore, the sequent P v Q, ~Q : P is semantically valid. Because we established the validity of this sequent semantically rather than syntactically we cannot use the syntactic *single* turnstile '⊢' to mark the relation of consequence involved here. Instead, we use the **semantic double turnstile** '⊨' to mark this notion of consequence. The double turnstile therefore represents the notion of **semantic consequence** in PL. Semantic consequence is defined as follows: a given formula of PL is a semantic consequence of some set of formulas of PL if and only if there is no interpretation which assigns T to each member of that set of formulas which does not also assign T to the given formula. Formally, we can subscript the first of our metalinguistic variables, namely 'A', to make a countable but possibly infinite set of such metalinguistic variables explicit, i.e. $\{A_1, A_2, A_3. \ldots\ldots\ldots A_n\}$.

We can now consider any such set as representing a set of premises in PL. Further, let 'B' represent any conclusion alleged to follow as a consequence from that set of premises. We can now represent the **general form of a sequent** as:

$\{A_1. \ldots\ldots A_n\} : B.$

In the present context we are specifically interested in the notion of semantic consequence represented by ' ⊨ '. So, we formally represent the relation of semantic consequence in PL as follows:

$\{A_1. \ldots\ldots A_n\} \models B$ *if and only if every interpretation which assigns* T *to each of the formulas* $\{A_1. \ldots\ldots A_n\}$ *also assigns* T *to formula* B.

Finally, where the sequent in question has no premises there will be no formulas to the left of the colon symbol to consider. In such cases, the test for semantic consequence simply reduces to a test for tautologousness of the formula on the right of the colon symbol (the conclusion). If the conclusion is always true under every interpretation then the sequent is certainly valid in PL. A given formula is a tautology, of course, only if the overall truth-value of the formula is T for *every* interpretation.

We have now arrived at two characterisations of the notion of logical consequence in PL, namely, the syntactical and the semantic. And it may well look as if we have two different answers to the fundamental question: when does one formula follow logically from some other formula or formulas in

PL? At this stage, it is natural to compare the two characterisations of consequence and, indeed, to wonder exactly how the two stand to one another.

In the last analysis, you may feel that the notion of semantic consequence, being so explicitly concerned with the preservation of truth across inference, exemplifies the notion of soundness of argument which we have been trying to capture since Chapter 1. But consider again, for example, the sequent P v Q, ~Q : P. We already knew both that P v Q, ~Q ⊢ P and that P v Q, ~P ⊢ Q (you should yourself have proved both sequents in Revision Exercise III in Chapter 3).[6] These syntactical results may well have led you to anticipate that P v Q, ~Q : P would indeed turn out to be semantically valid as well. In fact, you would have been perfectly justified in anticipating the semantic validity of this sequent just on the strength of its provability. Strikingly, the same holds true for *every* provable sequent of PL: every provable sequent is also a semantically valid sequent. In other words, all the syntactically valid sequents turn out to be semantically valid sequents anyway. This striking property of PL is known as **Soundness**. At this level, what is *Sound* is not some particular argument or sequent but rather the entire set of syntactic rules of inference. To prove Soundness of PL is just to prove that each and every rule of inference is *truth-preserving* and really can be relied upon to yield *only* valid sequents. A proof of Soundness is a guarantee that no rule will ever allow any semantically invalid step into a proof and that truth will always be preserved in every proper application of each and every rule. If no rule of proof allows anything invalid in and every application of every rule preserves validity then only what is valid will be provable.

Unsurprisingly, you will not be expected to be familiar with the nuts and bolts of a Soundness proof in your first-level course in formal logic. However, you may well be expected to know what the term means. Here it is important to distinguish clearly between uses of the terms 'sound' and 'Soundness'. Remember: to say that an argument is sound is just to assert that it is a valid argument with actually true premises. But to prove Soundness of PL is to demonstrate that every provable conclusion in the language is also a semantic consequence of the set of premises from which it is proved. Thus, Soundness ties together both our characterisations of logical consequence in PL, i.e. it ties '⊢' and ' ⊨ ' together. But note that Soundness ties these notions together, we might say, in one particular direction. At this point, we need new logical symbols to express ourselves clearly. If we adopt a **metalogical arrow**, '⇒', to express the connection between the two turnstiles and the symbol **S** for 'Sequent', we can represent Soundness as follows:

Soundness: $\vdash \boldsymbol{S} \Rightarrow \models \boldsymbol{S}$

As far as consequence in PL is concerned, Soundness is only half the story and reflects only half of the tremendous integrity between these two notions of consequence. In fact, PL has another, equally striking, complementary property known as **Completeness**. This property guarantees that every semantically valid sequent is also provable using the rules of inference. Therefore, there is no semantically valid sequent of which a proof cannot be constructed. Therefore, we know that everything we want in the system is in and that nothing has been overlooked or left out. Hence, the *Complete* in Completeness. Another common way to put the point is that every truth-functional tautology of PL is also represented by a theorem of PL. We will consider the point in this form again at the very end of this chapter. However we might put it, it should be clear that Completeness also ties the two notions of consequence together but in the *opposite* direction:

Completeness: $\models S \Rightarrow \vdash S$

Again, you will not be required to be familiar with the Completeness proof for first-level purposes, though, again, you may well be expected to know what the term means.

In conclusion, it should be clear that the two properties of Soundness and Completeness together exemplify a tremendous closeness and integrity between the proof-theory and the semantics of PL, i.e. between '$\vdash$' and '$\models$'. In virtue of those properties the 'two', seemingly very different, notions of logical consequence in fact turn out to be one.

Now study Box 4.4.

Guide to Further Reading

Space does not permit the inclusion of Soundness or Completeness proofs in the present text. For interested parties I include some historical background and pointers to texts which do contain these proofs. Note, however, that you really would be well advised to finish working through the present text before proceeding to another.

As noted in Chapter 1, modern formal logic begins with the work of the great German mathematical logician Gottlob Frege [1848–1925], Professor of Mathematics at Jena University in Germany from 1879 to 1918. Frege first published a rigorous formal system of logic as early as 1879 in his *Begriffsschrift*. This particular text contains a version of propositional logic but it also contains much more besides. In fact, it contains much of the logical machinery which we will go on to consider later in the text.

The point is often made that Frege's genius was lost on many of his

BOX 4.4

♦ A sequent is *semantically valid* if and only if there is no interpretation (i.e. no possible assignment of truth-values to the component formulas) under which all the premises of that sequent are true while the conclusion is false.

♦ To determine whether or not a given sequent is semantically valid, construct a *comparative truth-table* for the formulas constituting that sequent and consider whether there is any interpretation under which the premises are true while the conclusion is false (such a truth-table is said to be *comparative* simply because we *compare* the truth-values of the premises with the truth-value of the conclusion to determine semantic validity). If there is such an interpretation, the sequent is not semantically valid; if not, the sequent is semantically valid.

♦ Any interpretation under which the premises of a given sequent are true while the conclusion is false makes explicit an assignment of truth-values to the constituent formulas of that sequent which demonstrates the invalidity of that sequent. Any such interpretation is an *invalidating PL interpretation*, an *IPLI*, of that sequent.

♦ To generate an actual counterexample from an IPLI: replace each sentence-letter assigned F with a false natural language sentence, each sentence-letter assigned T with a true natural language sentence, and mimic the logical form of the original sequent exactly.

contemporaries. Although the point is a good one, it certainly does not apply to Bertrand Russell [1872–1970]. While a Fellow, and later Lecturer, of Trinity College, Cambridge [1895–1915], Russell became aware of the tremendous philosophical significance of Frege's work and, together with Alfred North Whitehead [1861–1947], set about developing his own system of formal logic. This work resulted in the publication of Russell and Whitehead's *Principia Mathematica*, in three volumes, between 1910 and 1913. Again, this text contains a version of propositional logic; and again much more besides. However, it was the American mathematician E.L. Post [1897–1954] who first proved *about* the system of propositional logic contained in *Principia* the metatheoretical result that the system was Complete, in our sense. In fact, Post's proof was the basis of his doctoral thesis for Columbia University in 1920.

Since 1920 a number of formal logicians have arrived at the same result in different ways. Perhaps the best known alternative version was presented by Leon Henkin in 1947. This version of the Completeness proof is considered to be somewhat simpler and quicker than many of its rivals, and the term 'Henkin proof ' is a commonplace of metatheoretical logical parlance.

More historical detail together with an outline of Post's proof[7] and a full Henkin proof of completeness[8] is given by Geoffrey Hunter in his *Metalogic*. Hunter's text represents a remarkable combination of rigour and accessibility at the metatheoretical level. It is therefore strongly recommended, though, sadly, it is presently out of print. However, interested parties will also find concise and very useful proofs of both the Soundness and Completeness of formal propositional logic in A.G. Hamilton's *Logic for Mathematicians*.[9] Interested parties should not be put off by the title. The text is quite readable and contains useful exercises. The relevant proofs can be found over pp. 27–45 of Chapter 2. Notably, Hamilton refers to the pair of results as establishing the *adequacy* of propositional logic. This is slightly non-standard. As we noted in Chapter 2, this term is usually used to describe a property of various sets of logical connectives (see below: Section XII). Finally, I again emphasise the importance of finishing at least the first part of the present text before proceeding to another.

Now try Exercise 4.3.

EXERCISE 4.3

1 Construct comparative truth-tables for the following sequents. For each sequent state whether it is semantically valid or not. If you find any to be semantically invalid make an IPLI explicit.

1. P v Q : P
2. P v ~Q, ~Q : ~P
3. Q : ~(~P & ~Q)
4. ~(P v ~Q) : ~P & Q
5. P → Q, Q → R : R → P
6. P → (Q → R) : Q → (P → R)
7. P & ~Q : ~(P → Q)
8. Q → P, P v Q : P v R
9. : ((~P → Q) → ~P) → ~P
10. : ~(P v ~Q) → (~P & Q)
11. ~R → Q : (P v Q) → (~R → P)

12. $\sim P \rightarrow (Q \vee R), \sim P \rightarrow \sim R : Q$
13. $P \,\&\, (Q \vee (Q \rightarrow R)) : (P \,\&\, Q) \vee ((P \,\&\, \sim Q) \vee (P \,\&\, R))$
14. $: (P \vee Q) \leftrightarrow \sim(\sim(Q \,\&\, \sim P) \,\&\, \sim(P \rightarrow Q))$
15. $(P \vee Q) \leftrightarrow (\sim R \vee S) : R \rightarrow (P \leftrightarrow \sim(R \,\&\, \sim S))$

VIII
Truth-Tables Again: Four Alternative Ways to Test for Validity

The truth-table test for semantic consequence has a mechanical straightforwardness which recommends it as a formal method for establishing the validity or invalidity of sequents of PL. Indeed, the method is sufficiently mechanical for machines to be programmed to carry out truth-table tests and to churn out results, we might say, unthinkingly. Further, the truth-table method contrasts with attempts to construct a proof for a given sequent just because, on a bad enough day, we may fail to construct a proof even if the sequent is ultimately provable. In contrast, the truth-table method offers a uniformly decisive test for validity which can be applied quite mechanically. As such, that method is known in formal terms as an **effective decision-procedure**, i.e. a procedure which, when followed correctly, always gives us a 'yes' or 'no' answer to our question about validity. However, the very mechanical nature of this test does not inspire a great deal of reflection about the notion of validity which we use that method to explore. So, before we become complete 'mechanicals', and in the hope of gaining a little more insight, let's consider four alternative methods for exploiting the truth-tables.

1 In the previous section we exploited truth-tables to test sequents for validity. There we noted the general form of a sequent:

$\{A_1. \ldots . A_n\} : B$

The truth-table tests we have employed so far simply involve taking a given sequent, constructing tables for premises and conclusion, and comparing the overall truth-values under every interpretation. However, the definition of validity allows at least one other approach. Suppose that the sequent we are dealing with really is semantically valid. If that is so, then there can be no interpretation under which the premises $\{A_1. \ldots . A_n\}$ are all true while the conclusion, B, is false. Therefore, for any valid sequent, the set of formulas consisting of the conjunction of each of the premises of that sequent together with the *negation* of the conclusion of that sequent must be strictly inconsistent.

Recall the original Blind Lemon Jefferson argument once more. Earlier, we noted that the argument could be formalised in PL as the sequent P, P → Q : Q. We know that this particular sequent is provable. So, by Soundness, we also know that it will turn out to be semantically valid. Note what happens when we run a truth-table test on the set of formulas consisting of the conjunction of each of the premises of that sequent with the negation of its conclusion:

	P	Q	(P	&	(P	→	Q))	&	~	Q
1.	T	T	T	T	T	T	T	**F**	F	T
2.	T	F	T	F	T	F	F	**F**	T	F
3.	F	T	F	F	F	T	T	**F**	F	T
4.	F	F	F	F	F	T	F	**F**	T	F
								m.c.		

Just as we would expect, the whole set of formulas is inconsistent. In other words, there is no way in which all of these formulas could possibly be true together. It follows that the sequent is semantically valid. Of course, if that set of formulas had not turned out to be strictly inconsistent then there would indeed have been at least one way in which those formulas could all be true together and therefore the sequent would have been shown to be semantically invalid.

2 Again, recall the general form of a sequent:

$\{A_1 . \ldots . A_n\}$: B

Once more, suppose that the sequent we are considering genuinely is a valid sequent. Finally, recall the truth-table definition of A → B. Remember that the conditional is only false in one case, namely, when its antecedent is true while its consequent is false. Now, given any valid sequent, we ought to be able to conjoin the premises of that sequent, take that conjunction as the antecedent of a conditional and then take the conclusion as the consequent to form a conditional which is tautologous. Why? Just because, if the sequent really is valid, then the premises cannot be true while the conclusion is false, i.e. the antecedent of the conditional we have constructed cannot be true while the consequent of the conditional is false. So, if the original sequent really is valid the conditional cannot possibly be false. Again, consider the sequent which represents the Blind Lemon Jefferson argument in PL:

	P	Q	(P	&	(P	→	Q))	→	Q
1.	T	T	T	T	T	T	T	**T**	T
2.	T	F	T	F	T	F	F	**T**	F
3.	F	T	F	F	F	T	T	**T**	T
4.	F	F	F	F	F	T	F	**T**	F
								m.c.	

Just as we would expect, the conditional we have constructed is a tautology and therefore the original sequent from which we constructed that conditional must be valid. Of course, had an F turned up in the main column there would indeed have been an interpretation under which the premises were true while the conclusion was false. In such a case we would have shown the original sequent to be invalid.

3 A little further thought about the truth-table for the conditional in PL quickly reveals that we can in fact conditionalise the entire sequent we are interested in and construct a nested conditional from it. That is, rather than form the conjunction of the set of premises as we did in (2) above, we could simply conditionalise those premises and take the resulting complex conditional as the antecedent of a more complex conditional with the conclusion as consequent.

Formally, given any sequent of the form $\{A_1 \ldots\ldots A_n\}$: B we can always construct a corresponding conditional:

$$A_1 \rightarrow (A_2 \rightarrow (\ldots A_n \rightarrow B))$$

Logically enough, such a conditional is known as the **corresponding conditional**. Now, if the original sequent is valid then its corresponding conditional must be a tautology. Recall our reasoning in (2) above: a conditional is false if, and only if, its antecedent is true while its consequent is false. For any valid sequent, however, the conditional which consists of the set of premises as antecedent and the conclusion as consequent cannot be false. If it were, then there would have to be a way in which all the premises were true while the conclusion was false. But, if the original sequent really is valid, that is impossible. When we construct a corresponding conditional it is just such reasoning which we exploit. Again, consider the sequent P, P → Q : Q. The corresponding conditional is just (P → ((P → Q) → Q)). The table for that formula is as follows:

	P	Q	(P	→	((P	→	Q)	→	Q))
1.	T	T	T	**T**	T	T	T	T	T
2.	T	F	T	**T**	T	F	F	T	F
3.	F	T	F	**T**	F	T	T	T	T
4.	F	F	F	**T**	F	T	F	F	F
				m.c.					

As we had anticipated, the corresponding conditional is indeed tautologous.

4 As you will undoubtedly have gathered from the exercises in this chapter, truth-tables can reach monstrous proportions of 16, 32, 64, 128 lines and worse, depending upon the number of sentence-letters involved. You may also have noticed how daunting it is to be faced with constructing such monsters and, indeed, how easy it is to get lost in a table of such proportions. So, it would be immensely valuable if we could discover some sort of short cut which might save our time and energy. Fortunately, just such a method is available. The **short-cut method** relies upon a good working knowledge of the truth-tables for the connectives so, if you have not yet learned the tables by rote, now is the time to do so.

In essence, our truth-table investigations into the validity of a given sequent really proceed by attempting to falsify that sequent. In each row of the table we construct a new interpretation and look to see whether that interpretation establishes the invalidity of the sequent. Only when every assignment fails to establish invalidity can we at last take the validity of the sequent to be established. Of course, if a sequent is invalid then there must be an assignment of truth-values under which the premises are all true *and* the conclusion is false. But each type of formula: negation, conjunction, etc., is only false under certain circumstances, i.e. when the unnegated formula is true, when one conjunct is false, and so on.

The short-cut method involves identifying the types of formula involved as premises and conclusion and the circumstances under which such formulas are true or false. We then go straight for the throat, as it were, and just consider that assignment which would bring the conclusion out as false and the premises out as true. So, we begin by simply writing T under the main connective of each premise and F under the main connective in the conclusion. In each case, we number the truth-values in order of assignment, writing each number immediately underneath the relevant truth-value. Next, we carefully consider the conclusion, note the type of formula involved and spell out the assignments to the sentence-letters which must hold if that formula really is going to turn out to be false. At each stage, we continue to number each new truth-value as we go.

Obviously, the assignment of truth-values to the sentence-letters in the conclusion will have implications for the truth-values of the same sentence-letters in the premises. So, we now look to see whether the premises really are true for such an assignment. In the best of all possible worlds, the premises will turn out to be true when the conclusion is false and we will have established the invalidity of a complex sequent with a one-line truth-table.

For example, consider the sequent R → Q, P v Q : P v R. First, assign T to each premise and F to the conclusion, numbering each assignment in order:

R	→	Q,	P	v	Q	:	P	v	R
	T			T				F	
	1			2				3	

The only way in which the conclusion, P v R, could be false is if both P and R are false. This insight gives us four more assignments which we list and number as follows:

R	→	Q,	P	v	Q	:	P	v	R
F	T		F	T			F	F	F
4	1		5	2			6	3	7

In turn, that assignment of truth-values has implications for the truth values of the complex formulas which form the premises of the sequent. Remember: both P and R are false. Now, because the antecedent of the conditional R → Q is false that conditional will be true whether its consequent is true or false. However, because the first disjunct of the second premise is false, the second disjunct must be true if the disjunction is to be true. Hence, we must assign T to Q. On just this assignment of truth-values the conclusion is indeed false while the premises are true. So, the sequent is invalid. Most importantly, a sequent involving three sentence-letters has been shown to be invalid by a truth-table which simply consists of one line. Finally, we assign T to Q to complete the table:

R	→	Q,	P	v	Q	:	P	v	R
F	T	T	F	T	T		F	F	F
4	1	8	5	2	9		6	3	7

What of valid cases, though? In every valid case you will find that, at some point down the line, you need to assign both truth-values simultaneously to one and the same sentence-letter! But that is clearly absurd. And that is surely just what we would expect: if the sequent really is valid then the attempt to show the sequent to be invalid should reduce to absurdity. For example, consider the following sequent: $Q \rightarrow P, P \vee Q : P \vee R$. First, assign F to the conclusion and T to the premises:

$Q \rightarrow P$,	$P \vee Q$	:	$P \vee R$
T	T		F
1	2		3

Again, the conclusion will only be false if both P and R are false, so:

Q	$\rightarrow$	P,	P	v	Q	:	P	v	R
	T	F		F	T		F		F
	1	4		5	2		3		6

Remember: both premises must come out true. Now, if $Q \rightarrow P$, is to be true Q must be false. But if $P \vee Q$ is also to be true Q must be true! Here is an absurdity. The attempt to show the sequent to be invalid has failed. What we learn is just that there is no assignment of truth-values to the sentence-letters which brings the premises out as true while the conclusion is false. Hence, the original sequent is indeed semantically valid.

The short-cut method outlined here goes by a number of names, e.g. 'the indirect method', 'the oblique method' or 'the quick-test method'. Occasionally, it remains nameless. A rose by any other name (or even no name at all) no doubt smells as sweet. The method should be a breath of fresh air after unwieldy truth-tables and use of it should enhance your understanding of the meanings of the connectives of PL. However, it is important to realise that the method doesn't always generate single, one-line tabulations. A little reflection on the nature of the formulas involved in our examples makes clear why these were guaranteed to result in one-line tables. All the formulas involved were either disjunctions or conditionals. But any such formula can only be false in one way. Hence, one particular interpretation of the conclusion is forced on us from the start. When we turn to the premises these again can only be false in one way and so, again, we are pushed in one particular direction. Of the remaining connectives, negation can also only be false in one way, namely, if the un-negated formula is true. However, any conjunction or biconditional can be false in more than one way, which opens up the possibility that, faced with formulas of these types, we might have to make more than one attempt.

In general, we will get a one-line table if either the conclusion can be false in only one way or the premises can all be true in only one way. If there is more than one way in which the conclusion can be false and it's not the case that the premises can only be true in one way, then we may well have to try again on a new line or even look at two new lines. Be that as it may, the method still generates short cuts, if slightly longer ones, and the ticklish cases are good mental exercise.

For example, consider the following sequent: P ↔ Q, Q ↔ P : P & Q. This time there are no fewer than three ways in which the conclusion might be false. So, the mere falsity of the conclusion doesn't push us in any one direction. Worse still, there are two ways in which each premise might be false. So the truth of the premises doesn't push us in any single direction either. But we can at least begin the test. After all, we know what we would like the end result to look like:

```
P ↔ Q, Q ↔ P  :  P & Q
  T      T          F
  1      2          3
```

But which move should we make next? In any such case, we may simply pick any sentence-letter, assign any value to it and begin to complete the table on that basis. For example, suppose that P is true, i.e.:

1.

```
P ↔ Q, Q ↔ P  :  P & Q
T T      T T     T F
4 1      2 5     6 3
```

Now, if P & Q is to be false, Q must be false. But if the premises are to be true Q must be true! Our first attempt ends in absurdity. But let's try again. Suppose that P is false:

2.

```
P ↔ Q, Q ↔ P  :  P & Q
F T      T F     F F
4 1      2 5     6 3
```

If the premises are to be true Q must also be false. But this time assigning F to Q will also ensure that P & Q is false, as the completed table shows:

3.

```
P ↔ Q, Q ↔ P  :  P & Q
F T F  F T F     F F F
4 1 7  8 2 5     6 3 9
```

Now we have, indeed, succeeded in showing the sequent to be invalid – but not at the first attempt. None the less, the short-cut method delivered the verdict in only two lines.

Box 4.5 summarises the four alternative ways to test for validity. Study it now.

BOX 4.5

♦ For any valid sequent, the set of formulas consisting of the *conjunction* of each of the premises of that sequent together with the *negation* of the conclusion of that sequent must be *inconsistent.* Hence, we can always conduct a truth-table to test for validity simply by forming just such a conjunction and testing it for consistency.

♦ For any valid sequent, the *conditional* consisting of the conjunction of the premises of that sequent as antecedent and the conclusion as consequent must be *tautologous;* just because, if the sequent really is valid, then the premises cannot be true while the conclusion is false. Hence, we can always conduct a truth-table to test for validity simply by forming such a conditional and testing it for tautologousness.

♦ Where the general form of a sequent is $\{A_1. \ldots\ldots A_n\}$: B the *corresponding conditional* is the formula:

$$A_1 \rightarrow (A_2 \rightarrow (\ldots A_n \rightarrow B))$$

For any valid sequent, the corresponding conditional must be tautologous. Hence, we can always conduct a truth-table to test for validity simply by forming such a conditional and testing it for tautologousness.

♦ The *short-cut method* exploits the fact that each type of formula is only false under certain circumstances, i.e. it involves identifying the types of formula involved as premises and conclusion and the types of circumstance under which such a formula can be true or false. We then consider just that assignment which would bring the conclusion out as false and the premises out as true, i.e. we write T under the main connective of each premise and F under the main connective in the conclusion; in each case numbering the truth-values in order of assignment and writing each number immediately underneath the relevant truth-value.

Next, consider the conclusion, note the type of formula involved and spell out the assignments to the sentence-letters which must hold if that formula really is going to get F. In each case, continue to number each new truth-value. In turn, that assignment has implications for the truth-values of the same sentence-letters in the premises. In the best of all possible worlds, the premises will turn out to be true when the conclusion is false and we establish the invalidity of a complex sequent with a one-line truth-table. If, however, there is more than one way in which the conclusion can be false and it's not the case that the premises can only be true in one way, then we may well have to try again on a new line or even look at two possible new lines.

With these four alternative ways of testing for validity clearly in mind it is time to attempt a few examples for yourself in Exercise 4.4.

EXERCISE 4.4

1 Test each of the following sequents for validity by constructing a truth-table which shows that the set consisting of the conjunction of the premises (if any) together with the negation of the conclusion is inconsistent.

1. ~(P v ~Q) : ~P & Q
2. P → Q, Q → R, P : R
3. ~Q → (~P → Q), ~Q : (~P → Q)
4. P → (Q → R), P, ~R : ~Q
5. : (P → Q) v (Q → R)

2 Test each of the following sequents for validity by constructing a truth-table which shows that the conditional whose antecedent is the conjunction of all the premises and whose consequent is the conclusion is tautologous.

1. P → Q, Q → R : P → R
2. P → Q, ~Q : P → R
3. P → (Q → R), P, ~R : ~Q
4. P, (Q & R) : (P & Q) v (P & R)
5. P ↔ Q, Q ↔ R : P ↔ R

3 Consider again sequents 1–5 of (1) above. Test each of those sequents for validity by constructing a truth-table which shows that the corresponding conditional for each sequent is tautologous.

4 Use the short-cut method to test the following sequents for validity. In each case, assign numerals to indicate the order of your interpretation.

1. P, ~(P & Q) : ~Q
2. P → (Q → R) : Q → (P → R)
3. Q → R : (~Q → ~P) → (P → R)
4. ~(P → Q), Q v (R & S) : R & S
5. (P & ~Q) v (Q & ~P) : P ↔ Q

IX
Semantic Equivalence

We can now explore another very important and extremely useful semantic notion, namely, that of **semantic equivalence**. In essence, semantic equivalence is truth-functional equivalence:

> Two or more compound formulas of PL are semantically equivalent if and only if for each and every assignment of truth-values to their component sentence-letters the overall truth-values of those formulas are one and the same.

To identify semantic equivalences we can exploit the truth-tables once again, i.e. to construct comparative truth-tables for the formulas we are interested in. This time, however, we are only interested in whether or not each formula takes the same truth-value under the same interpretation. If so, those formulas are semantically equivalent. If not, those formulas are not semantically equivalent. Now, semantically equivalent formulas don't just happen to have the same truth-values under every interpretation. Rather, they must have the same truth-values. The reason for this is precisely the same as the one we invoked to explain tautologousness. Just as a given formula is a tautology in virtue of its form so two or more formulas are equivalent in virtue of sharing the same form. In an important sense, two semantically equivalent formulas are really one and the same: they both express one and the same meaning in different ways. Hence, it should be no surprise that semantic equivalence between formulas exemplifies mutual or two-way semantic consequence, i.e. each formula is demonstrably a semantic consequence of the other. It follows that any two semantically equivalent formulas can be interchanged in the context of any sequent with a guarantee that the validity or invalidity of that sequent will remain wholly unaffected. Remember: if two compound formulas are semantically equivalent, then any interpretation which falsifies one falsifies the other and under any interpretation for which one is true so is the other. Hence, validity is always preserved under substitution of semantically equivalent compounds.

Investigating the particular semantic equivalences which hold in PL should again deepen your understanding of the logical connectives of PL. We use the connectives to form compound formulas and we define the meanings of those connectives just in terms of their truth-conditions. Identifying semantic equivalences between truth-functional compounds involving different connectives can be informative and, as we shall see later, practically useful. Moreover, semantic equivalences between compound formulas which involve distinct connectives underwrite some important logical laws. Among the most important of these are four laws due to the English mathematician and formal logician Augustus De Morgan in his *Formal Logic* [1847], as mentioned in Chapter 1. De Morgan's laws are particularly useful because they allow us to express any conjunction in terms of disjunction and negation and to express any disjunction in terms of conjunction and negation. Further, they also allow us to express negated conjunctions in terms of disjunction and negation and, again, negated disjunctions in terms of conjunction and negation. To express De Morgan's laws concisely I will use *double* semantic turnstile, i.e '⫤ ⊨ '. (Just as double syntactic turnstile indicates mutual or 'two-way' syntactic consequence so double semantic turnstile indicates mutual or 'two-way' semantic consequence.) Here are De Morgan's laws:

1. P & Q ⫤ ⊨ ~(~P v ~Q)
2. P v Q ⫤ ⊨ ~(~P & ~Q)
3. ~(P & Q) ⫤ ⊨ (~P v ~Q)
4. ~(P v Q) ⫤ ⊨ (~P & ~Q)

As you know, given the Completeness of PL, we could in fact replace the double semantic turnstile in each case with the double syntactic turnstile. It follows that the compound formulas involved in each law are interderivable.

To appreciate that De Morgan's laws exemplify semantic equivalences between compound formulas of just these forms is to understand why the compound formulas involved in each law should be interderivable. It is not just that we happen to be able to derive one compound from the other and vice versa in the case of each law, but rather that each formula conveys exactly the same meaning and has the same content as the other. As we might put it, each formula simply expresses the same point in a different way. In the case of each law, the formula on the right-hand side of the turnstiles re-expresses or, as we shall say, *rewrites* the formula on the left and vice versa. To prove that, in each case, a semantic equivalence is involved we must construct comparative truth-tables and compare overall truth-values under each interpretation. Remember: two formulas are semantically equivalent only if their overall truth-value is the same under every interpretation. For ease of comparison it is most convenient to put both formulas on a single line, as below, in the case of the first law:

1. P & Q : ~(~P v ~Q)

P Q	P & Q	~ (~ P v ~ Q)
T T	T T T	T F T F F T
T F	T F F	F F T T T F
F T	F F T	F T F T F T
F F	F F F	F T F T T F
	*	*

The asterisks mark the main columns which we are concerned to compare, and it is obvious at once from those columns that the overall truth-value is the same under each and every interpretation. Hence, De Morgan's first law does indeed exemplify a semantic equivalence.

Box 4.6 summarises semantic equivalence, and Exercise 4.5 contains some examples for you to try yourself.

EXERCISE 4.5

1 Show that each of De Morgan's laws exemplifies a semantic equivalence.

2 Represent the following arguments as sequents in PL, making your key explicit. In each case, show that the formulas involved are semantically equivalent:

(i) If Professor Cameron's car is in the car park then he is in his office. Therefore, if Professor Cameron is not in his office then his car is not in the car park.

(ii) If you eat your cake then you won't still have it. Therefore, you can't eat your cake and still have it.

BOX 4.6

♦ Two or more compound formulas of PL are *semantically equivalent* if and only if for each and every assignment of truth-values to their component sentence-letters the overall truth-values of those formulas are one and the same.

♦ To identify a semantic equivalence simply construct a comparative truth-table for the relevant formulas and consider whether or not each formula takes the same truth-value under the same interpretation. If so, those formulas are semantically equivalent. If not, then those formulas are not semantically equivalent.

(iii) Students love logic exams if and only if they are very enlightened. Therefore, if students do love logic exams then they are very enlightened students and if students are very enlightened then they do love logic exams.

(iv) The sun is shining and everything in the garden is coming up roses. So, it's not the case that either the sun is not shining or not everthing in the garden is coming up roses.

3 Use comparative truth-tables to determine whether or not the following pairs of formulas are semantically equivalent:

(i)	~P v ~Q	P → ~Q
(ii)	~(P → Q)	P & ~Q
(iii)	P → (P → Q)	P → Q
(iv)	P v Q	~(~P & ~Q)
(v)	P v (~~Q & R)	Q → (P & ~R)
(vi)	(P v Q) & ~(P & Q)	~(P ↔ Q)
(vii)	(P → Q) v ~(~R → S)	(~P v Q) v (R v S)

4 We can make further use of our metalinguistic variables to represent semantic equivalences between types of compound formulas in PL quite generally. Show that I have correctly identified the following pairs as semantically equivalent.

(i)	A → B	~A v B
(ii)	~(A → B)	A & ~B
(iii)	~(A & B)	~A v ~B
(iv)	~(A v B)	~A & ~B
(v)	A ↔ B	(A & B) v (~A & ~B)
(vi)	~(A ↔ B)	(A & ~B) v (~A & B)

X
Truth-Trees

You will be pleased to learn that the final semantic method which we will consider together in this chapter is a deceptively simple one which is remarkably easy to use. The method also has enormous potential which we will be able to exploit to our advantage both in this section and beyond. As we shall see, when the new method is combined with certain insights which we have already gained in the course of this chapter we will have arrived at a semantic method which we will still be able to exploit in the final chapter of this book.

The new method provides a straightforward way of testing formulas for **consistency**. The idea of consistency between sentences in a language is of fundamental importance in logic. In formal logic, however, the term 'consistency' is used in a number of importantly different senses. In the present context, the consistency of a set of well-formed formulas just implies that each and every well-formed formula in that set can be true at one and the same time, i.e.:

> A set of well-formed formulas of PL is consistent if and only if every member of that set can be true simultaneously.

It follows that a proof of consistency in this sense is a proof of the existence of a true interpretation of all the members of that set. Earlier in this chapter, we defined a true interpretation as a *model*. So, in the present context, we can equally well say that a proof of consistency is a proof of the existence of a model. The new method is precisely a test for consistency in that sense, i.e. it provides an answer to the question: could the formulas comprising a given set of formulas all be true together? In fact, the new method *guarantees* us an answer to just that question. If there is a true interpretation for all the formulas in the set then the new method will find it. In other words, the new method is an effective procedure for model-detection in PL.

The way in which the method proceeds to answer questions about consistency is by constructing what is known as a **consistency-tree** for a set of formulas. The tree format used is not entirely dissimilar to the format we exploited for the syntactical trees which we considered in Chapter 2. Like syntactical trees, consistency-trees are upside-down trees which break complex formulas down into their simple constituents. Indeed, consistency-trees can be more tree-like than their syntactical counterparts just because, when we construct a consistency-tree, we first list each formula on a separate line, one underneath the other. This procedure reflects the most obvious way in which the formulas might all be true together and gives a nice, trunk-like aspect to the inverted tree. Having listed the complex formulas we are interested in we then begin to break down or 'develop' each complex formula in terms of certain development rules. As ever, the development rules exploit the type of formula involved, i.e. conjunction, disjunction, etc.

The rule for conjunction simply requires us to write each conjunct on a separate line one below the other, like so:

```
1.        A & B
            |
            A
            B
```

In contrast, the rule for disjunction requires us to create two new branches, placing each disjunct on the end of a separate branch. This produces a pleasing, tree-like, branching effect:

```
2.        A v B
           / \
          /   \
         A     B
```

If we always develop every conjunction before we develop any disjunction then we will preserve the trunk-like effect for as long as possible and we will ensure that all the branching occurs at the top of the tree. Just as you would expect, that is exactly how we do proceed.

Having developed each formula in terms of that procedure we then carefully study the formulas in each branch, reading up from the tip of the branch back to the very beginning of the trunk. What we are looking for are contradictions among the formulas lying on that branch. If a branch does not contain any contradiction that branch is live and we mark that fact by writing a '✓' under that branch. However, if a branch does contain a contradiction then that branch is dead and we record that fact by writing an 'X' under it. If every branch dies then the tree is dead.

Although consistency-trees are easy to construct and produce pleasant-looking structures we should never lose sight of the fact that we use the trees to test for consistency. Indeed, each branch represents a different possible way in which all the formulas involved might be true together. So, for example, in the case of disjunction, our splitting the branch represents two possible ways in which the disjunction might be true, i.e. if either disjunct is true. But because there is only one way in which a conjunction can be true, namely, when both conjuncts are true, we do not split the branch. This represents the fact that there is only one possibility involved here.

In fact, each branch represents an attempt to assign truth-values to the component sentence-letters of the formulas so as to bring those formulas out as true simultaneously. Hence, each branch represents an interpretation. Moreover, because each branch represents an interpretation of the components of each formula, we know that when a branch dies that particular interpretation results in a contradiction, i.e. an *inconsistency*. If every branch dies then there is no interpretation which does not result in inconsistency. But, if that is so, then there is no way in which all the formulas being tested could all be true together. Therefore, that set of formulas is inconsistent. Conversely, if there is even one live branch then there is one way in which all the formulas being tested could be true together, i.e. a true interpretation or model. So, that set is consistent.

Let's consider some examples:

1. P & ~Q, P v Q, P

```
1.          P & ~Q
2.           P v Q
3.             P
4.             P              From line 1
5.            ~Q              From line 1
            /     \
6.        P         Q         From line 2
          ✓         × 5,6
```

Reading up the right-hand branch, we have a contradiction between Q on line 6 and ~Q on line 5. So, the right-hand branch is dead. However, there is no contradiction in the left-hand branch. So, we have a live branch on the left. Remember: each branch represents an attempt to assign truth-values to the sentence-letters composing the formulas in a way which makes all those formulas true. And a live branch represents an assignment which does not result in inconsistency. So, in this case there is an assignment which does not result in inconsistency and therefore this set is a consistent set.

Consider another case:

2. P & (Q v R), ~P v ~Q, ~R

```
1.             P & (Q v R)
2.              ~P v ~Q
3.                ~R
4.                 P               From line 1
5.               Q v R             From line 1
               /       \
6.            Q          R         From line 5
            /   \        × 3,6
7.        ~P     ~Q                From line 2
           ×      ×
          4,7    6,7
```

This time, all three branches are dead. So there is no interpretation under which all the original formulas could all be true together. It follows that this set is inconsistent.

Note carefully how each line containing one of the original formulas is developed as the tree progresses. In each case, I have annotated each new line with the number of the line developed at that point. Although this is a useful device for making explicit exactly which line is developed at which point, you may well find that you are not required so to annotate your trees in your particular course in formal logic. The same point holds for subscripting each contradiction with the line numbers of formulas which contradict one another. None the less, in the first few cases you attempt on your own, you may well find that these devices help you to keep track.

Before we move on, Exercise 4.6 gives a few examples for you to try yourself.

EXERCISE 4.6

1 Test the following sets of formulas for consistency using consistency-trees:

1. P & Q, R & ~S, P v S
2. P & Q, ~P v ~Q, ~Q
3. P & (Q v R), ~Q v ~R, ~R
4. (~P v ~Q) v R, ~P & ~Q, R
5. (~P v ~Q) v R, P & Q, ~R

XI More on Truth-Trees

At the outset of the last section I promised that we could further exploit the new method by combining it with some insights we had already gained in the course of the present chapter. It is now time to consider just how we might proceed in that respect. It is extremely useful to have an effective test for consistency for sets of formulas of PL. But aren't we interested, first and foremost, in validity rather than consistency? Perhaps. But the gap between these two fundamental notions is not as wide as it might first appear. In fact, the two notions are intimately related.

First, as we noted earlier, *a well-formed formula of PL is semantically valid if and only if the negation of that formula is inconsistent*. Second, as regards sequents, recall the first of the four alternative ways of using truth-tables to test for validity which we considered in Section VIII. That method involved testing the set of formulas consisting of the conjunction of all the premises of a sequent together with the negation of the conclusion of that sequent for consistency. If such a test resulted in inconsistency then we could conclude that the original sequent must be valid. The reasoning is simple: *a sequent is valid if and only if the set consisting of the conjunction of the premises together with the negation of the conclusion is inconsistent*.

Such a set is often referred to as a **counterexample set**. If the counterexample set to a sequent is consistent then there is an interpretation under which the premises of the original sequent are true while the conclusion is false. But, if that is so, the original sequent must be invalid. Conversely, if the counterexample set is inconsistent then there is no way in which the premises can be true while the conclusion is false, and we know that the original sequent is valid. So, consistency and validity are indeed intimately related, and this allows us to exploit consistency-trees to test sequents for validity. The consistency-tree test for validity in particular is known as the ***truth-tree* method**. But it is important to realise that we do *not* have two different kinds of tree here. In both cases, we simply have a consistency-tree. However, when we test for validity rather than testing the set of formulas composing the original sequent, we test that sequent's counterexample set for consistency.

Recall again the alternative truth-table test. This required us to conjoin all the premises with the negated conclusion. The first part of that requirement is already met by the consistency-tree method. That is precisely the point of listing all the original formulas at the start, i.e. in the trunk, just as we would break down a conjunction. So, all we have left to do is to substitute the negation of the conclusion of the sequent for the original conclusion and include it in the trunk together with the premises. In this way, we have designed a new and very elegant test for validity for sequents of PL, which is just as effective a procedure as truth-tables but is much less overtly mechanical and leads to less cumbersome diagrams.

Consider an old favourite: P v Q, ~P : Q. First, list the premises together with the negation of the conclusion. Next, break the complex disjunction down, keeping a watchful eye for contradictions:

```
1.        P v Q          Premise 1
2.         ~P            Premise 2
3.         ~Q            Negated conclusion
          /   \
         /     \
4.      P       Q        From line 1
        ×       ×
       2,4     3,4
```

In this case, both branches die in contradiction. Therefore, there is no way in which these formulas could be true together. The counterexample set is strictly inconsistent and so the original sequent must be valid.

Consider another example:

~P & Q, P v (Q & R) : R

```
1.      ~P & Q          Premise 1
2.      P v (Q & R)     Premise 2
3.       ~R             Negated conclusion
4.       ~P             From line 1
5.        Q
        /   \
6.      P   Q & R       From line 2
        ×     |
7.     4,6    Q
8.            R
              × 3,8
```

Again, both branches die. The counterexample set is shown to be inconsistent and, therefore, the original sequent is valid.

So far, we have only considered sets of formulas involving conjunction and disjunction. But now, we want to extend the truth-tree method to sets of formulas involving the other connectives. What's more, we also want to be able to develop the negations of complex formulas formed using any of the logical connectives of PL. In order so to extend the applicability of the method we require new **development rules** (for PL truth-trees), which are guaranteed to preserve truth across developments and so to preserve the validity or invalidity of the counterexample set. What might such new development rules look like? In fact, we have already answered this question. As we noted in Section IX, truth and validity are preserved across substitutions of semantically equivalent formulas. So, all we have to do is to make sure that we develop complex formulas in terms of semantically equivalent formulas. In the final question of Exercise 4.5, we used meta-linguistic variables to establish quite general semantic equivalences between forms involving precisely the remaining connectives together with negated cases for each and every connective. The new development rules we need simply exploit those equivalences. Hence, when disjunction is represented by splitting a branch and conjunction is represented by not splitting a branch, the equivalences we established in Exercise 4.5 generate the following set of development rules:

```
1.      A → B              2.      ~(A → B)
        /   \                         |
      ~A     B                        A
                                     ~B

3.     ~(A & B)            4.      ~(A v B)
        /   \                         |
      ~A    ~B                       ~A
                                     ~B

5.      A ↔ B              6.      ~(A ↔ B)
        /   \                      /      \
       A    ~A                    A       ~A
       B    ~B                   ~B        B
```

Finally, it is useful (and truth-preserving) to include a rule which allows us to eliminate double negations:

```
7.      ~~A
         |
         A
```

We are now in a position to apply the truth-tree method to sequents involving formulas of any level of complexity. But before we do we should consider a final example which illustrates an important point: if a tree has already branched and there is more than one surviving branch then, if we have to develop another branching formula, that formula should be developed over *every* surviving branch. This last example concerns one of a pair of logical laws known as **distributive laws** (to get the other in this pair, simply swap the premise with the conclusion in the sequent below):

P & (Q v R) : (P & Q) v (P & R)

```
1.              P & (Q v R)            Premise
2.          ~[(P & Q) v (P & R)]       Negated conclusion
3.               ~(P & Q)              From line 2
4.               ~(P & R)
5.                  P                  From line 1
6.               Q  v  R
               /        \
7.          Q              R           From line 6
          /   \          /   \
8.      ~P    ~Q      ~P      ~Q       From line 3
         ×     ×       ×       |
        5,8   7,8     5,8    /   \
9.                         ~P     ~R   From line 4
                            ×      ×
                           5,9    7,9
```

Now, every branch is dead. Therefore, the counterexample set is inconsistent and so we have shown that this distributive law is semantically valid. The tree also illustrates the point that branching formulas must be developed over all surviving branches. Hence, at line 8 we must develop the branching formula from line 3 over both existing branches. At line 9, however, only one branch remains live, so we need only develop the branching formula from line 4 over that live branch.

To date, we have not applied the truth-tree method to an invalid sequent. But, in fact, there is more to be gained from the truth-tree method when the sequent being tested is invalid. Certainly the method shows that the sequent is invalid, but it also highlights just those interpretations under which it is consistent to assert the truth of the premises and the falsity of the conclusion. For example, consider the following tree carefully:

P → Q, Q : P

```
1.     P → Q        Premise
2.      Q           Premise
3.      ~P          Negated conclusion
       /  \
4.   ~P    Q        From line 1
      ✓    ✓
```

This time, both branches are live. So, the counterexample set is consistent. Moreover, not only do we learn from the tree that the original sequent is invalid but we can also go on to identify the particular assignment of truth-values to the sentence-letters which will generate a counterexample, i.e. an IPLI of the sequent. And there is nothing new to do here! All the information we require is contained in the tree already. To make the IPLI explicit all we need do is list the sentence-letters involved in the sequent and then read up any live branch to identify the relevant assignment of truth-values to those sentence-letters. When the live branch contains an unnegated sentence-letter that letter on your list should get the value T. When the live branch contains a negated sentence-letter that letter should get the value F. In this case, the only sentence-letters involved are P and Q. Reading up the first or left-hand branch we can see that ~P holds there so P should be assigned F. However, Q is unnegated on that branch, so Q should be assigned T. And that is exactly the assignment of truth-values we require to construct the IPLI which is shown below.

IPLI: P: {F} Q: {T}

A little reflection shows that this assignment of truth-values to the sentence-letters of the original sequent does indeed generate true premises together with a false conclusion. Note, however, that we have two live branches here. In such a case, how do we know which to pick? The answer is that it doesn't matter. Either way, each live branch represents an IPLI. In the present case, we would arrive at the same assignment and the same IPLI. In other cases, as you will see, distinct live branches may represent distinct assignments. This simply reflects the fact that there is more than one IPLI for the sequent. And, of course, either will do the job.

In a sense, then, once we have constructed the tree we have already identified the IPLI. The IPLI is a free gift of the method. We simply read it off any surviving branch. Moreover, to get from the IPLI to an actual counterexample, simply substitute obviously false sentences for sentence-letters assigned F and obviously true sentences for sentence-letters assigned T. For exam purposes, it is advisable to choose clear-cut arithmetical sentences. Prime false sentences might be '0 = 1', '2 + 2 = 5' and so on. Prime true sentences might be '1 = 1', '2 + 2 = 4' and so on. In the present case we need a false sentence for P and a true sentence for Q. So, let P stand for '0 = 1' and let Q stand for '2 + 2 = 4':

Actual counterexample:

If 0 = 1 then 2 + 2 = 4	P → Q
2 + 2 = 4	Q
Therefore,	∴
0 = 1	P

Consider a final example:

P v Q, R → Q : R → P

```
1.           P v Q           Premise
2.           R → Q           Premise
3.          ~(R → P)         Negated conclusion
4.             R
5.            ~P             From line 3
             /   \
6.        ~R       Q         From line 2
        × 4,6    /   \
7.              P     Q      From line 1
             × 5,7    ✓
```

This time, the branch on the extreme right is live. Therefore, this set of formulas is also consistent and the original sequent is, therefore, invalid. Further, simply reading up the live branch reveals the following IPLI:

IPLI: P: {F} Q: {T} R: {T}

Finally, we proceed to construct a counterexample, again, by substituting actually true sentences and actually false sentences for the sentence-letters as indicated by the IPLI:

Either 0 = 1 or 1 + 1 = 2	P v Q
If 2 + 2 = 4 then 1 + 1 = 2	R → Q
Therefore,	:
If 2 + 2 = 4 then 0 = 1	R → P

Box 4.7 summarises the main points about the truth-tree method and restates both the **procedural rules** (for PL truth-trees) and the development rules (for PL truth-trees). The classic statement of the truth-tree method is due to Richard C. Jeffrey.[10] Famously, Jeffrey presents very helpful flow charts for checking the proper construction of truth-trees according to the rules. In the spirit of Jeffrey, I offer my own attempt at such a flow chart immediately prior to Exercise 4.7 (Figure 4.1). If that chart is useful, all credit is due to Richard C. Jeffrey. Study the contents of Box 4.7 *and* Figure 4.1 carefully before attempting Exercise 4.7.

BOX 4.7

The truth-tree method:

♦ The truth-tree method is used to test the *consistency* of sets of well-formed formulas of PL.

♦ A set of well-formed formulas of PL is *consistent* if and only if every member of that set can be true simultaneously. A proof of consistency is therefore a proof of the existence of a model.

♦ The method proceeds by constructing a *consistency tree* for a given set of formulas which breaks complex formulas down into their simple constituents via the following set of *development rules,* which simply exploit the kind of formula involved:

```
1.    A & B     2.   A v B     3.   A → B      4. ~(A → B)
        |            /   \          /    \           |
        A           A     B       ~A      B          A
        B                                           ~B

5.   ~(A & B)    6. ~(A v B)    7.   A ↔ B    8. ~(A ↔ B)    9. ~~A
       /   \           |             /   \        /    \         |
     ~A    ~B         ~A            A    ~A      A     ~A        A
                      ~B            B    ~B     ~B      B
```

♦ Each branch of a tree represents an attempt to assign truth-values to the sentence-letters composing the formulas in a way which makes all those formulas true. If any branch contains a contradiction then that branch is dead. A *dead* branch represents an assignment which results in inconsistency. A *live* branch represents an assignment which does not result in inconsistency. If every assignment results in inconsistency the original set is inconsistent. If there is any assignment which does not result in inconsistency the set is a consistent set.

To construct a *consistency tree,* observe the following *procedural rules*:

♦ List each formula on a separate line, one underneath the other.

♦ Consider each kind of formula carefully and develop every *conjunction* before any *disjunction.*

♦ Having developed each formula, carefully study the formulas in each branch, reading up from the tip of the branch back to the very beginning of the trunk. If a branch does contain a contradiction the branch is dead and we write an 'X' under it. If every branch dies then the tree is dead.

The ***truth-tree test for validity****:*

♦ A well-formed formula of PL is semantically valid if and only if the negation of that formula is inconsistent, and:

♦ a sequent is valid if and only if the set consisting of the conjunction of the premises together with the negation of the conclusion is inconsistent.

♦ Any such set is a *counterexample set.*

♦ If the counterexample set is consistent then there is an interpretation under which the premises of the original sequent are true while the conclusion is false. If that is so, the original sequent must be invalid.

♦ Conversely, if the counterexample set is inconsistent then the original sequent is valid.

♦ It follows that we can use the truth-tree method to test for validity simply by using that method to test a sequent's counterexample set for consistency.

To construct a truth-tree test for validity, observe two further procedural rules:

♦ Always develop every non-branching formula before any branching formula.

♦ Where a tree has already branched such that there is more than one surviving branch any other branching formula must be developed over *every* surviving branch.

Exercise 4.7 gives you the opportunity to construct some trees for yourself.

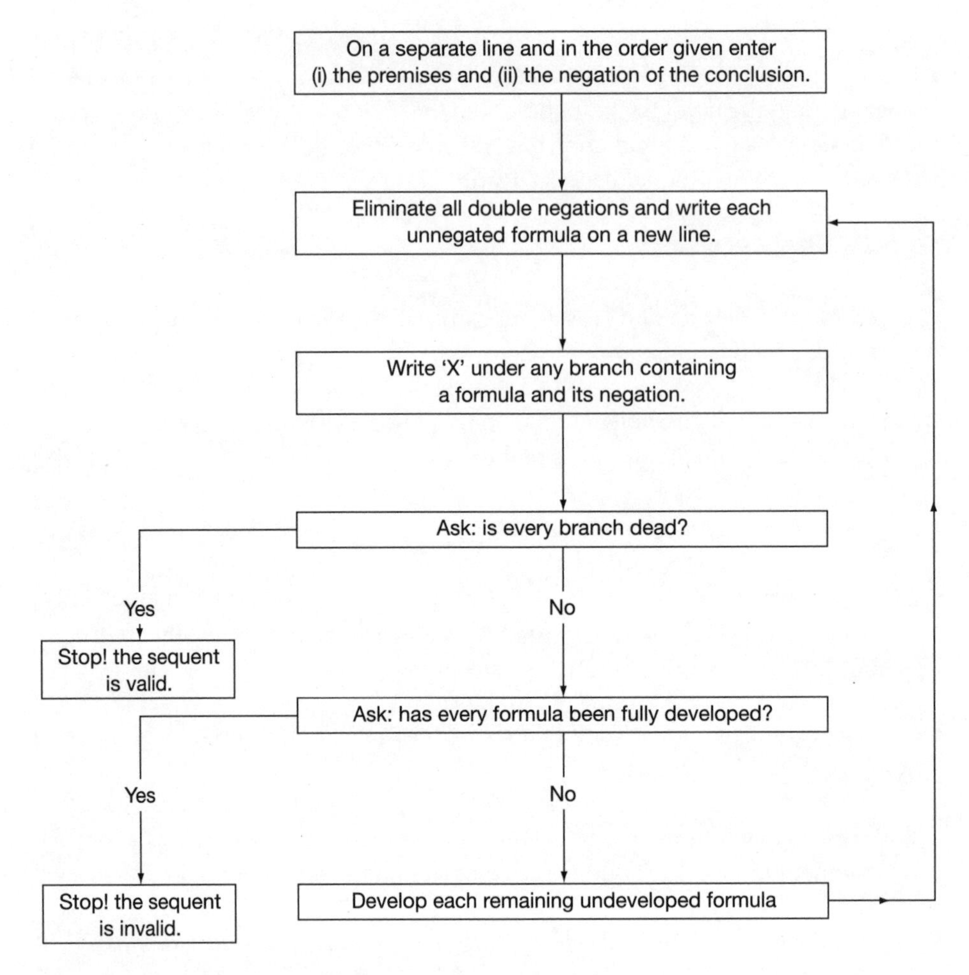

Figure 4.1 A Jeffrey-style flow chart for truth-trees

EXERCISE 4.7

1 Test the following sequents for validity using the truth-tree method. For any sequent which you find to be invalid give: (i) an IPLI of the sequent *and* (ii) an *actual* counterexample.

1. $P \to Q, \sim P : \sim Q$
2. $\sim P \to Q : Q \to P$
3. $P \to Q, Q \to R : P \to R$
4. $(P \to Q) \to P : P$
5. $\sim(P \vee \sim Q) : (\sim P \& Q)$

6. $: (P \vee P) \rightarrow P$
7. $: ((\sim P \rightarrow Q) \rightarrow \sim P) \rightarrow \sim P$
8. $(P \rightarrow Q) \rightarrow R : \sim R \rightarrow P$
9. $: (P \rightarrow Q) \rightarrow (\sim Q \rightarrow \sim P)$
10. $P \rightarrow \sim Q, \sim R \rightarrow P : Q \rightarrow R$
11. $: (P \vee \sim P) \rightarrow (Q \vee \sim (Q \vee R))$
12. $: (P \leftrightarrow Q) \leftrightarrow \sim (P \,\&\, \sim Q)$
13. $\sim R \rightarrow Q : (P \vee Q) \rightarrow (\sim R \rightarrow P)$
14. $(P \,\&\, Q) \rightarrow R, \sim P \rightarrow S : Q \rightarrow (R \vee S)$
15. $(P \vee Q) \,\&\, (R \vee \sim S) : ((\sim P \vee \sim R) \,\&\, (\sim P \vee S)) \rightarrow ((Q \,\&\, R) \vee (P \,\&\, \sim S))$

XII
The Adequacy of the Logical Connectives

Towards the end of Section II of Chapter 2, I promised to discuss the *adequacy* of the logical connectives at a later moment. Given what we have learned in the present chapter, that moment is now. So, what does the formal logician mean by the claim that the connectives of PL are *adequate*? Intuitively, what is meant is that not only do the five connectives represent the particular truth-functions with which you have become familiar during this chapter but, in fact:

> Each and every possible truth-function can be represented just in terms of those connectives.

This is a striking and very heartening fact about the connectives. You should find it heartening because it means that every truth-function which can possibly be expressed in PL will just reduce to some combination of the familiar connectives. You will also find it striking if it lets you feel something of the sheer expressive power of the connectives stretching throughout the deep structure of the language.

Certain facts about the complex formulas of PL may already have made you optimistic about the expressive power of the connectives. In particular, during our discussion of semantic equivalence, you will have picked up on the fact that different complex formulas are interdefinable. We explained this by saying that equivalent complexes exemplify the same forms in different ways. That possibility clearly suggests that we at least have more than the bare minimum of complex forms. Moreover, equivalence also exemplifies the fact that complex forms need not be wholly independent of one another. Indeed, we exploited just that fact in our discussion of the development rules for truth-trees in the last section when we used meta-linguistic variables to establish quite general equivalences between complex formulas. In fact, we obtained these results in virtue of the interdefinability of the connectives themselves. So, for example, we saw that any conditional formula can be rewritten in terms of disjunction and negation in a way which preserves truth and validity.

Let's explore the idea of adequacy a little more formally. When we first introduced truth-tables we noted that the number of rows in any table would just be 2^n where 'n' is the number of sentence-letters involved. And, as we noted in Section II of this chapter, because the principle of bivalence holds for PL, any given formula of PL can only be either true or false. There are only the *two* options. It follows that, for any table, there must be 2^{2^n} possible outcomes truth-value wise. In this way, we can get a complete picture of the number of possible truth-functions for any given number of sentence-letters. Let's pick an easy case, that of two sentence-letters. In this case, there will be 16 possible truth-functions, i.e. 2^{2^2} and so 16 possible columns. These will range from the case where the function gives T as value for every argument to the case where we have a function which gives F as value for every argument.

Further, because there are two sentence-letters (P and Q) and two truth-values there must be four rows of truth-value assignments, i.e. 2^2. Finally, because the list is exhaustive, it must include the truth-functions represented by our four binary connectives. Below is the complete table of truth-functions for the two sentence letters P and Q with the four familiar cases indicated:

P	Q	1	v	3	4	→	6	↔	&	9	10	11	12	13	14	15	16
T	T	T	T	T	T	T	T	T	T	F	F	F	F	F	F	F	F
T	F	T	T	T	T	F	F	F	F	T	T	T	T	F	F	F	F
F	T	T	T	F	F	T	T	F	F	T	T	F	F	T	T	F	F
F	F	T	F	T	F	T	F	T	F	T	F	T	F	T	F	T	F

In this context, the question about the adequacy of the connectives boils down to the question: can every single one of these truth-functions be expressed just in terms of those connectives? Obviously, we can certainly express four of the sixteen columns, but what of the other twelve? The short answer here is 'yes', we can indeed express the other twelve using only the familiar connectives. To show that this is so it suffices to establish semantic equivalences between the unfamiliar 'nameless' truth-functions for two sentence-letters and complexes constituted solely from the familiar ones. Here are some examples of such equivalences for the first four nameless cases. As an exercise, find equivalences for the remaining cases to complete the list:

1. (P v ~P) v Q
3. P v ~Q
4. P & (Q v ~Q)
6. (P & Q) v (~P & Q)
9. ?
10. ?

.

.

.

16. ?

It is crucial to realise that the complete list, even if correct, is nothing more than a useful illustration. It is certainly *not* a proof – that would require us to show that every possible truth-function for any number of sentence-letters could always be represented using the connectives. In effect, a proof would guarantee that given any nameless truth-function we can always construct a complex formula involving only the familiar connectives which will have exactly the same truth-table as the nameless truth-function. We certainly have not shown here that this is the case. None the less, it is indeed the case and has been proved to be so. In fact, formal proofs usually establish the adequacy of a subset of our set of connectives. Over pp. 64–6 of his *Metalogic*, for example, Geoffrey Hunter proves that the following set of three connectives {~, &, v} is sufficient to express every truth-function. Further, Hunter goes on to prove that the following sets consisting of only two connectives are also adequate: {~, →}, {~, & }, {~, v}. Hence my remark in Chapter 2 that our set of five connectives might be called a generous set!

Now, it is no accident that each of the adequate subsets we have considered included '~'. Negation is crucial to adequacy. A little reflection reveals why. Recall the complete table of truth-functions for two sentence-letters. The values of those functions range from the case in which they are true for every input to the case in which they are false for every input. But without negation no subset of the connectives could possibly result in a complete column of F's, i.e. without negation we cannot construct a formula which is strictly inconsistent. Hence, the set {&, v, →, ↔} is not adequate.

Nor are any of its subsets adequate. The crucial importance of negation is also illustrated by another striking fact about adequacy, namely, that a single binary connective alone is adequate. That connective represents not simple denial but **joint denial**, e.g. '*neither* . . . *nor* ---'. Obviously, negation is essential to joint denial. Strikingly, C.S. Peirce [1839–1914], an American physicist and philosopher whose work founded the tradition of American pragmatism, realised the adequacy of joint denial as early as 1880, though he did not prove that this was so. The proof was finally provided in 1913 by an American logician, Henry Sheffer [1883–1964],[11] for which Sheffer is justly famous.

In fact, following Sheffer, we can separate out two senses of joint denial, one of which involves '&' while the other involves 'v'. 'Neither . . . nor ---' clearly exploits 'v', i.e. this function is false if either disjunct is true. Therefore, the function will take the value T only if both disjuncts are false. Hence, this function is often alternately referred to as 'Not . . . and not ---'. Thus, where A and B are understood in the usual way and '/' represents 'neither . . . nor ---' we have the following truth-table:

A	B	A/B
T	T	F
T	F	F
F	T	F
F	F	T

Alternatively, we can use '&' to formulate joint denial in the sense of 'Not both A and B'. Obviously, this function will again take the value T when both A and B are false but it will also take the value T if just one of A or B is false. Hence, this function is often referred to as 'Not A or not B'. It is this particular sense of joint denial which is usually associated with Sheffer, and the symbol for this truth-function '|' is most commonly called **Sheffer's stroke**. Sheffer's stroke gives us the contrasting table:

A	B	A | B
T	T	F
T	F	T
F	T	T
F	F	T

Now, suppose we take Hunter's word that the set {~, v} is adequate. Could we then show that Sheffer's stroke is adequate? Strictly speaking, we should really first establish the adequacy of {~, v} ourselves. But Hunter is a capable and veritable logician and if we may take him at his word it is very easy to prove that Sheffer's stroke alone must be adequate too. As you might expect, the proof proceeds by finding semantically equivalent truth-functions for '~' and 'v' in terms of '|' alone. So, there are only two cases to worry about. Here are the relevant equivalences:

1 Negation:

(i)

A	~A
T	**F**
F	**T**

(ii)

A	A	|	A
T	T	**F**	T
F	F	**T**	F

2 Disjunction:

(i)

A	B	A	v	B
T	T	T	**T**	T
T	F	T	**T**	F
F	T	F	**T**	T
F	F	F	**F**	F

(ii)

A	B	((A	|	A)	|	(B	|	B))
T	T	T	F	T	**T**	T	F	T
T	F	T	F	T	**T**	F	T	F
F	T	F	T	F	**T**	T	F	T
F	F	F	T	F	**F**	F	T	F

Given these equivalences, we can now translate formulas of PL involving only '~' and 'v' into formulas which only involve Sheffer's stroke. For example, consider ~P v ~Q. We write the formula on a line and then replace each connective with an equivalence, one at a time, writing each replacement on a line below:

1. ~P v ~Q
2. (P | P) v ~Q
3. (P | P) v (Q | Q)
4. ((P | P) | (P | P)) | ((Q | Q) | (Q | Q))

Considering the obvious lengthening of formulas involved one might well wonder why we should be interested in Sheffer's stroke at all. It is a striking fact that *every possible* truth-function in PL can be represented in terms of a single binary connective but Sheffer's Stroke has also been considered to be of *philosophical* significance and practical importance. It's worth considering each point in turn.

Sheffer's Stroke plays a crucial role in Wittgenstein's early philosophy of language as exemplified in his *Tractatus Logico-Philosophicus.*[12] The *Tractatus* is a remarkable text in many respects, not least because it is ultimately composed of only seven sentences. That skeletal structure is fleshed out by remarks subordinated to each of the seven sentences by decimal numbering, but seven sentences constitute the heart of the enterprise. The central theme of the text concerns the nature of, and the connection between, thought, language and reality. Wittgenstein's own thinking on these topics is often difficult to discern, in part because he states the conclusions of his arguments before their premises. But we might reconstruct Wittgenstein's reasoning as follows.

Natural language contains sentences which can only be either true or false, sentences which have perfectly determinate meanings. How must thought, language and the world stand to one another to guarantee that language contains such sentences? In the last analysis, language must be able to represent reality so clearly as to ensure that what is said about reality can only be true or false. Determinacy of meaning therefore demands the existence of a fragment of language which can model and mirror reality precisely. For the Wittgenstein of the *Tractatus* those linguistic mirrors are the *elementary propositions*.[13] The clarity with which elementary propositions must represent reality requires that the linguistic building blocks from which those propositions are constructed are not complex signs but simple signs. For Wittgenstein, these simple signs are names. In turn, he thought, the meanings of these signs, the fundamental constituents of reality, must themselves be simple. Hence, he arrived at a picture of reality as ultimately composed of unalterable, self-subsistent *objects*. Logical investigation thus revealed not physical atoms but *logical atoms*.

These intellectual discoveries enabled Wittgenstein to explain linguistic representation as follows. Elementary propositions are composed of names but such propositions are not mere bundles of names. Rather they are precise structures of names. Such propositions reach out to reality not only via the naming-relation but also structurally, in terms of order-relations. Every elementary proposition reflects a possible structure of objects in reality but a true elementary proposition corresponds to an actual ordered set of objects, an *atomic fact*, and shares with that fact a common structure, a common *form*.

False elementary propositions do not share a structure with a fact. When false, there is nothing for the proposition to correspond to. Negated elementary propositions do not refer to non-existent facts. Rather, a negated elementary proposition is true when the unnegated elementary proposition is false. False propositions do have form, just as a possibility of structure in reality which is actualised in the fact of the proposition itself. Form is *shown* in the proposition itself; in the way in which the elements of an elementary proposition are combined we see what would be the case if it were true. Hence we can understand a proposition in advance of knowing its truth-value: we know what would be the case if it were true and what would be the case if it were false.

For present purposes, Wittgenstein's crucial insight was to realise that complex propositions could be formed from elementary propositions using truth-functional logical connectives. In turn, such complex propositions could themselves be combined, again, using truth-functional connectives and so on to ever greater levels of complexity. Each and every such complex proposition could now be shown to have a unique overall truth-value given the truth-values of the elementary propositions and definitions of the truth-functional connectives. (Wittgenstein independently invented the truth-table method for just that purpose.) Thus, each and every complex proposition of natural language is a truth-function of its elementary constituents. But every truth-function can be represented using Sheffer's stroke. As Wittgenstein puts the point, the general form of a truth-function is given by Sheffer's stroke. Moreover, that function gives us the general form of a proposition itself just because each and every possible complex proposition can be represented using only the stroke function. So, for the early Wittgenstein, Sheffer's discovery of the stroke function represents the discovery of the deepest element of the logical grammar of natural language.

In all honesty, the *Tractatus* has its fair share of critics, none more vicious or vociferous than the later Wittgenstein himself![14] But the *Tractatus* remains a landmark in the philosophy of language and is a particularly exciting text for students of formal logic. Moreover, the only way to appreciate the full force of Wittgenstein's later thought is to grasp first the nature of his early work. I can pursue the nature of the Tractarian philosophy of language no further here but I will offer some suggestions for further reading in the area.

Both Max Black's *A Companion to Wittgenstein's Tractatus* [1964] and Elizabeth Anscombe's *An Introduction to Wittgenstein's Tractatus* [1963] rightly remain classic commentaries. A clear and accessible exposition of much of the formal logic to be found in the *Tractatus* is given by H.O. Mounce in his *Wittgenstein's Tractatus: An Introduction* [1981]. Interested parties would also do well to read Peter Carruthers's excellent texts *Tractarian Semantics: Finding Sense in Wittgenstein's Tractatus* [1989] and *The Metaphysics of the Tractatus* [1990].

Finally, Sheffer's stroke also has some very valuable practical applications; as indeed do all of the truth-functions expressed by our familiar connectives. In fact, these functions form the basis of all electronic switching engineering and are therefore foundational in computer design. To appreciate the point it is only necessary to imagine that the variables A and B represent two switches which are connected in series to a battery:

[b] ________ / ________ / ________
A B

Electricity will flow through the circuit only if both switches are closed. But if we let T represent the fact that a switch is closed then electricity will flow only if both A and B are T. We might well call our circuit an '& circuit' just because it operates in exactly the same way as '&' does in PL, i.e. the two have a common character.

More generally, our truth-functional connectives can be used to characterise a set of circuits known as *logic circuits*. Suppose that a given circuit has two inputs and one output. Imagine that we are interested in voltage. Input voltage may be either high or low. Equally, output voltage may be either high or low. Suppose that output is high only if both inputs are high. Here, we have an '&' circuit. Consider a circuit where output is high only if one or both of the inputs is low. This time we have what electronic engineers refer to as a *nand* circuit, i.e. a 'not and' circuit. Now, label the inputs A and B and take T to mean 'high' and F to mean 'low'. The truth-table for Sheffer's stroke gives exactly the output values of the nand circuit. Again, the two have a common character.

What is common to both the logic of switching circuits and the logic of the connectives is the same basic algebraic structure. These algebraic structures are known as **Boolean algebras**, after the English mathematical logician George Boole, mentioned in Chapter 1, who first articulated the basic algebraic structures. Boole's 'algebra of classes' can readily be represented as a formal system. The reason for this is just that Boolean algebra is structurally isomorphic to formal logic. More precisely, all the key algebraic operations correspond exactly to logical operations. For example, the algebraic operation known as 'complementation' corresponds precisely to classical negation, the algebraic operation known as 'logical sum' has an exact counterpart in our sense of disjunction and so on.

Given these facts, it should be unsurprising that an algebraic perspective can give deep insights into the character of formal logic and indeed that algebraic logic is an area of interest in its own right.

Earlier in this chapter I referred to Gottlob Frege as the founding father of modern formal logic. You may now feel that the title really belongs to George Boole. Be that as it may, I hope that the foregoing discussion highlights something of the theoretical and practical significance of our set of connectives and the striking property of adequacy which belongs to that set. Formally, adequacy is an extremely important property and, like Soundness and Completeness, it is vital to the ambitions of the formal logician. The point is nowhere better put than by Hunter:[15]

> What do logicians want? The Holy Grail of logic would be a system or set of systems which caught *all* truths of pure logic. This nobody has yet found . . . So we ask: 'What do we want that we have some hope of getting?' Here an answer is: 'All truths of pure truth-functional logic' [and] there is a sense in which [PL] does catch all the truths of pure truth-functional logic . . . In the first place, the language . . . is adequate for the expression of any truth-function . . . Second, . . . every truth-functional tautology [can be] represented in a natural way by some theorem.

Now try Exercise 4.8, and then Examination 2.

EXERCISE 4.8

1 Consider again the complete table of truth-functions for two sentence-letters given above. Find semantically equivalent formulas for each of the nameless cases using the following sets of connectives:

(i) {~, v, &}

(ii) {~, v}

2 Represent the following formulas using only Sheffer's stroke. Show each substitution of a connective on a separate line.

(i) P v Q

(ii) ~P v Q

(iii) (~P v ~Q) v ~R

3 Given that {~, &} is an adequate set of connectives prove that {|} must also be an adequate set.

4 Given that {~, →} is an adequate set of connectives prove that {|} must also be an adequate set.

Examination 2 in Formal Logic

Answer every question.

1 *Study the following arguments carefully. Represent each argument as a sequent in PL. In each case make your key explicit. Test each sequent for validity using comparative truth-tables. Finally, construct a proof of any sequent which you find to be valid.*

1. If pigs had wings then pigs would fly and air traffic controllers would have nightmares. Therefore, if pigs had wings then pigs would fly.

2. Professor Plum was in the drawing room and Miss Scarlet was in the kitchen. If Professor Plum was in the drawing room and

the murder weapon was found in the drawing room then Professor Plum is in big trouble. Therefore, if the murder weapon was found in the drawing room then Professor Plum is in big trouble.

3. If it's not the case both that Professor Plum was in the study and Miss Scarlet was in the conservatory then the murderer was Reverend Green. But if it's not the case that Reverend Green was the murderer then Miss Scarlet was in the conservatory and Colonel Mustard was no doubt there too. So, if Reverend Green is not the murderer then Professor Plum was in the study and Colonel Mustard was there too.

2 *Test the following sequents for validity using the truth-tree method. For any sequent which you find to be invalid state: (i) an IPLI of the sequent* and *(ii) an actual counterexample to the sequent.*

1. R → Q, P v Q : P v R
2. : (~P & (P → Q)) → Q
3. : (P → Q) ↔ ~(P & ~Q)
4. P & (Q v R) : (P & Q) v (P & R)
5. (P v Q) → ~R : ((P & ~R) → ~R) & ((Q & R) → ~R)
6. (P & Q) → (R & S) : (P → (Q → R)) & (P → (Q → S))

3 *Show that the English language connective 'unless' is semantically equivalent to the PL connective* 'v'.

Notes

1 This claim is not entirely uncontroversial and raises questions which lie beyond the scope of this book.
2 See, for example, Benacerraf, Paul and Putnam, Hilary (eds), [1983], *Philosophy of Mathematics: Selected Readings*, second edition, Cambridge, Cambridge University Press. Part One of the text contains two papers (each) by Heyting and Brouwer. These are warmly recommended.
3 See, for example, Dummett, Michael, [1977], *Elements of Intuitionism*, Oxford, Clarendon Press. But also Dummett, Michael, [1978], 'The Philosophical Basis of Intuitionist Logic', in *Truth and Other Enigmas*, London, Duckworth.
4 Baker, G.P., and Hacker, P.M.S., [1984], *Language, Sense and Nonsense: A Critical Investigation into Modern Theories of Language*, Oxford, Blackwell, p. 171.
5 Luce, A.A., [1958], *Logic*, London, English Universities Press, p. 4.
6 See above, Chapter 3, Revision Exercise III, numbers 4 and 5.
7 Hunter, Geoffrey, [1971], *Metalogic: An Introduction to the Metatheory of Standard First-Order Logic*, London and Basingstoke, Macmillan. pp. 92–6.
8 *Ibid.*, pp. 105–16.

9 Hamilton, A.G., [1978], *Logic for Mathematicians*, revised edition, Cambridge, Cambridge University Press.
10 Jeffrey, Richard C., [1967], *Formal Logic: Its Scope and Limits*, New York, McGraw-Hill.
11 Sheffer, Henry M., [1913], 'A Set of Five Independent Postulates for Boolean Algebras, with Applications to Logical Constants', *Transactions of the American Mathematical Society*, XIV, pp. 481–8.
12 Wittgenstein, Ludwig, [1961], *Tractatus Logico-Philosophicus*, London, Routledge & Kegan Paul.
13 The expression *elementary proposition* translates Wittgenstein's German term *elementarsatz*. Although that term has come to be rendered in terms of English *proposition*, Wittgenstein has a highly distinctive view of the nature of both *satz* and *elementarsatz* which is not to be confused with the conception of a proposition discussed in Chapter 2.
14 See Wittgenstein, Ludwig, [1967], *Philosophical Investigations*, Oxford, Blackwell, Remarks 1–37.
15 Hunter [1971], pp. 93–4.

5
An Introduction to First Order Predicate Logic

5
An Introduction to First Order Predicate Logic

I
Logical Form Revisited: The Formal Language QL

Logic is the study of argument. The central problem for the logician lies in distinguishing good argument from bad argument, quite generally. As we have seen, the logician attempts to draw that distinction by exploiting the concept of validity. The claim of the formal logician in particular is just that the concept of validity can profitably be investigated in purely formal terms, i.e. that particular arguments are valid or invalid purely in virtue of the logical forms they exemplify. Above all, the primary task of the formal logician is to exhibit logical forms of argument. The programme of formal logic consists in the attempt to exhibit logical forms of argument and, further, to capture every valid form in a single formal language.

In Chapter 1 we noted certain possible limits to the success of that programme. The range of types of valid argument is a broad one which may not yield completely to the investigations of the formal logician. None the less, we were able to identify an impressive number of argument forms and to capture all of the valid ones in the formal language PL. Subsequently, we developed PL in two ways. First, we used rules of inference to exploit the different kinds of well-formed formula belonging to PL and arrived at a formal system for PL, a propositional calculus. Next, we went on to interpret the well-formed formulas of PL in terms of truth-values and to supplement our syntactic formal method of proof with formal semantic methods, i.e. truth-tables and truth-trees. Strikingly, we noted a perfect agreement between the syntax and the semantics of PL when we realised that, whatever method we used, we would arrive at an endorsement of exactly the same set of sequents.

For all the power, sophistication and remarkable formal integrity of PL, the sentence-letters which form its basic sentences are very limited instruments when it comes to representing the character of natural language sentences. In fact, the humble sentence-letter can only stand for a natural language sentence in its entirety, i.e. it can represent nothing of the **internal grammatical structure** of any such sentence. Indeed, PL is precisely the logic of those argument-forms whose validity does not depend in any way upon the internal grammatical structure of the atomic sentences which compose them. As a result, PL often fails to capture the form of natural language arguments in a satisfying way. Consider the following argument, for example:

1. All folk singers are groovy.
2. Arlo Guthrie is a folk singer.

Therefore,

3. Arlo Guthrie is groovy.

While the soundness of this argument is disputable, the intuitive validity of the argument is surely obvious. But note what happens when we try to formalise the argument as a sequent of PL. Because three distinct sentences are involved, three distinct sentence-letters must be used. Thus, the argument translates into PL as follows P, Q : R. But that sequent is obviously invalid! A valid argument may be an instance of a number of invalid argument-forms but the point is to show that it is also an instance of some valid argument-form. The problem here is that the validity of this particular argument depends not merely upon the relations between the sentences which make it up but also upon the internal grammatical structure of those sentences. But sentence-letters can represent nothing of the internal structure of a sentence. Therefore, we cannot accurately represent the form of this argument using sentence-letters. It follows that we will never be able to represent any argument of that form in PL accurately. Moreover, PL is inadequate for representing the form of *any* argument whose validity hinges on the internal structure of the sentences which make it up. Such forms certainly can be adequately represented formally but their representation requires a little more in the way of formal machinery than a language like PL provides. To that end, look closely at sentences 2 and 3 of the argument. Sentences of this type are **subject–predicate sentences**, i.e. sentences which attribute a property or properties to some individual person, place or object. In logical terms, such sentences predicate a property or properties of some particular subject. Thus, sentence 2 predicates 'being a folk singer' of the subject Arlo Guthrie while sentence 3 predicates 'being groovy' of that same subject. Further, because such sentences predicate a property of a single subject they are *singular sentences*.

To formally represent such sentences adequately, we need to be able to distinguish subjects from predicates in our notation and to express the subject–predicate relation formally. Therefore, we must introduce some new symbols which belong not to PL but to a new formal language which I will call **QL**, for **quantificational logic** (the meaning of this term will become clearer as we go). First, we will let lower-case italic letters from the beginning of the alphabet be names denoting individual subjects, e.g. we may simply use *a* to denote Arlo Guthrie, *b* to denote Blind Lemon Jefferson, and so on. Next, we identify upper-case italic letters from *F* onwards as predicate-letters. Hence, we may use *F* to represent the predicate ' . . . is a folk singer'. Again, for clarity, we will use these new symbols (and the sentences of QL which they compose) *autonymously*, i.e. we will let these symbols act as names for themselves and so keep our use of quotation marks to a minimum.

Note that while names are complete symbols, predicates are incomplete symbols, i.e. predicates contain a gap. Hence, when we write out the natural language expression which the predicate-letter represents we use a few dots to mark that gap or place, as in ' . . . is a folk singer'. Subject–predicate sentences are formed simply by plugging the gap in the predicate with a name. So, if *a* stands for Arlo Guthrie, and *F* represents the predicate ' . . . is a folk singer', the sentence 'Arlo Guthrie is a folk singer' is simply translated as *Fa*. (It may seem more natural to write *aF* here but later, when we consider relational expressions, you will see why following that convention could lead to confusion. Hence, the name should always be written after the predicate and not before it.)

Given these new elements of formal vocabulary, the conclusion: 'Arlo Guthrie is groovy' can easily be translated. As agreed, *a* stands for Arlo Guthrie. So, all we need to do is to pick a distinct predicate-letter to represent the new predicate ' . . . is groovy'. For that purpose, we may just use the next letter after *F* in the alphabet, i.e. we let *G* stand for the predicate ' . . . is groovy' and so formalise the conclusion as *Ga*.

So far in our formalisations we have followed the convention of picking letters for names and predicates which are identical with the first letters of the expressions we want to formalise; excluding the 'is' of predication, of course. This is often a convenient way of remembering what is representing what in any given context. But we are quite free not to follow that convention if we choose. For example, we could choose *b* as the formal name for Arlo Guthrie, or *c* and so on. Equally, we could have chosen *L* to represent '. . . is a folk singer', or *M* and so on. Although these practices may seem odd there is nothing in logic to forbid them. In fact, there will be occasions when we are forced to choose different symbols; for example, if we want to formalise two names which begin with the same letter or, indeed, two such predicates.

To complete the formalisation of the argument at hand, premise 1 must be formalised. This sentence is not a singular sentence but a *general sentence*, i.e. a sentence which makes a claim about *all* the members of a certain class, here, about the class of folk singers. To formalise this kind of general sentence, then, we need a way to capture formally what is said in an 'All' or, equivalently, an 'Every'-type sentence. This is done in three steps.

First, we adopt another new set of symbols known as *variables*. For that purpose, we will use lower-case italic letters from the end of the alphabet, i.e. x, y, z. Each and every such variable can simply be read as meaning 'thing'. Secondly, we adopt the following symbol $\forall$, upside-down 'A', which you can think of as formally expressing the term 'Every' or, better, the term 'Any'. Next, we place a variable, x, for example, alongside our upside-down 'A' like so: $\forall x$. The new expression which we have formed here is known as a **quantifier** and, logically enough, it allows us to talk about a given quantity of things, i.e. it enables us to *quantify* over sets of things. The use of a quantifier always carries with it an implicit recognition that there is a set of things over which we are quantifying, although on any particular occasion of use, that set may well be undefined or might even be thought of as including absolutely everything that we can think of as a thing at all. Now, because we will always use this particular quantifier to help us make assertions about *every* such thing or *any* such thing we will invariably use the new expression to indicate that we are quantifying over everything, over anything, quite universally. Hence, this quantifier is the **universal quantifier**. Simply, the expression indicates that we are quantifying universally, i.e. over every or any member of the set of things we are talking about.[1]

It is tempting to think of the expression $\forall x$ as simply meaning *everything*. But that is not quite true. For this expression cannot meaningfully occur just on its own, i.e. $\forall x$ is not itself a well-formed formula of QL. However, we can certainly use that expression to construct well-formed formulas. Keeping things intuitive, the quantifier must always be followed by a further expression which occurs in square brackets, i.e. the quantifier will be part of a formula whose overall shape will generally look like this:

$\forall x$ [. . .]

In an important sense, the square brackets are an integral part of the quantifier just because they indicate what it is that the quantifier applies to or, in Logicspeak, **governs**. Still keeping things intuitive, we can think of the expression we have just added to the quantifier as representing the **scope** of the quantifier, i.e. the square brackets indicate the scope of that universal quantifier. Hence, we can properly read formulas of this form as meaning, literally, 'For anything, x, . . . ' or 'Consider anything, x, . . . '. So, we do now have a way of expressing part of what is said in an 'All' or 'Every'-type sentence.

In contrast with subject–predicate sentences, to form general sentences in

QL with the universal quantifier, variables must always be used, i.e. rather than talking about specific named individuals in quantified formulas, we talk instead about any and every *thing*. The variable associated with the quantifier is said to range over the set of things we are talking about. That set which the variable ranges over is known as the **universe of discourse**, literally that world or universe about which we are talking or, more simply, as the **domain** where the domain is said to consist of **elements** (henceforth, I will use the simpler expression *the domain*). This is a crucially important concept to which we will return but, for the moment, it is sufficient to grasp that the term 'domain' refers to just that set of things which we are talking about in the general sentences we are concerned with and the quantified formulas which represent them in QL.

As noted, while use of a universal quantifier carries with it an implicit domain that domain may well be unspecified. In fact, it is common practice among formal logicians to leave the domain in just such an indefinite state. Further, I also said that the domain may be thought of as including absolutely everything that can be thought of as a thing. When the domain is construed in this way it is **unrestricted**. Logically enough, when the domain is confined to a particular kind of thing, such as human beings, animals, shapes, numbers, times, places or whatever, the domain is **restricted**, i.e. to things of just that kind. Hence, the domain has an inherent flexibility or relativity. Indeed, part of the beauty of this formal device consists in the sheer general applicability which derives from that flexibility. So, in QL, not only can we talk about whatever set of things we like but, given the universal quantifier, we can quantify over the entire set of things we've chosen to talk about.

For purposes of translation into QL, it is generally more intuitive to use a restricted domain and thus to take the variable as ranging over a definite kind of thing, i.e. just that kind of thing which is capable of having the property or properties we are interested in. Moreover, it is not the quantifier which gives the game away as regards the relevant kind of thing. Rather, that comes out in the predicates we want to use with the quantifier. For example, consider the first premise of the argument above: 'All folk singers are groovy.' Here, the presence of 'All' in initial position implies that the universal quantifier must be involved in any faithful translation. But the mere presence of 'All' does not tell us which kinds of thing are being talked about. To determine that fact we must look to the properties involved. In this case, the properties involved are those of being a folk singer and being groovy. The kinds of thing which can be groovy folk singers are human beings, persons. Therefore, the domain should be restricted to persons. Thus, any universally quantified formula which translates a general sentence in this context can be read as saying 'Consider every person . . . '. But what is implied about every person here? It is certainly not implied that every person is actually a groovy folk singer. Rather, premise 1 implies that those people who are folk singers are groovy. Hence, premise 1 is really a conditional, i.e. what it implies is that if a person is a folk singer then she or he is groovy. In fact, a universally quantified formula of this sort is really a **universal conditional**. Hence, we may paraphrase premise 1 as follows: 'Consider every person. If any person is a folk singer then that person is groovy.' In this case, then, premise 1 is a universally quantified conditional sentence about persons. Given that we are talking about persons quite generally here, we must use not a name but a variable throughout our formalisation of premise 1. The term 'person' occurs three times in the paraphrase of the premise: 'Consider every person. If any person is a folk singer then that person is groovy.' So, we make three uses of the same variable. Premise 1 can now be formalised as follows:

1.	$\forall x\ [Fx \rightarrow Gx]$	1. All folk singers are groovy.

And so we can complete the formalisation of the whole argument:

1.	$\forall x\ [Fx \rightarrow Gx]$	1. All folk singers are groovy.
2.	Fa	2. Arlo Guthrie is a folk singer.
	$\therefore$	Therefore,
3.	Ga	3. Arlo Guthrie is groovy.

Note carefully the presence of the three variables in the formalisation of premise 1. Further, note that although that whole expression is a well-formed formula, parts of the whole in isolation are not well formed. The universal quantifier just on its own, for example, is strictly meaningless. $\forall x$ must never occur other than as attached to some expression or other. What the universal quantifier is attached to, in this case $[Fx \rightarrow Gx]$, is a **matrix**. This is a useful notion just because the scope of a quantifier can be identified with its matrix, i.e. the square brackets are scope-indicating brackets.

Just as a universal quantifier always requires a matrix, so too a matrix always requires a quantifier, i.e. in isolation, the matrix is not a well-formed formula of QL. The key consideration here concerns the difference between names and variables. Names plug gaps in predicates and, in so doing, generate sentences (hence, predicates are **sentential functions**, i.e. predicates take names as arguments and give sentences as values). So, for example, premise 2 is certainly well formed and obviously meaningful. Further, we could properly write: $Fa \rightarrow Ga$. In this particular context, that singular conditional translates 'If Arlo Guthrie is a folk singer then Arlo Guthrie is groovy.' That is perfectly meaningful and the corresponding formal sentence is perfectly well formed. But the original matrix contains not names but variables. Variables must always be chaperoned by a quantifier and quantifiers must always have a matrix to look after. Romantically enough, neither can meaningfully stand without the other. Quantifiers and matrices belong together.

The fragment of new notation introduced here belongs not to the language of propositional logic but to the language of quantificational logic. In Chapter 1, we noted that specifying logical form is not really an all-or-nothing matter but, rather, is a matter of how deep we can dig into the structure of sentences and the arguments they compose with the formal vocabulary available to us. QL extends that formal vocabulary considerably and allows us to dig much more deeply into the internal grammatical structure of sentences. Moreover, the vocabulary of the new language is not confined to names, predicates, variables and the universal quantifier. Rather, it also inherits all of the vocabulary of PL, i.e. connectives and even sentence-letters, though these last are not always terribly useful in QL. Hence, just as arrow can be used with the universal quantifier to form universal conditionals so '&' can be used with that quantifier to form universal conjunctions, e.g.:

$$\forall x\ [Fx\ \&\ Gx]$$

And conjunctions of universally quantified formulas can also be formed, e.g.:

$$\forall x\ [Fx]\ \&\ \forall x\ [Gx]$$

Note the way in which the square brackets show up the differences in the scope of the quantifiers in these two examples. While the scope of the quantifier in the first case is the whole matrix, the scope of each quantifier in the second case is confined to a single conjunct. In that latter case, the main connective is '&'. Therefore, the scope of that connective is the entire formula. Hence, that formula is precisely a conjunction of universally quantified formulas. In any such case, each component formula (here, each conjunct) is a **sub-formula**. Hence, the complex formula $\forall x\ [Fx]\ \&\ \forall x\ [Gx]$ has two sub-formulas: $\forall x\ [Fx]$ and $\forall x\ [Gx]$.

Moreover, we can equally well form universal disjunctions, e.g.:

$$\forall x\ [Fx \vee Gx]$$

Equally, disjunctions of universal formulas can easily be formed, e.g.:

$$\forall x\ [Fx] \vee \forall x\ [Gx]$$

(again, note carefully the difference in the scope of the quantifiers in these cases).

Further, the connectives can also be used to connect subject–predicate formulas. Thus, conjunctions of subject–predicate formulas can be formed, e.g.:

(*Fa* & *Ga*)

And disjunctions of such formulas can be formed, e.g.:

(*Fa* v *Ga*)

And so on. Again, still more complex formulas can be formed by applying the connectives to these well-formed formulas in turn. At this stage, it is important to appreciate something of the potential for constructing different kinds of formula in QL. Exercise 5.1 should give you some idea of that potential.

EXERCISE 5.1

1 Construct universally quantified QL formulas with matrices whose main connective is one of the connectives: '&', 'v', '↔', '→' and '~'.

2 Use each of the connectives in turn to construct formulas such that each component formula involved, i.e. each sub-formula, is itself a universally quantified QL formula.

3 Use each of the connectives in turn to construct formulas each sub-formula of which is a subject–predicate QL formula.

II
More on the Formulas of QL

Exercise 5.1 should have begun to awaken your intuitions about the range of different kinds of formula which can be constructed in QL. The sheer extent of that range is testimony to the power, sophistication and articulateness of the new language. The key factor responsible for the increase in expressive power which we gain in QL consists in the introduction of formal vocabulary which allows us to represent the internal grammatical structure of natural language sentences for the first time. If you feel comfortable with that new vocabulary and found Exercise 5.1 quite straightforward then you may jump ahead to the next section. If elements of the new vocabulary still feel rather unfamiliar then you should work through the present section carefully.

In the discussion of formula-construction in PL, we considered a way of analysing complex PL formulas into their constituent parts by means of syntactical trees (for PL formulas). Those trees displayed the overall syntactic structure of a PL formula in diagrammatic form. Further, they did so by exploiting the contrast between compound PL formulas and atomic PL formulas. We can exploit a similar distinction between QL formulas to design a method for generating **syntactical trees for QL formulas**, i.e. for any given QL formula, the method can be used to analyse both the overall form of the formula and the form of each of its sub-formulas. Applying such a method to QL formulas is instructive in a number of important ways. I shall outline the bare bones of the method here. In later sections the usefulness of the method will be extended.

Much of the work required here was already completed when we set the method up for PL. For example, for purposes of tree-construction, we will still begin by simply designating the formula we are interested in as **F** (for clarity we will use the symbols introduced here autonymously). Hence, each tree will again begin with **F** on a line. Further, we still have just two kinds of connective, unary and binary, which we simply designate by **U** and **B** respectively. So, we can not only begin any tree by placing **F** on a line but we can also go on to rewrite any formula whose main connective is one of the familiar PL connectives. For example, consider the following QL formula: ∀*x* [*Fx*] & ∀*x* [*Gx*]. This formula is a complex conjunction, each conjunct of which is a universally quantified formula. Hence, we begin the syntactical tree by placing **F** on a line and then expand the formula simply by exploiting the first PL rewrite rule:

```
   F
 / | \
F  B  F
```

We know exactly which binary connective is involved here, namely, '&'. So, we can construct the following partial tree:

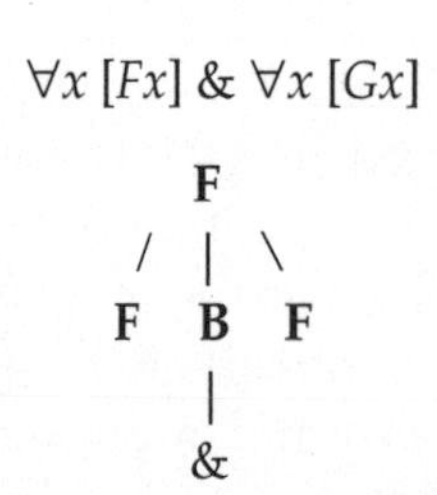

To complete the tree, we must develop each branch with an **F** on its tip. But these **F**s represent quantified formulas. Hence, at this point, we must supplement the existing stock of rewrite rules with a rule which allows us to rewrite quantified formulas. In fact, this is quite straightforward. Where **Q** denotes 'quantifier' and **M** denotes 'matrix', we add the following rewrite rule:

```
1.      F
       / \
      Q   M
```

Any quantified formula may itself be negated, i.e. the quantifier may be preceded by the unary connective '~'. Hence, where **U** denotes that unary connective, we add the following rule:

```
2.      F
       / \
     UQ   M
```

And now, using Rule 1 again, we can expand the tree still further:

$\forall x\ [Fx]\ \&\ \forall x\ [Gx]$

```
       F
     / | \
    F  B  F
   / \ | / \
  Q   M & Q  M
```

Now, every quantified formula must be developed completely. So, observe the following rule carefully: rewrite each and every quantifier, variable, predicate-letter, sentence-letter and name involved in each formula on a separate branch. Finally, to complete the tree, spell out just what kind of symbol is at the tip of each branch, i.e. write the name of the relevant element of vocabulary involved under each and every branch-tip, e.g.:

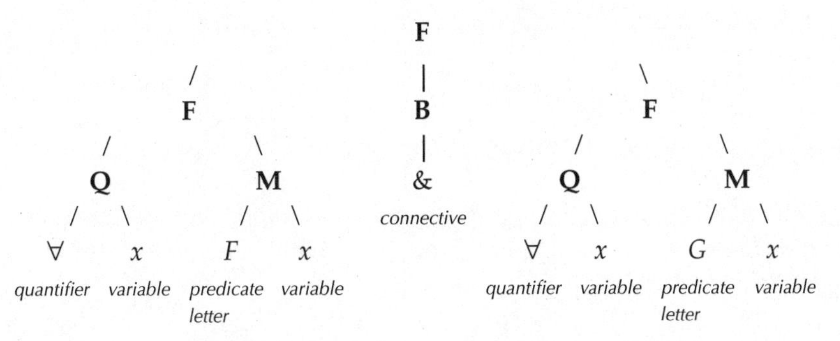

The vocabulary of QL includes subject–predicate formulas and so we require rewrite rules for these too. Hence, where **F** denotes 'formula', **S-P** denotes 'subject–predicate', **P-L** denotes 'predicate-letter', **N** denotes 'name' and **U** denotes the unary connective '~', the following rewrite rules should be used to develop subject–predicate formulas and their negations:

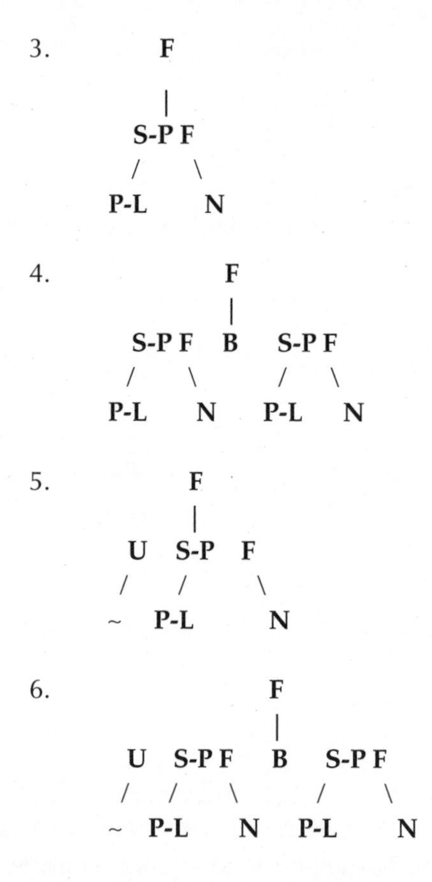

Exercise 5.2 gives you the opportunity to use syntactical trees to explore the formulas of QL a little further. Study Box 5.1 carefully before attempting the exercise.

BOX 5.1

Syntactical trees for QL formulas

♦ QL inherits all the vocabulary, grammar and formal apparatus of PL. Therefore, the following rewrite rules still apply (as do each of the other rules stated in Chapter 2):

```
1.   F        2.      F        3.    F
     |              / | \           / \
     A             F  B  F         U   F
```

♦ Further, because the vocabulary of QL is more extensive than that of PL we must observe the following procedure carefully:

♦ (i) Identify and distinguish quantifiers from matrices using the rules:

```
4.     F        5.      F
      / \              / \
     Q   M           UQ   M
```

Rewrite each quantifier and each associated variable on a new branch.

♦ (ii) Identify the main connective, if any, in each matrix. Matrices involving connectives may be conjunctive, disjunctive, conditional, biconditional or negated. But note that a negated matrix may negate a formula of any of these forms. Rewrite each and every element involved in each matrix specifying each predicate-letter, variable and connective involved on a separate branch.

♦ (iii) Use the following rules to rewrite subject–predicate formulas:

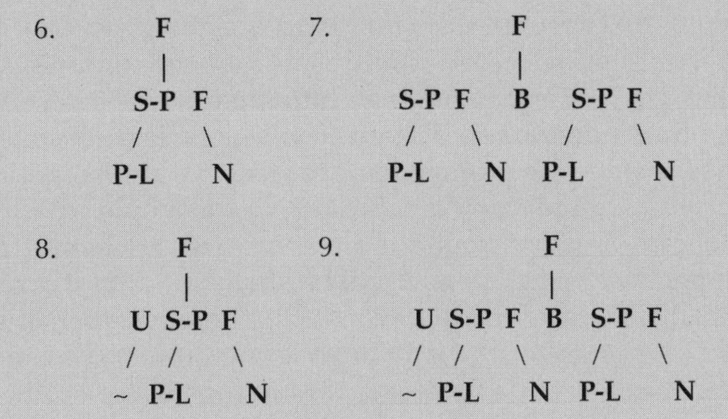

Again, specify each connective, predicate-letter and name involved on a separate branch.

♦ (iv) Finally, under each branch-tip describe the relevant element of the vocabulary of QL involved.

EXERCISE 5.2

1 Consider once more your answers to Questions 1–3 of Exercise 5.1 above. Construct syntactical trees for each and every formula you constructed when answering that exercise.

2 Consider the following formulas carefully:

(i) $\sim\forall x\ [Fx \rightarrow Gx]$

(ii) $\forall x\ [\sim(Fx \rightarrow Gx)]$

Indicate the scope of each quantifier and every connective in both cases.

III
The Universal Quantifier and the Existential Quantifier

Recall premise 1 of the argument we have been considering, i.e. the sentence 'All folk singers are groovy.' As noted, the logical grammar of this sentence is properly represented in QL by a *universal conditional*. And that is quite intuitive. After all, that premise does not imply that everyone is a groovy folk singer. In fact, it doesn't even imply that there are any folk singers. Rather, it implies that, for any individual, if that particular individual is a folk singer then that individual is also groovy. Moreover, this point holds quite generally, i.e. any sentence of the form 'All As are Bs' is properly translated into QL as a universal conditional. For example, consider the following sentence 'All trespassers will be shot.' This sentence does not imply that everyone will trespass and will be shot. Nor does it imply that anyone will actually trespass or that anyone will be shot. Rather, it is properly understood as a conditional, i.e. 'If anyone should trespass then he or she will be shot.' It follows that no universal conditional implies the actual existence of any trespasser or folk singer or of anything else for that matter. Therefore, universal conditionals do not generate any existential or, as philosophers say, **ontological commitments**. (Ontology is simply the study of what there is. So, ontological commitments are precisely existential commitments.) Again, this is highly intuitive, e.g. I might well want to say that 'All unicorns have a single horn' but I certainly do not want to imply the actual existence of any unicorn in the process (we will consider this point again later).

Universally quantified conditionals do not therefore imply the existence of anything. Rather, universal conditionals imply **singular conditionals**, e.g. the universal conditional 'All folk singers are groovy' implies the singular conditional 'If Arlo Guthrie is a folk singer then Arlo Guthrie is groovy.' And, indeed, it implies the singular conditional that 'If Bob Dylan is a folk singer then Bob Dylan is groovy.' And so on ad nauseam. Quite generally, then, a universal conditional implies all its instances. Thus, 'All trespassers will be shot' implies that 'If I trespass then I will be shot', that 'If you trespass then you will be shot' and so on. In this way, 'all' becomes 'every' or 'each and every', i.e. use of the universal quantifier indicates that something is being said about each (and every) particular thing of the relevant kind.

Therefore, the sense of 'all' which the universal quantifier expresses is not the **aggregative** sense of 'all', i.e. it is not the sense of 'all' involved in sentences such as 'All the tea in China will not induce me to stop reading this logic book' or 'All the riches of Arabia would not have persuaded her to stop loving him' or 'All the fans of the worst folk singer in the world could fit comfortably into a very small car.' Rather, the sense of 'all' or 'every' which the universal quantifier expresses is **distributive**, i.e. it distributes properties over each of its instances. Hence, a universal conditional does, indeed, imply singular conditionals for each and every one of its instances. But again, no singular conditional implies that anything of the relevant kind actually exists, e.g. the singular conditional 'If Arlo Guthrie is a folk singer then Arlo Guthrie is groovy' does not imply that Arlo Guthrie is a folk singer but just that if he is a folk singer then he is, indeed, groovy. Equally, the singular conditional 'If I trespass then I will be shot' does not imply that I am a trespasser or that there exist any trespassers at all.

Given that neither universal nor singular conditionals imply the actual existence of anything of any given kind, it is natural to ask: how do we make ontological commitments in QL? After all, part of the beauty of QL surely consists in that sense of being able to talk about actually existing sets of things and of being able to quantify over the members of those sets. Indeed, in these respects, QL has much of the feel of a natural language. So, how are ontological commitments to be expressed here? For this purpose, we must make explicit another expression which will be used precisely in order to make clear that ontological commitments are being made. The new expression is a second quantifier: the **existential quantifier**.

This time, the relevant symbol is not an upside-down 'A' but an 'E' written backwards, like so: $\exists$. Again, the quantifier must be combined with a variable, e.g. $\exists x$, and, again, that expression just by itself is not a well-formed formula of QL. It too always requires a matrix. To place the existential quantifier in initial position in any QL formula is to make the claim that there does indeed exist something of the kind indicated by the relevant predicate or predicates in the matrix. Hence, for purposes of translation, we can read the existential quantifier as implying that 'There is something . . . '. Before reading the matrix it generally helps to add 'such that . . . ', i.e. to read the existential quantifier as indicating 'There is something such that . . . '. The translation is then completed by filling in the dots with a translation of the matrix. Thus, while you can think of the universal quantifier as corresponding roughly in meaning to 'every', and, in combination with a variable, as meaning every*thing*, you might think of the existential quantifier as corresponding in meaning to 'some' and, in combination with a variable, as corresponding in meaning to some*thing*. But note that existentially quantified sentences are not singular or subject–predicate sentences. Rather, they are another kind of general sentence. The tell-tale sign here is just the use of variables rather than names. Hence, the variable ranges over the elements of the domain quite generally and although the presence of the existential quantifier does tell us that there is some element of the domain with a certain property, it does not tell us which element that is. Quine puts the point well when he notes that: 'variables of quantification . . . words like "something" . . . do not purport to be names at all; they refer to entities generally with a kind of studied ambiguity peculiar to themselves'.[2]

Our concern here is to exploit the potential of QL to represent the logical forms of sentences and, indeed, arguments. So, it is worth pausing to consider just which forms we can use the existential quantifier to express.

In the simplest possible case, basic existential sentences can be formed by combining the existential quantifier with a predicate and plugging the gap with a variable. Thus, we might represent the sentence 'There is a folk singer' or 'There exists a folk singer' in QL as:

$$\exists x\ [Fx]$$

In sharp contrast to universal conditionals, the presence of the existential quantifier does imply that there exists *at least one* instance of the kind of thing being talked about. So, in this case, we could equally well translate $\exists x\ [Fx]$ as 'There exists at least one folk singer' (note that the expression leaves it open whether there is more than one such thing, i.e. whether there are a few more folk singers out there).

Now, the existential quantifier should always be used when translating any sentence of the form: 'Some As are Bs.' Sentences of this form imply both that there is some A and that this A is also a B. Hence, such sentences are properly translated as **existentially quantified conjunctions**, e.g. the sentence 'Some folk singers are groovy' is properly translated into QL as $\exists x\ [Fx\ \&\ Gx]$.

BOX 5.2

♦ Any sentence of the form 'All As are Bs' should always be translated into QL as a *universal conditional*, i.e.:

$$\forall x\ [Ax \rightarrow Bx]$$

♦ Any sentence of the form 'Some As are Bs ' should always be translated into QL as an *existentially quantified conjunction*, i.e.:

$$\exists x\ [Ax\ \&\ Bx].$$

For purposes of translation, then, while we generally associate the use of the universal quantifier with a conditional matrix, we generally associate the use of the existential quantifier with a conjunctive matrix. Thus we arrive at two important rules of thumb for translation (see Box 5.2).

IV
Introducing the Notion of a QL Interpretation

Before we move on to practise translating natural language sentences into QL it is worth making explicit one particularly important notion which we have not yet considered in the present chapter. To that end, recall the notion of a *key* which we exploited for purposes of translation into PL. We used that notion to provide a useful picture of what was standing for what in each translation into that formal language. When translating into QL, it is equally important to spell out what is standing for what in terms of the formal vocabulary and, further, to make explicit exactly those things that we are using QL to talk about. The vocabulary of QL not only includes that of PL but also extends significantly beyond the limits of PL. Hence, in the context of QL, we replace the notion of a key with a new and fundamentally important notion, namely, that of a **QL-interpretation**.

You may recall the notion of an interpretation from our earlier investigation of the semantics of PL. There, the notion was defined as: *any assignment of truth-values to each of the component parts of a formula*. While that definition was adequate as regards PL, it cannot be adequate for QL. In QL we are no

BOX 5.3

To give a *QL-interpretation* is to give:

- your current choice of domain;

- a description of exactly how the elements of the formal vocabulary involved, e.g. names and predicates, are related to the elements of that domain, i.e. what each name and each predicate expression signifies in the context of that domain.

longer concerned with simple sentence-letters and so we cannot just assign truth-values to sentence-letters as component parts of complex formulas. In fact, no sentence of QL is ever true or false just in itself, absolutely, as it were. Rather, whether a formula of QL is true or false depends upon what it is that we take that formula to be about. In QL truth and falsity are relativised to the set of things we understand ourselves to be talking about, i.e. to the domain.

Moreover, the truth-value of a given QL formula depends not just on the choice of domain but also on how the elements of formal vocabulary involved, i.e. names and predicates, hook up with the elements of that domain. To specify a domain and to spell out precisely how all the elements of the formal vocabulary involved are related to the elements of that domain, i.e. what stands for what in any given case, is to give a QL-interpretation. As we shall see, a QL-interpretation gives us all the truth-conditions we require for each and every sentence of QL. What I have said here is nothing more than an intuitive explanation, i.e. it is not a formal definition of a QL-interpretation. For present purposes, however, the account in Box 5.3 is adequate.

Setting up a QL-interpretation is thus a matter of spelling out a particular choice of domain and making clear what each element of vocabulary stands for so as to make sense of translations of sentences into QL. With a little practice at constructing QL interpretations, you will quickly become familiar in a practical sense with some of the most important semantic concepts for QL. We will cash in on that familiarity later. But now consider the following recipe for making a QL-interpretation carefully.

To construct a QL-interpretation follow 1–3 below:

1. First, write out a big '*I*' (for *interpretation*), with something of a Germanic flourish, like so:

   $\mathfrak{I}$

2 Next, consider the properties involved and select a domain. Write: **D**: (for **domain**) and then spell out a choice of domain, putting that choice in curly brackets to indicate that it is a set of things. Examples might include the set of human beings, i.e. **D**:{**human beings**}; or the set of brand-name fishfingers in the local supermarket, i.e. **D**:{**brand-name fishfingers in the local supermarket**}; or the set of cats, i.e. **D**:{**cats**}, and so on. Remember also that the choice of a domain should be dictated by the predicates involved in the particular context of translation, i.e. the domain should consist of all and only those things capable of possessing the relevant properties. Hence, a domain of human beings is an appropriate choice for translations involving folk singers, philosophers and florists while a domain of cats is appropriate to translations involving predicates like '... is a tabby', '... is a tortoise-shell', etc.

3 Finally, list the elements of formal vocabulary involved, i.e. all the names and predicates, making clear what is to stand for what in this particular interpretation.

By way of an example, let's construct a QL interpretation for the sentences composing the argument we considered earlier:

1. All folk singers are groovy.
2. Arlo Guthrie is a folk singer.

Therefore,

3. Arlo Guthrie is groovy.

Here, the things capable of having the relevant predicates are just persons or human beings. Hence, I write an 'I' for interpretation and then state that the domain for this interpretation is the set of human beings, i.e.:

$\mathfrak{I}$ **D**: {**human beings**}

In this case, two predicates and one name are involved. So, I select and list the elements of formal vocabulary I require, spell out exactly what each element stands for and complete my interpretation as follows:

F: . . . is a folk singer

G: . . . is groovy

a: Arlo Guthrie

Given this interpretation, each natural language sentence can be translated into QL. And so the whole argument can be represented thus:

1. $\forall x\,[Fx \rightarrow Gx]$
2. Fa

$\therefore$

3. Ga

Finally, in any translation into QL always try to represent as much of the internal structure of the original sentence as possible. In practice, this frequently means resisting the temptation to lump distinct predicates together. For example, the sentence:

Gina is a gallant and green-fingered gardener

involves not one but three distinct predicates. Each distinct predicate should be translated by a distinct predicate-letter. Therefore, where a denotes Gina, this sentence is not properly translated simply using the predicate-letter G, i.e. it is not properly translated as Ga. Rather, an interpretation which separates out each predicate should be constructed, e.g.:

$\mathfrak{I}$ **D**: {**human beings**}

G: . . . is gallant

H: . . . is green-fingered

I: . . . is a gardener

a: Gina

In this case, all three predicates apply to the subject, i.e. Gina. Given the interpretation, we can construct the following three subject–predicate formulas:

$Ga \quad Ha \quad Ia$

In QL as in PL, the logical connectives can be used to characterise relevant

BOX 5.4

For purposes of translation into QL always observe the following rules:

- Any sentence of the form 'All As are Bs' must always be translated as a *universally quantified conditional*, i.e.:

  $\forall x\ [Ax \rightarrow Bx]$

- Any sentence of the form 'Some As are Bs' must always be translated as an *existentially quantified conjunction*, i.e.:

  $\exists x\ [Ax\ \&\ Bx]$

- *Never* translate any sentence of the form 'All As are Bs' as:

  $\forall x\ [Ax\ \&\ Bx]$

or any sentence of the form 'Some As are Bs' as:

$\exists x\ [Ax \rightarrow Bx]$

relationships between sentences. In this case, it is natural to use '&' to conjoin the formulas and complete the translation as:

$((Ga\ \&\ Ha)\ \&\ Ia)$

With this point and the notion of a QL-interpretation clearly in mind, you can now usefully attempt some translations for yourself. However, before you do Exercise 5.3, consider the contents of Box 5.4 carefully.

EXERCISE 5.3

1 Translate the following English sentences into QL. In each case make your QL-interpretation explicit and try to represent as much of the structure of each sentence as possible.

(i) Some florists are greengrocers.

(ii) All greengrocers are happy people.

(iii) There is at least one folk singer who is also groovy.

(iv) Sandy Denny is a folk singer.

(v) Both Sandy Denny and Julie Felix are folk singers.

(vi) All folk singers who are groovy play guitar.

(vii) Either all folk singers are groovy or some folk singers are dreadful.

(viii) Some folk singers are also florists and all greengrocers are groovy.

(ix) All folk-singing florists are happy, generous and interesting people.

(x) If all folk-singing florists are groovy greengrocers then some groovy greengrocers are fearless fire-fighters.

(xi) Sandy Denny is a folk singer if and only if Julie Felix is a flamenco dancer.

(xii) If everyone is a folk-singing flamenco dancer then Sandy Denny is a folk-singing flamenco dancer.

(xiii) Sandy Denny, Julie Felix and Tom Paxton are either folk singers or flamenco dancers.

(xiv) Either someone is a folk-singing flamenco dancer or everyone is either a folk singer or a flamenco dancer.

(xv) Everyone is a folk singer if and only if it is not the case that if Tom Paxton is a folk singer then Julie Felix is a flamenco dancer.

V
Valid and Invalid Sequents of QL

In Chapter 1, I defined a valid argument as an argument such that if the premises are true then the conclusion must be true. It followed that an argument is invalid if it is possible that its premises are true and its conclusion false. I then went on to define a valid form of argument as an argument-form all of whose substitution-instances are valid arguments. Consequently, an invalid form of argument was defined as a form which had any invalid substitution-instance. These definitions have been faithfully preserved throughout the text and, to date, they have been perfectly adequate. In QL, however, the notion of truth is relativised to the notion of a QL-interpretation and so it is useful to restate the definitions of validity and invalidity in a way which preserves those earlier intuitions but also takes account of the new-found relativity of the notion of truth in QL.

For example, consider the following QL sequent $\exists x\ [Fx]\ :\ \forall x\ [Fx]$. A little reflection should quickly tempt us to conclude that this sequent is invalid, i.e. that the premise can be true when the conclusion is false. And, indeed, if we did know that the premise was true and the conclusion false then we would know that the sequent is invalid. But, on the basis of our present state of information, we do not yet know that this is so. To generate the truth-values which each sentence should receive we must construct a QL-interpretation. Consider the following two interpretations:

1. $\mathfrak{I}$ **D**: {**Bill Clinton**}

   F: . . . is president of the USA

Under this interpretation, $\exists x\ [Fx]$ and $\forall x\ [Fx]$ are both true. Hence, this interpretation fails to show that the sequent in question is invalid. Now consider another interpretation:

2. $\mathfrak{I}$ **D**: {**human beings**}

   F: . . . is president of the USA

This time, the premise $\exists x\ [Fx]$ is plainly true; *some* human being is president of the USA. But the conclusion, $\forall x\ [Fx]$, is plainly false. It is certainly not the case that every human being is president of the USA! Hence, on this interpretation we do show that the sequent is invalid. Therefore, we must indeed reformulate the definition of validity in a way which preserves our intuitions but also takes account of the relativity of the notion of truth in QL. Validity and invalidity in QL are defined as follows:

1. A sequent of QL is valid if and only if, for every domain, there is no possible interpretation under which all the premises are true and the conclusion is false.

Therefore:

2. A sequent of QL is invalid if and only if, for some domain, there is a possible interpretation under which all the premises are true and the conclusion is false.

Finally, to complete the definition of validity for QL:

3. A formula of QL is valid if and only if for every domain that formula is true under every possible interpretation.

It follows from these definitions that to construct an interpretation under which the premises of a given sequent are true and the conclusion is false is to construct a counterexample to that sequent. Hence, interpretation 2 above is a counterexample to the sequent $\exists x\ [Fx] : \forall x\ [Fx]$. Formal methods for generating counterexamples to QL sequents are given in Section II of Chapter 7. At this stage, the matter of constructing counterexamples can be left to intuition, art and imagination. Before attempting Exercise 5.4, however, note the following complementary approach to establishing the validity of QL sequents carefully.

Consider the following argument:

1. If all formal logicians are generous then Paul is generous.
2. Paul is generous.

Therefore,

3. All formal logicians are generous.

Consider the following interpretation:

$\mathfrak{I}$ **D**: {**human beings**}

F: . . . is a formal logician

G: . . . is generous

a: Paul

Premise 1 is a conditional. Given the presence of 'all', the antecedent of that conditional should be translated as a universal conditional. The consequent of the conditional is a subject–predicate sentence. Thus we arrive at the following translation: 1. $\forall x\ [Fx \rightarrow Gx] \rightarrow Ga$. The second premise is identical with the consequent of the conditional we have just formed, i.e. 2. Ga. Finally, the conclusion is identical with the antecedent of the conditional which translates premise 1, i.e. $\forall x\ [Fx \rightarrow Gx]$. Therefore, this argument exemplifies the fallacy of affirming the consequent which we considered in Chapters 1 and 3. It follows that the argument is invalid. Hence, even though the language of quantificational logic was used to translate, the invalidity of the argument is established by showing that it is an instance of an invalid PL form. This practice is known as **shallow analysis**. Keep the possibility of shallow analysis in mind when attempting Exercise 5.4, but first study Box 5.5.

BOX 5.5

Validity and invalidity in QL

♦ A sequent of QL is valid if and only if there is no possible interpretation under which all the premises are true and the conclusion is false.

♦ A sequent of QL is invalid if and only if there is a possible interpretation under which all the premises are true and the conclusion is false.

♦ A formula of QL is valid if and only if that formula is true under every possible interpretation.

EXERCISE 5.4

1 Represent the following arguments as sequents of QL. In each case, make your QL-interpretation explicit and state whether you consider the argument to be valid or not.

(i) 1. Someone is groovy.

Therefore,

2. Everyone is groovy.

(ii) 1. All florists are generous.

2. All generous people are happy.

Therefore,

3. All florists are happy.

(iii) 1. Some greengrocers are folk singers.

2. Some folk singers are haberdashers.

Therefore,

3. Some greengrocers are haberdashers.

(iv) 1. All philosophers are absent-minded.

2. Some philosophers are also logicians.

Therefore,

3. Some logicians are absent-minded.

VI
Negation and the Interdefinability of the Quantifiers

QL inherits all of the formal vocabulary of PL. Therefore, the unary connective '~' is as much a part of QL as of PL. As ever, negation plays a uniquely important role in the formal language and allows us to increase the expressive power of QL still further. For example, consider the following sentence: 'There are no folk singers.' Where *F* represents the property of being a folk singer, this sentence implies that there are no *F*s, i.e. that nothing is *F*. Neither quantifier just on its own can enable us to express the idea of there being nothing which is *F*. Given negation, however, the quantifiers can easily be used to express sentences of that form. The most natural way to translate is to represent the sentence as a **negated existential generalisation**, e.g. as $\sim\exists x\ [Fx]$. To paraphrase: 'It is not the case that there exists anything which is *F*.' It follows that nothing is *F*. And note just how clearly the point is made by this QL sentence. It entails that there does not exist any element of the domain with the property in question. In other words, QL formalisations of sentences with 'Nothing' in initial position will not mislead us into supposing that there does exist something with the property in question (compare and contrast premises 1 and 2 of the first argument we considered on p. 3 of this text). Once again, the application of quantificational logic to natural language provides genuine insight into the logical grammar and structure of natural language sentences and a clearer understanding of the consequential character of such assertions, i.e. of the nature of the logical consequences of assertions of that kind. Thus the application of quantificational logic to natural language is a paradigm of **logical analysis**.

While it is natural to use the existential quantifier to tackle the translation of 'nothing', we could equally well have exploited the universal quantifier for the same purpose here. To do so, we connect the negation symbol not with the quantifier but with the predicate within the matrix, i.e. $\forall x\ [\sim Fx]$. In logical terms, the negation is now internal rather than external. To paraphrase: 'Consider anything, *x*: it is not the case that *x* is *F*.' Again, it follows that there does not exist anything which is *F*, i.e. no element of the domain is *F*. Moreover, negation can be used to establish equivalences between the quantifiers, i.e. **quantifier-equivalences**, quite generally. For example, to assert that everything has a given property is to imply that it is not the case that there is anything which lacks that property. Therefore, for any QL formula, given negation, any occurrence of a universal quantifier can be replaced by an occurrence of an existential quantifier in terms of the following equivalence:

1. $\forall x\ [\ldots] = {\sim}\exists x\ [{\sim}(\ldots)]$

Equally, to assert that there does exist something which has a given property is to imply that it is not the case that everything in the domain lacks that property. Therefore, given negation, any occurrence of the existential quantifier can be replaced with a universal quantifier in terms of the following equivalence:

2. $\exists x\ [\ldots] = {\sim}\forall x\ [{\sim}(\ldots)]$

Finally, where *A* and *B* stand for any arbitrary predicate-letters, we can go on to exploit negation still further to represent formally sentences of the form: 'No *A*s are *B*s', i.e. nothing is both *A* and *B*. Again, it is natural to combine negation with the existential quantifier for this purpose. Thus, we can simply write: ${\sim}\exists x\ [Ax\ \&\ Bx]$, i.e. 'nothing is both *A* and *B*'. Alternatively, we could use the universal quantifier for the same purpose by constructing a universal conditional of the following kind: $\forall x\ [Ax \rightarrow {\sim}Bx]$, i.e. consider anything, *x*: if *x* has the property *A* then it does not have the property *B*. (Note that even though negation is involved here, universal sentences remain firmly associated with the conditional while existential sentences remain firmly associated with conjunction.)

We have now considered how to formalise a number of different forms of sentence in QL and noted an important equivalence between the quantifiers which negation enabled us to express. At this stage, it is useful to have a summary of these sentence-types and their formal translations. So, to recap, study Box 5.6.

BOX 5.6

◆ Something is *A*:	$\exists x\ [Ax]$	*or*: ${\sim}\forall x\ [{\sim}Ax]$
◆ Nothing is *A*:	${\sim}\exists x\ [Ax]$	*or*: $\forall x\ [{\sim}Ax]$
◆ Everything is *A*:	$\forall x\ [Ax]$	*or*: ${\sim}\exists x\ [{\sim}Ax]$
◆ Some *A*s are *B*s:	$\exists x\ [Ax\ \&\ Bx]$	*or*: ${\sim}\forall x\ [{\sim}(Ax\ \&\ Bx)]$
◆ All *A*s are *B*s:	$\forall x\ [Ax \rightarrow Bx]$	*or*: ${\sim}\exists x\ [Ax\ \&\ {\sim}Bx]$
◆ No *A*s are *B*s:	${\sim}\exists x\ [Ax\ \&\ Bx]$	*or*: $\forall x\ [Ax \rightarrow {\sim}Bx]$

Translation into QL involves a deeper concern with the internal form and structure of sentences than translation into PL. And that new emphasis is quite natural. After all, we motivated the need for a new language precisely on the basis of the inability of PL to express arguments whose validity hinged on the internal structure of the basic sentences which composed them. Hence, when translating into QL, first look very carefully at the type of natural language sentence to be translated. Next, try to assimilate the sentence type to one of the forms listed on the left-hand side in Box 5.6 and select the appropriate formalisation from the list on the right. Finally, when each sentence has been translated, use the colon to express the set of formal sentences as a sequent of QL.

In an important sense, translation into QL requires a two-tiered concern with form. Before we can go on to consider the logical form of the argument, we must first settle the question of the logical forms of the sentences which compose it. Translation into PL included something of that two-tiered approach, of course, but the powerful new vocabulary of QL simultaneously makes the concern with sentence-formalisation even more important and provides the means to do much more justice to such questions. Box 5.6 is intended to provide some useful guidelines in this respect but, as ever, there is no algorithm for translation into the formal language and, as you know, many formalisations are equivalent. As time goes on, you will develop a feel for the sentence-types illustrated and, hopefully, for a few other types as well. Exercise 5.5 will allow you to let practice make perfect.

EXERCISE 5.5

1. Translate the following English sentences into QL. In each case make your interpretation explicit and try to represent as much of the structure of each natural language sentence as possible:

   (i) Unicorns do not exist.

   (ii) There is no such thing as a free lunch.

   (iii) Everything is beautiful.

   (iv) All logic students are geniuses.

   (v) No folk singers are grunge fans.

   (vi) Everyone is either a folk fan or a jazz person.

   (vii) Everyone is either a fan of traditional folk music or is not a person of taste.

   (viii) No folk-singing logic students are also happy-go-lucky haberdashers.

(ix) Everyone who is both a folk fan and a jazz person is either a person of taste or an eccentric.

(x) Either it is not the case that some florists are greengrocers or it is the case that no haberdashers are ironmongers.

2. Represent the following arguments as sequents of QL. Again, make your interpretation explicit and state whether you consider each argument to be valid or not.

(i) Someone is a folk-singer and someone is groovy.
Therefore,
Someone is a groovy folk singer

(ii) All formal logicians are generous.
All generous people are happy.
Therefore,
All formal logicians are happy.

(iii) Some veggieburgers are wholesome and tasty.
Some veggieburgers are cheeseburgers.
Therefore,
Some cheeseburgers are wholesome and tasty.

(iv) All fire-fighters are fit and fearless.
No folk singers are fit and fearless.
Therefore,
No fire-fighters are folk singers.

(v) Either someone is a florist or everyone is a greengrocer.
Not everyone is a greengrocer.
Therefore,
Someone is a florist.

(vi) Either someone is a florist or everyone is a greengrocer.
Everyone is not a greengrocer.
Therefore,
Someone is a florist.

VII
How to Think Logically about Relationships: Part One

Completing Exercise 5.5 should have shown you something of the expressive power of QL. But there is at least one form of natural language sentence which cannot yet be represented in that language, namely, sentences which express **relations** between things. Sentences involving relations such as: 'loves', 'hates', 'is the sister of', 'is the father of', 'is taller than' and so on are a commonplace of ordinary discourse. As developed so far, however, the form of such sentences cannot be represented in QL. Therefore, the form of any argument consisting of such sentences cannot be represented accurately either.

The existing formal vocabulary does include predicate-letters which allow us to ascribe properties to things but the problem here is that no predicate-letter can express any relation which might hold *between* things. Relations, however, precisely do hold between things. This can be seen more clearly by giving examples. Consider the following examples based on the list given in the previous paragraph:

Paul loves Zebedee.
Zebedee hates mice.
Tiffin is the sister of Zebedee.
Descartes is the father of modern philosophy.
Peter is taller than Paul.

In all these cases, a relation holds between exactly two things. In a sense, then, my original list blurred the logical form of these relations, which is more clearly expressed as follows:

'. . . loves ---'
'. . . hates ---'
'. . . is the sister of ---'
'. . . is the father of ---'
'. . . is taller than ---'

Looked at in this way, all these relations involve two gaps, i.e. a verb links two gaps which are plugged by terms identifying things which are related to each other in that way. In Logicspeak, gaps are more widely known as *places* which are *filled* rather than plugged. Henceforth, we will adopt this terminology. In fact, the examples given here all involve the simplest kind of relation, i.e. two-place or **dyadic relations** (and that is precisely why no predicate-letter is adequate to translate any relation; predicate-letters only involve one place and are as such monadic). Moreover, three-place or triadic relations can also be represented in QL (we will consider these in the next chapter) as, indeed, can four-place relations and so on. Hence, quite generally, relations of any number, polyadic relations, can in principle be represented in QL.

Now, because the relations we are considering here hold between at least two things, we must have a way of representing that two things are involved and that some relation holds between them. Again, it helps to give an example. Consider the first sentence on the list given above: 'Paul loves Zebedee.' How am I to express my enduring affection for my cat in QL?

First, the expression *R* can be used to represent any given relation, i.e. *R* for *relation*. When a further, distinct relation is involved we will use *S* and so on. Hence, in the present context, we can think of *R* as formalising: ' . . . loves --- '. To construct a well-formed relational formula in QL it only remains actually to fill the places, i.e. plug the gaps. In the case in point, the places are filled by the names 'Paul' and 'Zebedee' respectively, i.e. by the pair of names ⟨Paul, Zebedee⟩. Note that we use a special kind of bracket for such pairs. The vocabulary of QL already includes names and so, where *a* stands for 'Paul' and *b* for 'Zebedee', I simply fill the first place with *a* and the second with *b* to complete the translation (expressions which fill places in relations are generally called *terms* in the relation). It may seem intuitive to try to reflect the grammatical form of relational sentences in natural language in QL here, i.e. by writing:

$$aRb$$

But, in fact, we will write both names after the relational expression like so:

$$Rab$$

This convention is useful just because it extends much more easily to cases of three-place relations, four-place relations and so on. Hence, we read the formal expression *Rab* as asserting that *a* stands in, or, better, *bears* the relation *R* to *b*.

For purposes of translation, the way in which places are filled in formal relational expressions is crucially important, i.e. the sense of the resulting translation is sensitive to the *order* of the terms. In slightly more formal terms, then, a QL interpretation assigns to relational expressions **ordered pairs** of names of elements of the domain. Hence, our new brackets indicate that any such pair is ordered. For example, both *Rab* and *Rba* are equally well formed. But note that reversing the order of the names affects the sense of the translation. In the present context, for example, *Rab* translates the obvious truth that 'Paul loves Zebedee' into QL while *Rba* translates the much less obviously true claim that 'Zebedee loves Paul.' Moreover, filling both places with the same name also produces formulas which are well formed though again the sense of the translation alters. So, if I had a greater affection for myself than I do, I might plug both gaps with *a* to form:

$$Raa$$

This formula translates 'Paul loves Paul' into QL. Finally, I can equally well formalise the indisputable truth that 'Zebedee loves Zebedee' as:

Rbb

Further well-formed QL formulas can easily be formed using relational expressions together with the quantifiers, and, indeed, new and more complex kinds of quantified formula can be formed in this way. As ever, quantified formulas concern the domain of quantification. Hence, variables rather than names must be used in such cases. Remember: no variable can occur meaningfully without a quantifier to govern or *bind* it. Hence, in QL every variable is a *bound variable* and, conversely, there are no *free variables* in QL, i.e. no variable occurs free from a quantifier. To represent formally the simplest kinds of general sentence involving relations, only one quantifier need be used. For example, consider the certain untruth that 'Everyone loves Paul.' Because a relation is involved here there will be two places to fill.

Clearly, the sentence involves a name which will certainly fill one of the places. However, the presence of 'Everyone' in initial position indicates that the sentence is properly translated using the universal quantifier, which in turn requires the use of a variable. That variable should also fill a place. But which expression should fill which place? A little reflection quickly reveals that the first place should be filled with the variable while the second should be filled with the name, i.e. where *R* represents ' . . . loves --- ', *a* stands for 'Paul' and *x* is the variable, 'Everyone loves Paul' is properly translated into QL as:

$\forall x\ [Rxa]$

To paraphrase: 'Consider every single person, each person stands in the relation of loving to Paul.' Again, note the sensitivity of the sense of the translation to the order of the terms. The QL formula:

$\forall x\ [Rax]$

is perfectly well formed but rather than translating 'Everyone loves Paul' this formula translates the very different claim that 'Paul loves everyone.' To paraphrase: 'Consider every person, Paul stands in the relation of loving to each person.' Using the same interpretation, we can also easily formalise the more plausible claim that 'Somebody loves Paul' as:

$\exists x\ [Rxa]$

To paraphrase: 'There exists someone such that that person loves Paul.' Again, using the same interpretation, we can formalise the (hopefully untrue) claim that 'Nobody loves Paul' most naturally as:

$\sim\exists x\ [Rxa]$

Given the interdefinability of the quantifiers, the same claim can also be translated as:

$\forall x\ [\sim Rxa]$

It should now be easy to see how we might turn the tables here, as it were, to represent formally sentences such as 'Paul loves/someone/no one.' Again, all we need do is switch the order of the terms (spell out for yourself the QL translations of these cases, i.e. 'Paul loves someone' and 'Paul loves no one').

When the formal vocabulary of QL is supplemented with relations a new level of formal logic is reached. In fact, the addition of relations makes some very important differences which we will consider in more detail in Chapter 7. Hence, it is useful to mark the distinction between these different levels of formal logic. Earlier, I described expressions which involve only one place, such as the predicate-letters of QL, as *monadic* expressions. Therefore, I will refer to that fragment of QL which does not include relations, but confines itself to monadic predicates and to the methods for establishing the validity and invalidity of sequents within that fragment, as **monadic quantificational logic** or, more simply, as **monadic QL**.

When monadic QL is supplemented with formal vocabulary adequate to the task of representing relations which might be dyadic, triadic and so on, the result is **polyadic quantificational logic** or, again, more simply, **polyadic QL**.

In this section we have examined a number of different forms of sentence involving relations. So, consider Box 5.7 carefully, and then do Exercise 5.6.

EXERCISE 5.6

1 In Box 5.7 I deliberately do not spell out the possible (equivalent) formalisations which can be made using the alternate kind of quantifier together with negation in cases 3–8 above. Spell these out for yourself in order to familiarise yourself with the alternative kinds of formalisation which are possible in such cases. (If need be, you might like to refer back to Section VI for a refresher course on negation and the interdefinability of the quantifiers at this point.)

BOX 5.7

- 1 *a* loves *b*: Rab
- 2 *b* loves *a*: Rba
- 3 Everyone loves *a*: $\forall x\ [Rxa]$
- 4 Someone loves *a*: $\exists x\ [Rxa]$
- 5 No one loves *a*: $\sim\exists x\ [Rxa]$
- 6 *a* loves everyone: $\forall x\ [Rax]$
- 7 *a* loves someone: $\exists x\ [Rax]$
- 8 *a* loves no one: $\sim\exists x\ [Rax]$

VIII
How to Think Logically about Relationships: Part Two

To date, the sentences involving relations which we have considered have been of a relatively simple kind, e.g. in translation, we have only ever had to have recourse to a single quantifier in any given case. Equally, the matrices we have been considering have been as simple as possible, i.e. we have not made use of any of the logical connectives in forming any matrix. As the natural language sentences to be translated become more complex, however, we may well want to use more than one quantifier and, indeed, to use connectives when constructing matrices. Let's look at the latter possibility first.

For example, consider the sentence 'Paul loves every folk singer.' It may not be immediately obvious to you how to go about formalising this sentence, so let us first consider what the sentence implies. Well, if Paul loves every folk singer, it follows that everything which has the property of being a folk singer is loved by Paul, i.e. everything which is a folk singer stands in the relation of being loved by to Paul. Hence, we have identified a particular predicate and a specific relation and so we can construct an interpretation. First, we select the relevant domain, so as to specify the set of things capable of having the property in question and standing in the relevant relation. Obviously, these are human beings. Hence, we can construct the following interpretation:

$\mathfrak{I}$ **D: {human beings}**

Further, one predicate, one relation and one name are involved. Note that we adopt the convention of listing relations first, predicates second and names last. Thus, the interpretation can be completed as follows:

R: . . . loves ---

F: . . . is a folk singer

a: Paul

Recall the paraphrase given above: everything which is a folk singer stands in the relation of being loved by to Paul. Given 'everything' in initial position here, it is natural to exploit the universal quantifier in the translation, i.e. to represent the sentence as a *universal conditional*. So, we may begin translating as follows:

$\forall x\,[Fx \rightarrow \ldots]$

To paraphrase, we state 'Consider any person, if that person is a folk singer then . . . '. To complete the translation we must decide how to formalise the consequent here, i.e. how to formalise the claim that Paul loves any such person. But that is quite straightforward. We can assert that Paul loves any such thing as follows:

Rax

Hence, the formalisation is completed as:

$\forall x\,[Fx \rightarrow Rax]$

To paraphrase: 'Consider any person, if that person is a folk singer then Paul loves that person.'

Now, let's look at another example, i.e. the sentence 'Paul does not like some folk singers.' Given the presence of 'some' here it is natural to exploit the existential quantifier in translation this time, i.e. to form an *existentially quantified conjunction*. Using the same interpretation, then, we translate 'Paul does not like some folk singers' as:

$\exists x\,[Fx \,\&\, {\sim}Rax]$

Given the quantifier-equivalences, however, we could construct an alternative translation, i.e.:

$$\sim\forall x\ [Fx \rightarrow Rax]$$

As ever in translation, note that universally quantified formulas remain firmly associated with the conditional while existentially quantified formulas remain associated with conjunction. Further, note that the sense of the resulting translation is sensitive to the order of terms in the relation, e.g. under the present interpretation the QL formula:

$$\forall x\ [Fx \rightarrow Rxa]$$

translates 'Every folk singer loves Paul' – a pleasant if unrealistic thought! Naturally, the complexity of the required matrix increases with the complexity of the natural language sentence to be translated. Here, we should keep in mind the important rules of thumb and standard associations we have observed to date, e.g. a sentence such as 'Paul loves all groovy folk singers' should be translated as a universal conditional with a complex antecedent:

$$\forall x\ [(Fx\ \&\ Gx) \rightarrow Rax]$$

IX
How to Think Logically about Relationships: Part Three

In the previous section we considered relations within complex matrices for the first time and noted how the complexity of the matrix increased with the complexity of the natural language sentence to be translated. In the present section I want to draw attention to the fact that the complexity of the formal sentence as a whole will also increase when we translate still more complex kinds of natural language sentence, i.e. sentences which involve **multiple generality**. Invariably, formalisation of these types of sentence will require the use of more than one quantifier. For example, consider the sentence: 'Everyone loves everyone.' The obvious choice of domain here is again persons or human beings and we have only one relation to worry about in this context. Hence, we arrive at the following very simple interpretation:

$\mathfrak{I}$ **D**: {**human beings**}

R: . . . loves _ _ _

Given the presence of 'Everyone' in initial position, it is natural to exploit the universal quantifier for purposes of translation. But note that there are two occurrences of that term in the sentence. So, for the first time, any proper translation will require two distinct quantifiers. For if we make use of only one quantifier, then, given that no variable can occur free in any well-formed QL formula, we will only be able to use one variable. Hence, that approach to translation would generate a formula of the form:

$\forall x\, [Rxx]$

This formula is perfectly well formed. But, under the current interpretation, it translates the sentence 'Everyone loves themselves'! In order to express the sense of the sentence 'Everyone loves everyone', then, we require two distinct variables and, therefore, two distinct quantifiers to bind those variables. Only given two distinct variables can we formally open up the possibility that one thing bears a relation to another rather than to itself. Hence, 'Everyone loves everyone' is properly translated as follows:

1. $\forall x\, \forall y\, Rxy$

Strictly, however, this is not a well-formed QL formula, for we have omitted the scope-indicating square brackets which are an integral part of any quantifier, and so we must supply them. Moreover, note that each quantifier has a distinct scope. The scope of the first quantifier includes the second quantifier, the relation and both variables, even though that quantifier only actually binds the first variable coming after *R*, i.e. *x*. In contrast, the scope of the second quantifier is identical only with the expression *Rxy*. Hence, we enter the scope-indicating square brackets to arrive at the following well-formed QL formula:

1.′ $\forall x\, [\forall y\, [Rxy]]$

Finally, note that reversing the order of the variables within the matrix does not affect the sense of the resulting translation, i.e. under the present interpretation:

1.″ $\forall x\, [\forall y\, [Ryx]]$

still translates the natural language sentence: 'Everyone loves everyone.'

Next, consider the sentence 'Someone loves someone.' Given the presence of 'Someone' in initial position it is natural to exploit the existential quantifier in order to translate. Again, however, in order to express the sense we want here we must choose two distinct variables and two distinct quantifiers to govern them. Otherwise, adopting the single variable approach will lead us to construct a formula of the following form:

$\exists x\ [Rxx]$

Again, this is a well-formed formula but again it is not an adequate translation. Under the present interpretation, the sense of this QL formula is just that someone stands in the relation of loving to herself or himself, i.e. 'Someone loves herself or himself.' So, we must again use two distinct quantifiers and two distinct variables in order to express the sense we want here. Further, we should again be careful to include each pair of scope-indicating brackets as appropriate. Here is the correct translation:

2. $\exists x\ [\exists y\ [Rxy]]$

Again, note that reversing the order of the variables within the matrix does not affect the sense of the translation, i.e. under the present interpretation:

2.′ $\exists x\ [\exists y\ [Ryx]]$

still translates the natural language sentence: 'Someone loves someone.'

Finally, we should consider some of the kinds of natural language sentence whose translation into QL generates formulas which combine both quantifiers. For example, consider the plausible claim that 'Everyone loves someone.' Here, the presence of both 'Every' and 'some' indicates that we will need both quantifiers. Further, the presence of 'Everyone' in initial position makes clear that this is a universally quantified sentence, i.e. that the order of the quantifiers will be as follows:

$\forall x\ \exists y \ldots$

We can complete the formalisation by including appropriate brackets and making clear that x bears R to y:

3. $\forall x\ [\exists y\ [Rxy]]$

But note the importance of the order of the variables in this case. For the first time, reversing the order of the variables in the matrix does alter the sense of the translation. Under the current interpretation:

3.′ $\forall x\ [\exists y\ [Ryx]]$

translates the very different claim that for each individual in the domain there exists someone who loves him or her, i.e. this QL formula translates the sentence: 'Everyone is loved by someone', which, of course, has a very different sense from the sentence 'Everyone loves someone.' We can see both this point and some further subtlety involved in translating natural language sentences into QL by considering possible translations of Russell's alleged counterexample to the cosmological argument, i.e. translations of the argument that:

4. Everyone has a mother.

Therefore,

5. Someone is the mother of everyone.

How might we represent this argument as a sequent of QL? Consider the premise carefully. Note that where the domain consists of human beings and R represents the relation '. . . is the mother of - - -' the QL formula:

$\forall x\ [\exists y\ [Rxy]]$

does not translate the premise in a satisfying way. The sense of this particular formula is as follows: consider each person, x, there exists some person, y, such that x is the mother of y, i.e. 'Everyone is the mother of someone.' Again, that sentence certainly is not synonymous with 'Everyone has a mother.' In order to get the sense we want all we need do is reverse the order of the variables in the matrix like so:

4. $\forall x\ [\exists y\ [Ryx]]$

To paraphrase: consider each person, x, there is some person, y, such that y is the mother of x. It follows that everyone has a mother.

Next, consider the conclusion, i.e. the sentence 'Someone is the mother of everyone.' The presence of 'Someone' in initial position suggests that, this time, we require an existentially quantified QL formula whose scope includes a universally quantified formula. So, the formula we now require switches the order of the quantifiers given in 4 above. Moreover, that is all that is required here, i.e. if we simply switch the order of the quantifiers in 4 we arrive at a QL formula which expresses the sense of the conclusion exactly. Hence, we formalise 5 as:

5. $\exists y\ [\forall x\ [Ryx]]$

To paraphrase: 'There exists some particular individual who stands in the relation of being the mother of to every individual.' It follows that

BOX 5.8

- Everyone loves themselves: $\forall x\ [Rxx]$
- Someone loves themselves: $\exists x\ [Rxx]$
- Everyone loves someone: $\forall x\ [\exists y\ [Rxy]]$
- Someone loves everyone: $\exists x\ [\forall y\ [Rxy]]$
- Someone is loved by everyone: $\exists x\ [\forall y\ [Ryx]]$
- Everyone is loved by someone: $\forall x\ [\exists y\ [Ryx]]$

someone is the mother of everyone. Hence, we arrive at the following QL sequent:

$$\forall x\ [\exists y\ [Ryx]] : \exists y\ [\forall x\ [Ryx]]$$

A little reflection quickly reveals that this sequent is certainly intuitively invalid. Further, because the inference from premise to conclusion involves reversing or switching the order of the quantifiers this particular sequent is an example of what is often described as the *quantifier switch fallacy*. Such then is the form of Russell's counterexample to the cosmological argument which we considered in Chapter 1. Hence, Russell is alleging that this is exactly the form of the cosmological argument. Is that argument really of the form we have made clear here? The answer to that particular question must be decided by the reader.

We have considered a number of different kinds of sentence involving multiple generality so study the contents of Box 5.8 carefully.

X
How to Think Logically about Relationships: Part Four

In the previous two sections we considered some of the roles relations can play in complex matrices and some kinds of sentence which involve multiple generality. In the present section I outline certain kinds of sentence whose translation into QL will involve both multiple generality and complex matrices. For example, consider the sentence: 'All folk singers love every bluesman.' We will use the following interpretation to translate:

$\mathfrak{I}$ **D: {human beings}**

R: . . . loves ---

F: . . . is a folk singer

G: . . . is a bluesman

Given the presence of 'All' in initial position in the sentence in question, it is natural to exploit the universal quantifier here and so to form a conditional, i.e. to begin to translate as follows:

1. $\forall x\ [(\ldots \rightarrow \ldots)]$

But note that we need to represent formally *both* the initial 'All' *and* the subsequent 'every'. It follows that we need to include two occurrences of the universal quantifier when we form the relevant conditional and that in turn suggests that the particular conditional we require will be nested, i.e. a conditional within a conditional. Hence, we require something of the following general form:

1.′ $\forall x\ [(\ldots) \rightarrow \forall ?\ [(\ldots \rightarrow \ldots)]]$

Further, we want to talk about, and distinguish between, two kinds of thing: folk singers and bluesmen. Therefore, we will need to use two distinct variables to distinguish between the two kinds of thing involved before we can assert the relation of loving between them. It follows that the translation will require two distinct quantifiers. For note that, if we make use of only one quantifier, then, given that no variable can occur free in any QL formula, we will only be able to use one variable. Hence, the nested conditional must involve two distinct quantifiers:

1.″ $\forall x\ [(\ldots) \rightarrow \forall y\ [(\ldots \rightarrow \ldots)]]$

Let's think carefully about the nested conditional we want to form here. What we want to say is surely this: consider each person, *x*, if any person is a folk singer then, consider each person, *y*, if any person is a bluesman then the first person stands in the relation of loving to the second. Hence, the requisite conditional asserts that if *x* is a folk singer and *y* is a bluesman then *x* loves *y*. The formalisation is completed as follows:

1.‴ $\forall x\ [Fx \rightarrow \forall y\ [(Gy \rightarrow Rxy)]]$

Again, note carefully that unless we use two distinct quantifiers and two distinct variables we cannot express the relation between the two classes involved, i.e. the QL formula:

$\forall x\ [Fx \rightarrow (Gx \rightarrow Rxx)]$

translates the sentence: 'Every folk-singing bluesman loves himself.'

In general, the point to remember is just this:

> unless we use two distinct quantifiers and two distinct variables we cannot express a relation holding universally between members of two classes.

Moreover, it follows that where three classes are involved we require three distinct variables and, therefore, three distinct quantifiers, where four classes are involved we require four distinct variables and, therefore, four distinct quantifiers, and so on.

Now consider the following sentence: 'Some folk singers love some bluesmen.' This time, given 'Some' in initial position, it is natural to exploit the existential quantifier. Again, there must be two occurrences of the quantifier in the formal translation, two distinct variables, and, again, we can simply list the quantifiers before forming the appropriate conjunction:

2. $\exists x\ [\exists y\ [(Fx\ \&\ Gy)\ \&\ Lxy)]]$

Equally, we could place the second quantifier inside the conjunction and closer to the variable it binds, as follows:

2.′ $\exists x\ [Fx\ \&\ \exists y\ [(Gy\ \&\ Lxy)]]$

In general, we should always be as sensitive as possible when considering the meaning of the natural language sentence to be formalised. For example, consider the sentence: 'Every folk singer loves some bluesman.' In this case, the presence of both 'Every' and 'some' indicates that an accurate translation should involve both quantifiers. But note that, just as it stands, the sentence in question is ambiguous. It could mean either that:

(i) Every folk singer is such that there is some bluesman whom he or she loves.

or that:

(ii) Some bluesman is such that he or she is loved by every folk singer.

QL provides all the formal resources required to express clearly either sense and, as you might expect, the difference comes out, in part, at least, in the order of the quantifiers. To capture the first sense we try to express the following claim: consider any person, x, if that person is a folk singer then there is some bluesman, y, that x loves. Clearly this is a universal conditional. But note the form of the consequent: ' . . . there is some bluesman . . . '. Hence, the consequent must be prefixed with the existential quantifier. So, we construct a universal conditional whose antecedent is an existentially quantified formula. Under the current interpretation, we can complete the formalisation of (i) as follows:

(i) $\forall x\, [Fx \rightarrow \exists y\, [Gy \,\&\, Lxy]]$

In sharp contrast, sentence (ii) implies that there is some bluesman whom every folk singer loves. In this case, then, the formalisation should begin with the existential quantifier. So, we would anticipate a conjunction. But note that one of the conjuncts should be a universally quantified formula about every folk singer. Therefore, that conjunct should have the form of a

BOX 5.9

- *A* loves all *B*s: $\forall x\, [Bx \rightarrow Rax]$
- *A* loves all *B*s which are *C*: $\forall x\, [(Bx \,\&\, Cx) \rightarrow Rax]$
- All *A*s love every *B*: $\forall x\, [\forall y\, [(Ax \,\&\, By) \rightarrow Rxy]]$

  or:

  $\forall x\, [Ax \rightarrow \forall y\, [By \rightarrow Rxy]]$
- Some *A*s love some *B*s: $\exists x\, [\exists y\, [(Ax \,\&\, By) \,\&\, Rxy]]$

  or:

  $\exists x\, [Ax \,\&\, \exists y\, [By \,\&\, Rxy]]$
- For every *A* there is some *B* whom *A* loves: $\forall x\, [Ax \rightarrow \exists y\, [By \,\&\, Rxy]]$
- Some *B* is loved by every *A*: $\exists x\, [Bx \,\&\, \forall y\, [Ay \rightarrow Ryx]]$

conditional. Finally, because the formalisation involves a mixture of kinds of quantifier, we should be careful about the order of the variables in the matrix featuring the relation. Under the current interpretation, then, (ii) is properly translated as follows:[3]

(ii) $\exists x\,[Gx\;\&\;\forall y\,[Fy \rightarrow Lyx]]$

Ambiguities such as these are a commonplace of natural language and so we should always be as sensitive as possible to the sense of the sentence we are trying to formalise. Again, only practice makes perfect. Study the contents of Box 5.9 carefully before attempting Exercise 5.7.[4]

EXERCISE 5.7

1 Use the interpretation provided to translate the following English sentences into QL:

$\mathfrak{I}$ **D**: {**human beings**}

M: . . . is the mother of ---

F: . . . is the father of ---

W: . . . is the wife of ---

H: . . . is the husband of ---

S: . . . is a sister of ---

B: . . . is a brother of ---

L: . . . loves ---

a: Alice

b: Bill

c: Carol

d: Derek

(i) Everyone loves Derek.

(ii) Everyone loves themselves.

(iii) Nobody loves anybody.

(iv) Everybody loves somebody.

(v) Somebody loves everybody.

(vi) Bill loves Alice and Carol and Carol's mum.

(vii) Alice, Carol, Derek and Derek's uncle all love Bill.

(viii) Derek loves Alice's grandfather.

(ix) Carol loves Bob's sister-in-law.

(x) Either Alice loves Derek's grandmother or Carol loves Alice's sister-in-law.

(xi) Grandparents are parents too.

(xii) Derek loves Bob's paternal grandfather.

(xiii) No father is anyone's mother.

(xiv) All sisters are siblings.

(xv) All parents love their children.

2 Represent the following arguments as sequents of QL. In each case, construct a QL-interpretation for your translation. Consider, intuitively, whether each particular natural language argument is valid or not:

(i) Everyone loves someone.
Therefore,
Someone is loved by everyone.

(ii) Every event has a cause.
Therefore,
Some event is the cause of every event.

(iii) Everyone is taller than someone.
Therefore,
Someone is taller than everyone.

(iv) Every folk fan likes the Amazing Blondel.
Some bluesmen are folk fans.
Therefore,
Some bluesmen like the Amazing Blondel.

(v) Every formal logician admires every German philosopher who wrote a logic text.
Gottlob Frege is a German philosopher.
Gottlob Frege wrote a logic text.
Therefore,
Every formal logician admires Gottlob Frege.

(vi) Every cat hates all mice.
Some dogs like some mice.
Therefore,
All cats hate some dogs.

(vii) Some florists respect all greengrocers.
No florists respect any hitmen.
Therefore,
No greengrocers are hitmen.

(viii) Every florist likes a nice flower.
The rose is a nice flower.
Therefore,
Every florist likes the rose.

3 Translate the following sentences into QL. In each case, construct a QL-interpretation for your translation.[5] *Hint*: use a sentence-letter to translate the sentence mentioning Newton's Law in (viii) (no such law is properly understood simply as a universal generalisation).

(i) Every object attracts every other.

(ii) The brighter the day the better the day.

(iii) Every cat loves a master who feeds him generously.

(iv) Every proton is heavier than any electron.

(v) Every mutation results from a change in some gene.

(vi) If everybody loves Arlo Guthrie then nobody hates Blind Lemon Jefferson.

(vii) Whenever Rover sees a cat, he barks at it.

(viii) If a body decelerates, then if Newton's Second Law holds, there exists some force acting upon it.

(ix) No one has been fooled by everyone.

(x) No one has fooled everyone.

(xi) Everyone has either fooled someone or been fooled by someone.

(xii) Everyone who has fooled someone has fooled herself.

(xiii) Everyone who has been fooled by someone has fooled herself.

(xiv) Everyone has both fooled someone and been fooled by her.

(xv) Someone has fooled someone if and only if it is not the case that no one has fooled anyone.

XI
Formal Properties of Relations

It is important to appreciate that not all relations are of the same kind. This is intuitive and is easily shown. Being the parent of, for example, is the kind of relation which one might bear to others but which one could not bear to oneself! In contrast, 'being exactly the same person as' is a relation which one can only bear to oneself. Further, 'being as tall as' is the kind of relation one both bears to oneself and to others. Formal logicians have had considerable success in identifying and categorising different types of relation in terms of their formal properties. The present section is intended to provide a basic map of the conceptual geography of the area rather than any detailed consideration of the issues. Although many of the issues surrounding the different types of relation which can be identified are genuinely fascinating, any detailed consideration would take us straight into pure mathematics generally and set theory in particular. That would be no bad thing but it is strictly beyond the current remit. None the less, we should at least be aware of the different kinds of relation that there are, their formal properties, and how such properties combine.

In addition to the utility of these notions for the pursuit of pure mathematics, they are also very helpful when considering certain problems belonging to **philosophical logic** and, indeed, to the philosophy of mathematics. In fact, the theory of relations belongs to an area of formal logic known as the **theory of definition**. This is an important area in its own right and interested parties would do well to consult Haskell B. Curry's *Foundations of Mathematical Logic*, [1976], New York, Dover, Chapter 3, Section C, for more information on it. Further, certain elements of the theory of definition may well have genuinely momentous applications in the philosophy of mathematics. So, it is doubly important to have at least something of a picture of what goes on in this area of formal logic.

To that end, let's begin by considering that kind of relation which, like 'being the same person as', one can bear to oneself. Such relations are known in Logicspeak as **reflexive** relations. In general, a relation R is reflexive if and only if it is such that everything bears that relation to itself. In slightly more formal terms, then, a relation is reflexive if and only if, for any and every a, a stands in that relation to itself, or, as I shall put it, if and only if a bears R to itself, i.e. if, and only if $\forall x\ [Rxx]$. In contrast, a relation is **irreflexive** if and only if it is *not* the case that a thing can bear that relation to itself. That is, if and only if, for any and every a, a does not bear R to itself. Formally, then, a relation is irreflexive if, and only if $\forall x\ [{\sim}Rxx]$. Finally, a relation is said to be **non-reflexive** if and only if it is both the case that some object, or objects, does bear that relation to itself and that some other object, or objects, does not bear that relation to itself, i.e. if and only if ('iff' for short):

$$\exists x\ [Rxx]\ \&\ \exists x\ [\sim Rxx]$$

Reflexivity is one of three cardinally important defining features of relations. The next kind of relation to consider is like that of 'being as tall as'. As noted, one can bear that relation both to others and to oneself. But here the important point is that if it's true to say that you are as tall as me then it must also be true to say that I am as tall as you. This property of relations is known, naturally enough, as **symmetry**. Symmetrical relations are a commonplace of ordinary discourse, e.g. 'as tall as', 'as long as', 'equidistant from' and so on. For any relation R and any individuals a and b, R is a symmetrical relation if and only if whenever a bears R to b, b also bears R to a. Formally, a relation R is symmetrical iff:

$$\forall x\ [\forall y\ [Rxy \rightarrow Ryx]]$$

In contrast, a relation is said to be **asymmetrical** just in case, for every a and b, a's bearing R to b means that b does not bear R to a, i.e. iff:

$$\forall x\ [\forall y\ [Rxy \rightarrow \sim Ryx]$$

Further, a relation R is **non-symmetrical** if it is both the case that there is some a that bears R to some b when that b bears R to a and that there is some c which bears R to some d while d does not bear R to c, i.e. iff:

$$\exists x\ [\exists y\ [Rxy\ \&\ Ryx]]\ \&\ \exists x\ [\exists y\ [Rxy\ \&\ \sim Ryx]]$$

Finally, a relation is **anti-symmetrical** if when it is both the case that a bears R to b and that b bears R to a it follows that a and b are in fact identical objects. So, for example, consider the relation: 'is a number not greater than'. If a is not greater than b but b is not greater than a, then a and b must be one and the same number. Presently, we have no way of representing the idea of being one and the same thing as in QL, i.e. no way of representing **identity** in QL. Hence, we cannot fully formalise this particular kind of relation here. In the following section, however, we will supplement QL precisely with a way of talking about identity. So, you might return to this particular task at the end of the next section.

And so we come to the last of the triumvirate of important properties of relations mentioned earlier. Again, we can use 'is taller than' by way of illustration. Suppose that you are taller than Blind Lemon Jefferson and that I am taller than you. It follows at once that I too am taller than Blind

Lemon Jefferson. This important property of certain relations is known as **transitivity**. Consider any three objects going by the names of *a*, *b* and *c*: if *a*'s bearing *R* to *b* and *b*'s bearing *R* to *c* means that *a* must also bear *R* to *c*, then the relation is transitive. So, a relation *R* is transitive iff:

$$\forall x\ [\forall y\ [\forall z\ [(Rxy\ \&\ Ryz) \rightarrow Rxz]]]$$

In contrast, a relation is said to be **intransitive**, intuitively enough, if and only if:

$$\forall x\ [\forall y\ [\forall z\ [(Rxy\ \&\ Ryz) \rightarrow {\sim}Rxz]]]$$

Finally, a relation is **non-transitive** if, again, there are both cases in which *a* bears *R* to *b*, *b* bears *R* to *c* and *a* bears *R* to *c* *and* cases where *a* bears *R* to *b* and *b* bears *R* to *c* but *a* does not bear *R* to *c*. Formally, then, a relation is non-transitive if and only if:

$$\exists x\ [\exists y\ [\exists z\ [(Rxy\ \&\ Ryz)\ \&\ Rxz]]]\ \&\ \exists x\ [\exists y\ [\exists z\ [(Rxy\ \&\ Ryz)\ \&\ {\sim}Rxz]]]$$

The triumvirate of important properties of relations which we have spelled out here is actually more useful for logical purposes than might appear to be the case. In fact, much of what formal logicians seek to do in their investigations concerning particular relations is to determine which of those properties, or which combination of those properties, characterise the relation under investigation. Moreover, the logician can easily go on to define new kinds of relation precisely in terms of different combinations of those properties. For example, relations which have the properties of being reflexive, transitive and anti-symmetrical are known as **partial ordering relations**. This notion brings the domain of the relation sharply into focus just because what a partial ordering relation partially orders is (some of) the elements of the domain. The domain, of course, is just a set and sets, of course, consist of members. So, a partial ordering relation literally establishes an order among some of the members of a set. In turn, the formal logician can go on to exploit the notion of a partial ordering relation to define further order relations. For many philosophers and logicians, however, another kind of relation is of even more interest. This kind of relation combines the three fundamental properties of being reflexive, transitive and symmetrical and is known as an **equivalence relation**. Mathematically, equivalence relations have an extremely useful and significant effect on the elements of the domain. Again, our earlier example of 'being as tall as' provides a useful illustration.

Let the domain be the set of human beings and suppose that God sorts out the domain into sub-groups by reference to the relation of 'being as tall as', grouping everyone of the same height together and drawing circles around each group. Note that everyone will be in some group; even if one person is the only one of that height she or he is still as tall as herself or himself (the relation is reflexive). And no one will be standing in more than one group or wavering between groups.

In Logicspeak, God would have created a series of mutually exclusive **partitions** of the domain, i.e. he will have split the original set into neat, independent subsets. Each such subset is known as an **equivalence class** of the domain (with respect to the relation in question). Further, consider the members of any equivalence class. These will all be the same with respect to height, i.e. each member is the same height as every other. In just that respect, then, all the members of each equivalence class 'coalesce' and, for many formal purposes, they are interchangeable. Again, relations of this kind obviously suggest themselves as useful in mathematics. But there may well be an even more important use for equivalence relations and equivalence classes in the philosophy of mathematics. Here, unfortunately, we can do nothing more than hint at this use and provide interested parties with a guide to further reading. None the less, the possible significance of that use is sufficiently momentous to merit a hint.

As we have seen, modern formal logic was born in the work of Frege and of Russell and Whitehead. Equally, the Wittgenstein of the *Tractatus* also played an important part in that development. In the work of all these authors, the development of logic was inextricably bound up with an extremely ambitious project known as **logicism**. Logicism is nothing less than the attempt to show that mathematics itself can ultimately be reduced to and derived from the principles of logic alone. The heart of the logicist endeavour is to show that arithmetic can be derived from logical principles and, in both Frege's and Russell's attempts to derive arithmetic, equivalence classes play a crucial role. Interested parties can find the traditional logicist programme explained and critically discussed in Stephan Korner's *The Philosophy of Mathematics: An Introductory Essay*, [1960], London, Hutchinson University Library, especially Chapters II and III. Those who wish to pursue the debate beyond the confines of that text would do well to consult the volume on *Philosophy of Mathematics: Selected Readings*, edited by Paul Benacerraf and Hilary Putnam, [1983], second edition, Cambridge, Cambridge University Press. But note also that the logicist programme has recently undergone something of a revival. The prize represented by the possible success of this particular programme is indeed a glittering one and interested parties can find the new logicism #3 outlined in Crispin Wright's *Frege's Conception of Numbers as Objects*, [1983], Aberdeen, Aberdeen University Press (note especially Chapter 3 of that volume).

Study the contents of Box 5.10 carefully before attempting Exercise 5.8.

BOX 5.10

- A relation R is *reflexive* iff: $\forall x\ [Rxx]$
- A relation R is *irreflexive* iff: $\forall x\ [\sim Rxx]$
- A relation R is *non-reflexive* iff: $\exists x\ [Rxx]\ \&\ \exists x\ [\sim Rxx]$
- A relation R is *symmetrical* if and only if: $\forall x\ [\forall y\ [Rxy \rightarrow Ryx]]$
- A relation R is *asymmetrical* iff: $\forall x\ [\forall y\ [Rxy \rightarrow \sim Ryx]$
- A relation R is *non-symmetrical* iff: $\exists x\ [\exists y\ [Rxy\ \&\ Ryx]]\ \&\ \exists x\ [\exists y\ [Rxy\ \&\ \sim Ryx]]$
- A relation R is *transitive* iff: $\forall x\ [\forall y\ [\forall z\ [(Rxy\ \&\ Ryz) \rightarrow Rxz]]]$
- A relation R is *intransitive* iff: $\forall x\ [\forall y\ [\forall z\ [(Rxy\ \&\ Ryz) \rightarrow \sim Rxz]]]$
- A relation R is *non-transitive* iff: $\exists x\ [\exists y\ [\exists z\ [(Rxy\ \&\ Ryz)\ \&\ Rxz]]]$ $\&\ \exists x\ [\exists y\ [\exists z\ [(Rxy\ \&\ Ryz)\ \&\ \sim Rxz]]]$

EXERCISE 5.8

1 Explain intuitively and represent formally what is the case when a relation has the property of being:

(i) reflexive

(ii) symmetrical

(iii) irreflexive

(iv) transitive

(v) non-symmetrical

(vi) non-transitive.

2 Which combination of the properties listed in 1 above defines an equivalence relation?

3 Consider the following relations carefully. List as many of the formal properties of each relation as you can.

(i) . . . is as happy as ---

(ii) . . . is the sister of ---

(iii) . . . is the mother of ---

(iv) . . . loves ---

(v) . . . is the same length as ---

(vi) . . . is a number not less than ---

XII
Introducing Identity

We come now to the final element which we will introduce to the formal vocabulary of QL, namely, *identity*. Notably, the addition of identity makes no difference to the purely formal properties of QL.[6] However, it does significantly increase the expressive power of the language in a number of important respects. Naturally enough, the equals sign '=' is used to represent identity in QL and that sign is read as meaning 'is'. But this 'is' emphatically is not the same 'is' as the 'is' of predication. When we take a one-place predicate such as '. . . is a folk singer' or '. . . is groovy' and fill the place with a name such as 'Bob Dylan', for example, we are attributing a property to an individual, i.e. we say of Bob Dylan that he has the property of being a folk singer, being groovy or whatever. In sharp contrast, to use ' = ' we must fill two places, i.e. the general form of an identity statement is:

' . . . = --- '

Hence, identity statements are formed using two terms and ' = ', i.e. identity is a dyadic relation. Moreover, an identity statement does not simply attribute a property to an object. Rather, it *identifies* whatever is referred to by each term. Hence, '=' allows us to express the idea of *being the very same object as* in QL. So, we should, indeed, contrast the 'is' of predication with the 'is' of identity. In truth, this distinction is an important one, and failure to draw the distinction explicitly can lead to serious philosophical errors. Therefore, supplementing QL with the 'is' of identity is a very valuable and useful addition to the language.

Let's consider carefully the different kinds of sentence that identity can be used to express in QL.

First, because we can always fill both places in any identity statement with one and the same name, identity allows us to construct a vast number of extremely boring and uninformative sentences each of which must be true just in virtue of the form of sentence involved, e.g. I can truly assert that 'Bob Dylan is Bob Dylan', that 'Blind Lemon Jefferson is Blind Lemon Jefferson', that 'Arlo Guthrie is Arlo Guthrie' and so on, ad nauseam. But an important point underlies this possibility, namely, that *everything is identical with itself*. In terms of its formal properties, then, identity is reflexive. This point is captured formally in QL by the *law of identity*:

$\vdash \forall x\, [x = x]$

In the next chapter we will consider how to prove the law of identity. Here, we need only note that it is provable.

Identity can also be used to construct more interesting and informative sentences which might or might not be true. For example, it is a well-known fact that Bob Dylan's real name is Robert Zimmerman. Moreover, Bob Dylan has used a number of other names during his career, e.g. in a well-known film, he used the name 'Alias', amusingly enough. Now, you may well not have known either of these facts. So, in sharp contrast to the first kind of identity statement, the 'is' of identity can be used to construct true sentences which genuinely are informative. For example, I might assert that:

1. Bob Dylan is Robert Zimmerman.

or that:

2. Bob Dylan is Alias.

In both cases, what I have asserted is true just because each term in each identity, 'Bob Dylan', 'Robert Zimmerman' and 'Alias', refers to one and the same human being. Again, you might well not have known that either of these identity statements did hold and so those statements might be genuinely informative for you.

I noted above that identity is a reflexive dyadic relation. But note carefully that if it's true that Bob Dylan is Robert Zimmerman then it's equally true that Robert Zimmerman is Bob Dylan. Hence, identity is symmetrical. Further, if it's true that Bob Dylan is Robert Zimmerman and it's true that Robert Zimmerman is Alias then it must be true that Bob Dylan is Alias. Hence, identity is transitive. Summing up, then, given that identity is reflexive, symmetrical and transitive, identity is an equivalence relation.

Before we go on to consider the usefulness of identity for translation in QL it is helpful to note a further point: formal logicians tend not to use the names of folk singers to illustrate the nature of identity statements. Instead, you will find the literature dominated by two examples. The first is due to

Frege and concerns the Morning Star. In fact, the Morning Star is the same heavenly body as the one which is known as the Evening Star (both these expressions refer to the planet Venus). Again, you might well have thought that the first two expressions referred to distinct heavenly bodies and so, again, a sentence identifying the Morning Star with the Evening Star might well be genuinely informative. The second standard example was given prominence by Quine and concerns the great Roman philosopher Cicero. Cicero's proper name was in fact 'Marcus Tullius Cicero' and so he also came to be known as 'Tully'. Again, we can form a true and genuinely informative identity statement to the effect 'Cicero is Tully.'

Obviously, it is also useful to be able to represent sentences asserting **non-identity** in QL, e.g. 'Bill Clinton is not identical to Bob Dylan.' And this double-underlines the usefulness of the identity symbol for purposes of translation into the language QL. So far, we have had no way of representing any such sentence. Now, however, we can easily represent this sentence (and any sentence of that form) in QL. First, choose any two distinct names, say, a and b to stand for Bill Clinton and Bob Dylan, respectively. Next link the two using the identity symbol and enclose the resulting identity statement in brackets thus:

$$(a = b)$$

Finally, simply prefix the whole with a negation sign:

$$\sim(a = b)$$

This formula can now be paraphrased as: it is not the case that a is identical with b. Hence, both identities and non-identities are expressible in QL. Moreover, the availability of the identity symbol opens up new possibilities for translating sentences which attribute properties to specific named individuals. For example, suppose that we wanted to translate the sentence 'Bob Dylan is a glamorous and groovy folk singer.' Consider the following interpretation:

$\mathfrak{I}$ **D**: {**human beings**}

F: . . . is glamorous

G: . . . is groovy

H: . . . is a folk singer

a: Bob Dylan

Without identity, we could translate as follows:

1. $(Fa \mathbin{\&} Ga) \mathbin{\&} Ha$

But the addition of identity to QL enables us to translate the same sentence in a rather more sophisticated way, i.e.

1.′ $\exists x\ [((Fx\ \&\ Gx)\ \&\ Hx)\ \&\ (x = a)]$

To paraphrase: there exists someone who is a glamorous, groovy folk singer and that person is identical with Bob Dylan.

Finally, in conjunction with the existential quantifier, the identity symbol makes it easy to translate sentences which simply assert that particular individuals, identified by name, exist. For example, consider the sentence: 'Bob Dylan exists.' Given the interpretation above, we can translate as follows:[7]

$\exists x\ [x = a]$

To paraphrase: there exists someone who is Bob Dylan.

Consider the contents of Box 5.11 carefully before proceeding to Exercise 5.9.

BOX 5.11

♦ The introduction of the symbol ' = ' for *identity* to the language QL allows us to represent a sense of 'is' which is distinct from the 'is' of predication.

♦ The 'is' of identity is a two-place relation ' . . . is --- ' which is reflexive, transitive and symmetrical. Therefore, identity is an *equivalence relation.*

♦ The identity symbol can be used to construct QL sentences asserting either *identity*, e.g.:

$(a = b)$

or *non-identity*, e.g.:

$\sim(a = b)$

♦ In conjunction with the *universal quantifier*, identity can be used to express formally the *law of identity*, i.e. $\vdash \forall x\ [x = x]$ and/or universal sentences asserting universal non-identity, e.g. $\forall x\ [\sim(x = x)]$.

♦ In conjunction with the *existential quantifier*, identity can be used to translate sentences which assert that some specific named individual exists, e.g. $\exists x\ [x = a]$ and to translate sentences which attribute a property or properties to some specific named individual, e.g. $\exists x\ [Fx\ \&\ (x = a)]$.

EXERCISE 5.9

1 Translate the following English sentences into QL. In each case, make your interpretation explicit:[8]

(i) Blind Lemon Jefferson exists.

(ii) The Morning Star is the Morning Star.

(iii) The Morning Star is the Evening Star.

(iv) In fact, the Morning Star and the Evening Star are just the planet Venus.

(v) Everything is identical with itself.

(vi) Nothing is identical with itself.

(vii) Something is not identical with itself if and only if it is not the case that everything is identical with itself.

(viii) Everyone has fooled herself only if she has fooled someone else.

(ix) Everyone has fooled herself only if she has been fooled by someone else.

(x) Everyone has fooled himself only if everyone has fooled someone else.

2 Represent the following arguments as sequents of QL. In each case, make your interpretation explicit and state whether you consider each argument to be valid or invalid:

(i) The Morning Star is the Evening Star.
The Evening Star is the planet Venus.
Therefore,
The Morning Star is the planet Venus.

(ii) If the Evening Star is the Morning Star and the Morning Star is Venus then the Evening Star is Venus.

(iii) Wittgenstein is a German philosopher.
No German philosophers are badly behaved.
Therefore,
Either Wittgenstein is not badly behaved or Wittgenstein is not a German philosopher.

(iv) No one is her or his own father.
Paul is Zebedee's father.
Therefore,
Paul is not identical with Zebedee.

(v) Every folk singer is either groovy or is in fact Dr Strangely Strange.
Therefore,
If Dr Strangely Strange is groovy then every folk singer is groovy.

(vi) Every folk singer is groovy.
Therefore,
If someone is a folk singer and someone isn't groovy they are certainly not one and the same person.

XIII Identity and Numerically Definite Quantification

The addition of identity to the formal vocabulary of QL allows us a degree of precision as regards expressing the quantity of things we might talk about which, up to now, has been beyond our reach. Of course, we can easily assert that there is something, x, which has a particular property, say F, simply by exploiting the existential quantifier, e.g. if we wanted to formalise, 'There exists a folk singer', then given the interpretation in the previous section, we could simply write:

$$\exists x\ [Fx]$$

However, as we noted earlier, the existential quantifier carries with it a sense of there being *at least one* such thing, i.e. it leaves open the possibility that there is more than one such thing. But suppose that we did want to rule that possibility out, i.e. that we wanted to assert that there is only one folk singer or, in other words, *exactly one* folk singer. Just as it stands, the existentially quantified formula above is not adequate to represent such a claim formally. It does tell us that there is at least one such thing. But in order to express the claim that there is exactly one such thing we must also be able to express the assertion that there is *at most one* such thing. Now, if it is both the case that there is at least one F and that there is at most one F then it follows that there is exactly one F. Therefore, we must find a way of expressing the claim that there is at most one F. And this is precisely where identity can help.

Intuitively, we need to assert the following: consider everything in the domain: if anything, x, has the property F and anything, y, has the property F, then y is identical with x. Note very carefully that any such assertion is properly represented not as an existentially quantified formula but rather as a universally quantified formula. Therefore, that formula should have not a conjunctive matrix but a conditional matrix. As a universal conditional the formula will leave open the possibility that, in fact, there are no *F*s. And that possibility is precisely what is left open when we claim in natural language that, for example, '*At most* there exists one folk singer.' Hence, we represent the claim that there is at most one *F* by the following QL formula:

$$\forall x\ [\forall y\ [(Fx\ \&\ Fy) \rightarrow (x = y)]]$$

Finally, to capture the sense of 'There is exactly one *F*' in QL we can simply conjoin the claim that there is at least one *F* with the claim that there is at most one *F*. It follows that there is an *F* and that nothing distinct from that *F* is also *F*, i.e. there is only one and, therefore, *exactly* one *F*. Formally, then:

1. $\exists x\ [Fx]\ \&\ \forall x\ [\forall y\ [(Fx\ \&\ Fy) \rightarrow (x = y)]]$

To paraphrase: there exists something which is a folk singer and, for anything else in the domain, if it has the property of being a folk singer then it is the very same thing as the first. But note carefully that exactly the same sense is captured by the following rather leaner QL formula:

1.′ $\exists x\ [Fx\ \&\ \forall y\ [Fy \rightarrow (x = y)]]$

Moreover, this is not the only way in which we can make the point in QL. The following formula again captures the same sense, perhaps even more clearly:

1.″ $\exists x\ [Fx\ \&\ {\sim}\exists y\ [Fy\ \&\ {\sim}(y = x)]]$

Informally, 1″ precisely represents the claim that something is *F* and that nothing else exists which is also *F*.

Further, this technique can be extended to as many *F*s as we like. For example, suppose that we want to represent formally the assertion that there are at least two *F*s. This time, it won't do to write simply:

$$\exists x\ [\exists y\ [Fx\ \&\ Fy]]$$

Why not? Well, what guarantee have we that these are not one and the same thing? How do we know that we are not just referring to the same element of the domain with each conjunct? In short, we don't. But again identity makes it easy to express the claim that these are not identical. All we need do is add a further conjunct which asserts that *x* and *y* are not identical as follows:

$$\exists x\ [\exists y\ [(Fx\ \&\ Fy)\ \&\ {\sim}(x = y)]]$$

Now, while this formal sentence does express the claim that there are at least two *F*s it does not follow that there aren't more than two *F*s, i.e. it does not follow that there are at most two *F*s. In order to express the claim that there are exactly two *F*s we need to assert both that there are at least two *F*s and that there are at most two *F*s; only then will it follow that there are exactly two *F*s. To that end, we can easily exploit the technique of adding on a clause to the effect that anything else in the domain which is *F* is simply identical with one or other of our original two *F*s, formally:

2. $\exists x\, [\exists y\, [((Fx \,\&\, Fy) \,\&\, {\sim}(x = y)) \,\&\, \forall z\, [Fz \rightarrow ((z = x) \vee (z = y))]]]$

Again, this is not the only way in which we can make this point in QL. The following formal sentence captures the same sense, again, perhaps even more clearly:

2.′ $\exists x\, [\exists y\, [((Fx \,\&\, Fy) \,\&\, {\sim}(x = y)) \,\&\, {\sim}\exists z\, [Fz \,\&\, ({\sim}(z = x) \,\&\, {\sim}(z = y))]]]$

In principle, then, there is no limit to the number of things with a given property or properties that we can talk about formally in QL. Therefore, we can exploit identity to quantify in a numerically definite way, quite generally. This kind of quantification, **numerically definite quantification**, is perhaps the greatest single increase in expressive power which the addition of identity to QL bequeaths us.

Once you have considered the contents of Box 5.12 carefully you can put what you've learned here into practice in Exercise 5.10.

BOX 5.12

Numerically definite quantification

♦ In conjunction with the existing vocabulary of QL we can exploit identity to quantify in a *numerically definite* way. Note carefully, however, that each of the locutions 'at least', 'at most' and 'exactly' have distinct senses.

♦ In order to assert that there are 'at least n *F*s' not only must you assert the relevant number of positive existentially quantified sub-formula;s, e.g.:

$\exists x\, [\exists y \ldots [((Fx \,\&\, Fy) \ldots$

but you must also explicitly rule out the possibility that any of the elements of the domain involved are identical, e.g.:

$\ldots \,\&\, {\sim}(x = y) \,\&\, \ldots)$

♦ In order to assert that there are 'at most n *Fs*' not only must you assert the relevant number of positive universally quantified sub-formulas, e.g.:

$\forall x\ [\forall y \ldots [((Fx\ \&\ Fy) \ldots \rightarrow \ldots$

but you must also explicitly rule out the possibility that any other element of the domain has the property in question, e.g.:

$\forall z\ [Fz \rightarrow ((z = x)\ \text{v}\ (z = y) \ldots$

♦ Note very carefully that a formal sentence whose sense is simply 'At most there are n *Fs*' leaves open the possibilities that there are no *Fs*, that there are n – 1 *Fs*, that there are n – 2 *Fs*, etc. Thus, for example, the QL sentence which translates 'At most there are 2 *Fs*', i.e. the QL formula:

$\forall x\ [\forall y\ [(Fx\ \&\ Fy) \rightarrow \forall z\ [Fz \rightarrow ((z = x)\ \text{v}\ (z = y))]]]$

not only leaves open the possibility that there are 2 *Fs* but also leaves open the possibility that there is 1 *F* and, indeed, that there are, in fact, no *Fs*.

In order to assert that there are exactly n *Fs*:

♦ Assert the relevant number of positive existentially quantified sub-formulas, e.g.:

$\exists x\ [\exists y\ [((Fx\ \&\ Fy) \ldots$

♦ Explicitly rule out the possibility that any of the elements of the domain involved are identical, e.g.:

$\ldots\ \&\ {\sim}(x = y)\ \&\ \ldots)$

♦ Explicitly rule out the possibility that any other element of the domain has the property in question, e.g.:

$\ldots\ \&\ \forall z\ [Fz \rightarrow ((z = x)\ \text{v}\ (z = y) \ldots$

EXERCISE 5.10

1 Monotheists are those people who believe that there is only one God.[9] Polytheists believe that there is more than one God. Further, different kinds of polytheist may well posit different numbers of Gods. The

following sentences might well come from an exchange of views between monotheists and polytheists (one of whom is obviously rather confused). Translate these sentences into QL:

(i) There is at least one God.

(ii) It's just not true that there is one and only one God.

(iii) There are exactly two Gods.

(iv) There are at least three Gods.

(v) At most, there are two Gods.

(vi) Either there are no Gods or there is exactly one God or there are exactly two Gods.

(vii) There are three, and only three, Gods.

(viii) There are at least four Gods.

(ix) There are at most four Gods.

(x) There are exactly four Gods.

2 Recall the discussion of the formal properties of relations in Section X. There, we noted that a relation *R* is anti-symmetrical if and only if when it is both the case that *a* bears *R* to *b* and that *b* bears *R* to *a* it follows that *a* and *b* are in fact identical objects. Represent this kind of relation formally in QL in the manner in which other properties of relations were represented in Section X.

XIV
Russell #1: Names and Descriptions

The addition of identity to the formal vocabulary of QL enables us to quantify in a numerically definite way for the first time. As we shall see in the present section, the same technique also enables us to represent formally an aspect of logical grammar which was first uncovered by Bertrand Russell. Moreover, this is not simply a matter of increasing the expressive power of QL. Rather, it allows us to appreciate a paradigm case of logical analysis which contributes much to our understanding of the nature of a particular kind of natural language expression, namely, descriptions. Russell's analysis of descriptions, the **theory of descriptions**, is one of the most valuable contributions ever made to formal and philosophical logic and, indeed, is thought by many to be the paradigm of logical analysis. To appreciate fully the nature of that theory it is useful to make clear a little of Russell's own philosophical position as regards logic and language first.

Russell shared with the Wittgenstein of the *Tractatus* a very straightforward view about how it is that names are meaningful, i.e. both authors maintained that the meaning of a name is identical with the object to which the name refers. For Russell in particular, to know the meaning of a name is just to be acquainted with the object to which that name refers. But this is only one aspect of Russell's philosophy and, in his *Lectures on the Philosophy of Logical Atomism* of 1918, for example, we find that view embedded in the wider context of a philosophy of language and a particular view about the nature of philosophy itself. The primary task of philosophy here is precisely logical analysis. Analysis ends, Russell holds, when we arrive not at physical atoms but at *logical atoms*, i.e. when we arrive at knowledge that reality is composed of simple particulars, qualities and relations.

As a language, QL is particularly well adapted to hooking up with what, in Russell's view, reality ultimately is. Names are meaningful just as a result of referring to the simple particulars which are the ultimate constituents of reality. Predicates represent the properties or qualities of particulars and relational expressions allow us to represent relations which might hold between particulars. Thus, Russell argues, reality consists of facts, i.e. concatenations of simple particulars. Language consists of propositions. Propositions attempt to describe facts. The simplest kind of fact consists in the possession of a quality by a particular, or the sharing of a relation by two particulars, or three and so on. All such basic facts Russell terms *atomic facts*. Logically enough, those propositions which describe atomic facts are *atomic propositions*. Moreover, atomic propositions consist of names which have meaning simply by denoting objects. Hence, real names or, as Russell puts it, **logically proper names**, mean objects. Russell's account of names might well seem an intuitive one but it does generate a significant philosophical headache when it comes to a certain kind of identity statement. Here, as you know, there are two types to consider:

Type 1. $a = a$ e.g. 'Elvis Presley is Elvis Presley.'

Type 2. $a = b$ e.g. 'Elvis Presley is the King of rock'n'roll.'

If the terms in these identities are logically proper names then all true identity statements must be uninformative, i.e. the only content they can have is 'a is identical with a.' For Type 1 identities this analysis is absolutely correct. The problem is that Russell seems to be committed to exactly the same analysis for Type 2 identity statements. But that cannot be correct. Certainly, I learn nothing when I am informed that Elvis Presley is Elvis Presley. But suppose I don't know that Elvis Presley is the King of rock'n'roll and you tell me that he is. Here I do learn something. Hence, there is a discrepancy in informativeness here. But how is that discrepancy possible given Russell's analysis of logically proper names? According to Russell, the answer is just that although the 'is' involved is the 'is' of identity and not that of predication it is not the case that two logically proper names are involved here. For Russell, the expression 'the King of rock'n'roll' is not a name at all. Rather, it is a description. Further, names and descriptions do not contribute to the meanings of sentences in the same way. In essence, Russell's theory of descriptions is an analysis of precisely how it is that descriptions do contribute to the meanings of propositions in their own special way.

To that end, Russell first distinguishes two types of description, namely, definite and indefinite. **Definite descriptions** are prefixed by the definite article: '*The* present Prime Minister', '*The* man in the street', etc. **Indefinite descriptions** are prefixed by the indefinite article: '*A* man . . . ', '*A* philosopher . . . ', etc. Thus, the expression 'the King of rock'n'roll' is not a name but a definite description. The theory of descriptions is precisely a theory of definite descriptions. More generally, Russell's claim is that for any $a = b$ type identity statement where b is a phrase of the form: 'the so and so', b is not a logically proper name but a definite description. Further, he argues, definite descriptions cannot have meaning in virtue of picking out objects just because there need not actually be anything in the world which corresponds to the description. Therefore, it is always possible to deny the existence of anything so described quite meaningfully, e.g. 'The King of rock'n'roll does not exist', 'The present King of France does not exist', 'The greatest finite number does not exist' and so on. Finally, Russell promises, any such description can be analysed out of the sentence in which it occurs and replaced, **salva veritate** (i.e. such that the truth-value of that sentence remains unchanged). Hence for any problematic Type 2 identity, analysis can make the apparently nominative expression 'b' disappear while preserving the truth-value of the original sentence. Further, the theory of descriptions provides all the machinery that such analysis requires.

So, what is Russell's analysis of a sentence such as 'Elvis Presley is the King of rock'n'roll '?

According to Russell, the use of any sentence containing a definite description entails that the described thing exists, i.e. that there exists one and only one such thing. Hence, in the present case, it is entailed that there is exactly one thing in the world which is the King of rock'n'roll. Given the discussion of numerically definite quantification in the previous section, we know that any sentence asserting the existence of exactly one thing can be understood in terms of there being at least and at most one such thing. Thus, if we let F represent the predicate: '... is the King of rock'n'roll' the first part of what is entailed can be represented in QL by the existentially quantified formula:

1. $\exists x\ [Fx]$

So far so good. But what is entailed is not just that there is at least one such thing but also that there is at most one such thing. Given our earlier discussion, we can easily formalise the latter claim in the usual way:

2. $\forall x\ [\forall y\ [(Fx\ \&\ Fy) \rightarrow (x = y)]]$

To paraphrase: 'For any x and any y, if x is King of rock'n'roll and y is King of rock'n'roll then x is identical with y.'

To complete the analysis we conjoin the two QL formulas as follows:

3. $\exists x\ [Fx]\ \&\ \forall x\ [\forall y\ [(Fx\ \&\ Fy) \rightarrow (x = y)]]$

The same sense can also be captured by the rather leaner QL formula:

3.′ $\exists x\ [Fx\ \&\ \forall y\ [Fy \rightarrow (x = y)]]$

Either way, as promised, the definite description 'the King of rock'n'roll' has now been completely analysed in a way which makes clear that no name whatsoever is involved. Therefore, the logical grammar of that expression certainly is not that of a logically proper name. And now it is a short step to state the full analysis of the original problematic Type 2 identity: 'Elvis Presley is the King of rock'n'roll':

$$\exists x\ [Fx\ \&\ \forall y\ [(Fy \rightarrow (x = y))\ \&\ (x = \text{Elvis Presley})]]$$

Moreover, sentences ascribing properties to the King of rock'n'roll such as, for example, 'The King of rock'n'roll is an American boy' can easily be translated: simply repeat the analysis of the definite description and complete the translation by adding a conjunct to represent the relevant property, i.e. repeat the formalisation of 'The King of rock'n'roll':

$$\exists x\ [Fx\ \&\ \forall y\ [Fy \rightarrow (x = y)]]$$

And, where G stands for the predicate '... is an American boy', conjoin that predicate as follows:

$$\exists x\ [Fx\ \&\ \forall y\ [Fy \rightarrow ((x = y)\ \&\ Gx)]]$$

Whether a sentence containing a definite description attributes a property to the thing described or not, the use of any such sentence entails that the thing described exists. But the crucial point is this: it does not follow that there does exist any such thing in the world. Indeed, the King of rock'n'roll is a case in point here. Some (rather misguided) people might want to say that the King of rock'n'roll does exist, of course, but consider a favourite example of Russell's:

> The present King of France is bald

This sentence is certainly meaningful. The grammatical subject of the sentence is 'The present King of France' and use of the sentence certainly entails the existence of the thing described, i.e. the present King of France. But, presently, there is no King of France. Therefore, the sentence is false. Hence, the expression 'The present King of France' cannot contribute to the meaning of the sentence as a logically proper name. If the grammatical subject of the sentence were a logically proper name then it could not fail to refer to an existing object. Indeed, the very idea of a logically proper name which does not refer to an existing object is contradictory. But there is no logical impropriety whatsoever involved in the idea of a sentence containing a definite description as grammatical subject failing to refer to an existing object. Hence, definite descriptions cannot contribute to the meanings of sentences in the same way that logically proper names do.

Finally, we can deepen the contrast between names and descriptions by considering the ways in which sentences containing such terms can turn out to be false. For example, consider any subject–predicate sentence involving a logically proper name as grammatical subject. Clearly, any such sentence will be false only if the referent of the logically proper name lacks the property attributed to it by the predicate. Otherwise, the sentence will be true. However, a sentence containing a definite description may be false in one of three ways. First, if the thing described does not exist the sentence containing the description will be false. This takes care of the present King of France precisely because there isn't one. Second, if there is more than one such thing, then, again, the sentence containing the definite description will be false, e.g. if there were two or three present Kings of France, any sentence about the one and only King of France would be false. Third, a sentence containing a definite description is false if the described entity does exist but lacks the property ascribed to it in the sentence. For example, supposing that there did presently exist a hirsute King of France, the sentence 'The present King of France is bald' would still be false but, this time, in virtue of the lack of baldness rather than the lack of existence. Therefore, Russell takes the problem of Type 2 identities to be solved.

The theory of descriptions remains a paradigm, perhaps the paradigm, of philosophical analysis but it is not entirely uncontroversial. For example, Peter Strawson has proposed an alternative account of sentences containing definite descriptions. Surely, Strawson argues, the sentences constituting Russell's analysis are not entailed by the original sentence but are implied or presupposed by it. On Strawson's view, where what is presupposed does not actually exist the original sentence is not false but *neither true nor false*. Thus, while 'The present King of France is bald' comes out false for Russell just because there is now no present King of France, for Strawson, if there is now no such king, that assertion is neither true nor false. Plausible as these suggestions may seem, there is a price to be paid for accepting them, i.e. the failure of the principle of bivalence. Hence, we either accept truth-value gaps, as it has been put, or recognising a third truth-value: 'neither true nor false'. Moreover, Russell's theory has been plausibly defended against Strawson's criticisms by Stephen Neale in his *Descriptions*, Cambridge MA, MIT Press [1990].

An even more radical departure from Russell's approach to Type 2 identities is clear in the work of Frege. To understand such sentences properly, Frege argues, we do not need to distinguish between names and definite descriptions at all. Rather, we need to distinguish between the reference of such expressions and their senses. The reference is, indeed, the object referred to but it is the sense which fully discloses the meaning of the expression. For Frege, sense is that which fixes reference. Sense is a way of presenting the referent, a criterion for identifying the referent. According to Frege, then, there is more to the meaning of a name than is dreamt of in Russell's philosophy.

Moreover, on Frege's view, it is precisely in terms of the sense–reference distinction that the informativeness of Type 2 identity statements should be explained. For example, consider the sentence:

The Morning Star is the Evening Star.

Certainly, both 'The Morning Star' and 'the Evening Star' refer to the same object, i.e. to the planet Venus. But they do so in rather different ways. Each expression presents that object differently and, as it were, shows us a different route to the object (think of the two different senses as two different telescopes trained on the same star. We see the same object but by different means). So, when we do learn something from a Type 2 identity statement, what we learn is that an unfamiliar route can be taken to a familiar object. Therefore, the discrepancy in content is at the level of sense rather than reference. Frege's distinction is upheld and vigorously defended by the contemporary logician and philosopher Michael Dummett. Further, as is clear from his *Philosophical Investigations*, Remarks 39–59, Wittgenstein abandoned his early Russellean view in favour of a position more like that of

Frege. In sharp contrast, Russell's fundamental idea that reference is constitutive of meaning for proper names is persuasively promoted and defended in the work of another eminent contemporary logician and philosopher, Saul Kripke. Do names really have sense or is their meaning exhausted by reference? And how exactly are Type 2 identity statements properly understood? These questions remain definitive of the contemporary philosophy of language but we can do no more justice to them here. Interested parties can find

BOX 5.13

Russell on names and descriptions

♦ According to Russell, the meaning of a *logically proper name* is identical with the object to which that name refers.

♦ In contrast, *definite descriptions* cannot have meaning simply in virtue of picking out objects because there need not actually be anything in the world which corresponds to any such description.

♦ However, sentences containing *definite descriptions* do *entail* that the described thing exists.

♦ Hence, a correct formal analysis of any sentence containing a definite description in QL must have the overall form of an *existentially quantified formula*, e.g.:

$$\exists x\,[Fx]$$

♦ Further, sentences containing definite descriptions entail that *exactly one* such one thing exists.

♦ Hence, a proper formal analysis of a sentence containing a definite description in QL should result not simply in an existentially quantified formula but rather in a *numerically definite* existentially quantified formula, e.g.:

$$\exists x\,[Fx\ \&\ \forall y\,[(Fy \rightarrow (y = x))]]$$

♦ When a sentence containing a definite description also involves the *ascription of a property* to the thing described, repeat the analysis and conjoin the relevant predicate, e.g.:

$$\exists x\,[Fx\ \&\ \forall y\,[(Fy \rightarrow (y = x))\ \&\ Gx]]$$

Frege's paper 'On Sense and Reference' together with papers by Russell, Strawson, Dummett, Kripke and other important figures in the field in the Oxford Readings volume on *Meaning and Reference*, [1993], edited by Adrian Moore, Oxford, Oxford University Press.

Exercise 5.11 will allow you to put into practice what you have learned about names and descriptions in the present section. First, study Box 5.13.

EXERCISE 5.11

1 The following sentences all involve definite descriptions. Translate these sentences into QL in the manner of Russell. In each case make your interpretation explicit and try to reflect as much of the structure of each sentence as you can:

(i) The King of rock'n'roll is dead.

(ii) The King of the blues was a genteel Delta bluesman.

(iii) The Blind Lemon Jefferson album on the turntable is deeply groovy.

(iv) The head of the philosophy department was a cool, calm and collected character.

(v) Paul preferred the Greatest Hits album that Bob Dylan had recently released.

(vi) Everyone rated the bluesman who rated everyone.

XV
Russell #2: On Existence

In the previous section we considered one philosophical headache which arose from Russell's view of just how it is that logically proper names are meaningful, i.e. the problem of explaining the informativeness of Type 2 identity statements. In this, the final section of the present chapter, we consider a related headache arising from the same source. Again, Russell's remedy is given in terms of the theory of descriptions. Moreover, in the solution to this corollary headache, the value and importance of that theory can be clarified and double-underlined.

This time, the headache concerns attributions and denials of *existence*. Famously, Russell argued that 'exists' is far from being an ordinary predicate such as '... is a folk singer' or '... is groovy'. Indeed, Russell denied that existence could be properly predicated of individuals at all. Why? Consider again his account of logically proper names: either a name denotes an individual or it is meaningless. Suppose that 'Mr N.N.' is a logically proper name in just that sense and suppose I assert that:

Mr N.N. exists.

If that assertion is meaningful at all then what I have asserted must be true, i.e. if you know that 'Mr N.N.' is a logically proper name then you must know that there exists an individual to which that name refers. Worse still, by parity of reasoning, the assertion:

Mr N.N. does not exist

if meaningful, is self-contradictory.

Moreover, Russell argues, not only can existence not simply be predicated of individuals but to try to reason to the existence of an individual invariably involves a fallacy, e.g. to reason that:

1. Men exist.
2. Socrates is a man.

Therefore,

3. Socrates exists.

is no better than reasoning that:

1. Men are numerous.
2. Socrates is a man.

Therefore,

3. Socrates is numerous.

Hence, existence cannot be meaningfully applied to individuals denoted by logically proper names in singular sentences. But what of general sentences and existence? Here again we should be wary, for no sentence of the form 'All *A*s are *B*s' ever implies that there exist any *A*s. But universal sentences are not the only kind of general sentence. Existential generalisations can equally well be formed and these do have ontological implications. What, then, is Russell's analysis of the logical character of existential sentences?

In essence, Russell's view is that existential sentences predicate one particular property of other properties. More specifically, Russell holds that an existential sentence predicates the property of *having at least one instance* of a given property. The point can be seen more clearly by separating out three different kinds of predicate. First, certain predicates always result in false sentences no matter which name we use to plug the relevant gap, e.g. 'x is not identical with itself' is false for every value of the variable x. In other words, the property of not being self-identical has no instances whatsoever. Equally, certain predicates always result in true sentences no matter which name we use to plug the relevant gap, e.g. 'x is identical with itself' is true for every value of the variable x. Hence, each and every thing is an instance of that predicate. Finally, certain predicates result in a true sentence for at least one argument, e.g. 'x is a woman' generates a true sentence for at least one value of the variable x. In other words, the predicate 'x is a woman' has at least one instance.

On Russell's view, to make an existence-claim is to assert that a predicate is of the third kind, i.e. what we mean when we make an existence-claim is precisely that, for the relevant predicate, there is at least one value of the variable which generates a true sentence. Hence, to claim that men exist is precisely to claim that the predicate 'x is a man' has at least one instance, i.e. that there is at least one value of the variable x which generates a true sentence. As Quine famously put it, 'to be is to be the value of a variable'.[10]

In one sense, the foregoing should be unsurprising. Ontological commitments are properly expressed in QL by positive existentially quantified formulas and, indeed, any such formula is true if and only if there is at least one value of the relevant variable which makes it so. But I draw attention to Russell's analysis here not to emphasise the logical grammar of talk about what there is but to highlight the logical grammar of talk about what there isn't. For Russell may well have hit upon a solution to a philosophical puzzle which is as old as philosophy itself. This puzzle is often described as the **problem of non-being**, or the 'problem of Plato's beard'. The problem is just this: how can we possibly talk meaningfully about what does not exist? For example, if I assert that the present King of France does not exist doesn't the very meaningfulness of that assertion commit me to the existence of what I want to say does not exist? And, if not, what exactly am I making that assertion about?

The first point to make here is simply that the expression 'the present King of France' is not a name but a description. Given the theory of descriptions, any sentence in which that expression occurs can be analysed to reveal that no names whatsoever are involved in that expression. Therefore, the meaningfulness of the sentence is entirely independent of the existence of any King of France and so my use of that sentence does not commit me to the existence of any such king. But when I deny that such a king exists what then am I talking about? Given Russell's analysis of existence, it is quite clear that what I am talking about here is the predicate '*x* is the present King of France.' More precisely, what I am asserting is that there is no value of the variable for that predicate which generates a true sentence. Thus, Russell not only clarifies the logical character of talk about what exists, he also clarifies the logical character of talk about what does not exist. Hence, Plato's beard is trimmed.

But what of sentences which deny existence using singular nouns rather than descriptive phrases, e.g. ordinary names such as 'Pegasus' or 'Santa Claus'? On Russell's view, the meaningful use of a logically proper name does presuppose a referent object. So, don't we become entangled in Plato's beard if we assert that Pegasus does not exist or that Santa Claus does not exist?

Certainly, on Russell's view, logically proper names do presuppose referents. But few if any of the names of ordinary language are ever logically proper in his sense. Indeed, in the *Lectures on the Philosophy of Logical Atomism* Russell surmises that the demonstratives 'this' and 'that' which are generally used simply to refer to objects are the nearest approximations to logically proper names that a language such as English can provide. Moreover, Russell writes:

> when Adam named the beasts, they came before him one by one, and he became acquainted with them and named them. We are not acquainted with Socrates and, therefore, cannot name him. When we use the word 'Socrates' we are really using a description . . . 'The Master of Plato' . . . 'The philosopher who drank the hemlock' or 'The person whom logicians say is mortal', but we certainly do not use the name as a name in the proper sense of the word.[11]

So, ordinary proper names such as 'Pegasus' and 'Santa Claus' are nothing other than truncated descriptions; proxies for expressions such as 'the winged horse belonging to Bellerophon' or 'the owner of Rudolph the Red-Nosed Reindeer'. But these expressions are definite descriptions. Hence, again, any sentence in which any such ordinary name occurs can equally well be analysed in terms of the theory of descriptions to reveal that no logically proper name is involved. Therefore, the meaningful use of any sentence involving ordinary proper names such as 'Pegasus' or 'Santa Claus' certainly does not commit the speaker to the existence of Pegasus or of Santa Claus.

Finally, Russell's theory is given a small but useful embellishment by Quine. Suppose that, for a given ordinary proper name, there is no obvious associated description at hand. In any such case, Quine suggests that we expand the singular noun into a singular description; a practice which has come to be known as **Quining** singular nouns. For example, if we can think of no description associated with 'Pegasus' we can simply rewrite the singular noun 'Pegasus' as the singular description 'The Pegasizer'. The latter expression can now be analysed in terms of the theory of descriptions, e.g. where the relevant predicate is '*x* Pegasizes.' And, again, the meaningful use of singular nouns in sentences denying existence does not commit the speaker to the existence of what she or he seeks to deny. Thus, Russell and Quine provide not only a clear criterion of ontological commitment but also a way of making sense of ontological disagreements, i.e. a way of understanding how, when I assert that something does not exist, I do not commit myself to the existence of the thing in the process. If I want to number Pegasus among my ontological commitments I will assert that, for some value of the variable, the predicate '*x* Pegasizes' generates a true sentence. If you do not want to include Pegasus among your ontological commitments you may simply claim that my assertion is false. The beauty of Quining, then, is precisely that we can deny the existence of anything we choose without the act of denial committing us to the existence of the very thing we want to deny.

As you may have realised, the technique of Quining can be applied to ordinary names quite generally, i.e. to real names such as 'Blind Lemon Jefferson' just as much as to dubious names such as 'Santa Claus'. Of course, formal logic cannot prove that Santa Claus does not exist any more than it can prove that Blind Lemon Jefferson does exist. But that is eminently right. Formal logic is not natural science. And ontological commitments are a matter of cultural and personal belief; perhaps, even, a matter of individual taste. To paraphrase Quine, those with a taste for the aesthetics of desert landscapes will deny the existence of many more things than those who appreciate the beauty of densely populated cityscapes. However, what formal logic can do is provide a way of making sense of the logical character of the kinds of discussion which will inevitably arise between individuals of such different tastes. In the process, it might also have solved a problem which puzzled philosophers for more than 2,000 years.

Again, Russell's analysis is not entirely uncontroversial and, again, there is more to this particular debate than we can do justice to here. Russell's analysis of existence together with an explanation of the theory of descriptions can be found in his 'Lectures on the Philosophy of Logical Atomism', in *Logic and Knowledge: Essays 1901–1950*, [1984], ed. Robert Marsh, London, George Allen & Unwin, especially Lectures V and VI. Quine's account of the logic of ontological disagreements can be found in Chapter 1 of his *From a Logical Point of View*, [1963], New York and Evanston, Harper & Row, which is entitled 'On What There Is' (though it might be more aptly titled: 'On What There Isn't'). Useful commentary on Quine's criterion (and its limits) is provided by Christopher Hookway in his *Quine*, [1988], Oxford Polity Press. For more of the flavour of the controversy here see Stephen Read's '"Exists" is a Predicate', *Mind*, 89 [1980].

Examination 3 will allow you to put into practice much of what you have learned in this chapter.

Examination 3 in Formal Logic

Answer every question.

1 *Translate the following sentences into QL in the manner of Russell or Quine. In each case, make your interpretation explicit:*

(i) The President of the United States is Bill Clinton.

(ii) The President of the United States is male.

(iii) The purple people-eater is a monster.

(iv) Santa Claus exists and he is a charming fellow.

(v) Santa Claus does not exist and neither does Pegasus.

(vi) Flubjub does not exist.

2 *Represent the following arguments as sequents of QL. Translate each definite description in the manner of Russell. In each case, make your interpretation explicit:*

(i) The man in the iron mask is a bore.
No one likes a bore.
Therefore,
No one likes the man in the iron mask.

(ii) Alice sat the logic exam that Professor Frege had recently devised.
The happiest student in the room was the one who passed with flying colours.
Alice passed with flying colours.
Therefore,
Alice was the happiest student in the room.

3 The following argument has a certain infamy due to the fact that prior to the development of QL logicians found the argument difficult to formalise. *Construct an interpretation and represent the argument as a sequent of QL:*

1. All horses are animals.

Therefore,

2. All horse's heads are animal's heads.

4. *Use the interpretation provided below to translate the following argument:*

1. Nothing is better than eternal happiness.
2. A cheese sandwich is better than nothing.

Therefore,

3. A cheese sandwich is better than eternal happiness.

$\mathfrak{I}$ **D: {human beings and their possessions}**

B: . . . is better than ---

C: . . . is better off than ---

D: . . . has ---

a: a cheese sandwich

b: eternal happiness

Notes

1 In the attempt to reduce proliferation of brackets as far as possible, I do not enclose quantifiers in brackets in what follows.

2 Quine, W.V.O. [1963], *From a Logical Point of View*, New York, Harper & Row, Ch. 1, p. 6.

3 Note very carefully that we cannot faithfully render (ii) as: '$\exists x\ [Gx\ \&\ \forall y\ [Fy \rightarrow Lxy]]$'. Under the current interpretation, this translates as: 'Some bluesman loves every folk singer.'

4 The idea for Question 1 was inspired by an exercise designed by Mates, Benson, [1972], *Elementary Logic*, second edition, New York, Oxford University Press. See Ch. 5, Ex. 1.

5 I am indebted to Barker, Stephen F., [1957], *Induction and Hypothesis: A Study of the Logic of Confirmation*, Ithaca NY, Cornell University Press, for the ideas for examples (i), (iii) and (iv) and to John Slaney for the kind of example involved in (v). Examples (ix)–(xv) are based on a set of illustrations provided by Tennant, Neil, [1978], *Natural Logic*, Edinburgh, Edinburgh University Press. See Ch. 3. 3.8, pp. 33–4.

6 As we shall see in Ch. 7, monadic QL is Sound, Complete and Decidable *with or without* identity and, again, QL beyond the monadic fragment is Sound, Complete and Undecidable, again, with and without identity.

7 I am indebted to Alexander Broadie for this suggestion.

8 The idea (and the credit) for examples (viii)–(x) is due to Tennant, [1978], Ch. 3, p. 34.

9 The idea for this particular exercise was inspired by certain examples used by Morse, Warner, [1973], *Study Guide for Logic and Philosophy*, second edition, Belmont CA, Wadsworth, Ex. 9–4, (p. 102).

10 Quine, [1963], Ch. 1, p. 15.

11 Russell, Bertrand, 'Lectures on the Philosophy of Logical Atomism', in: Robert C. Marsh (ed.), [1984], *Logic and Knowledge Essays 1901–1950*, London, George Allen & Unwin, p. 201.

6
How to Argue Logically in QL

6
How to Argue Logically in QL

Introduction: Formal Logic and Science Fiction

To date, we have considered QL as a formal language into which we can translate natural language arguments in a way which clearly shows up both the logical form of those arguments and the internal, logical structure of the sentences which compose them. I hope that you have also seen something of the power and sophistication of QL as a tool for *logical analysis*, e.g. in Russell's theory of descriptions, and, therefore, something of the philosophical insight which can derive from the application of formal logic to natural language. So far, we have been content simply to represent natural language arguments as sequents of QL and to consider questions of their validity or invalidity either purely intuitively or via *shallow analysis*, i.e. as yet, we have no formal method in terms of which we can demonstrate the validity of sequents of QL just as such. As ever, the focus of the formal logician's concern is precisely that possibility of characterising valid reasoning purely formally not merely at the level of PL but, equally, at the level of QL. How are we to approach the question of formal methods for QL? For PL, the first formal method of demonstration we designed was a syntactical one, i.e. we introduced a set of rules of inference for PL in terms of which the notion of *proof-in-PL* was defined. In formal terms, we supplemented the formal language PL with a *deductive apparatus* and so arrived at a *formal system* of propositional logic. In this chapter, we adopt a similar strategy as regards QL, i.e. we introduce a set of rules of inference so as to develop a formal system of quantificational logic.[1]

As a formal language, QL inherits all of the formal vocabulary of PL. Moreover, as a formal system, QL inherits the entire deductive apparatus of PL. So, all the familiar rules of inference from premise-introduction to RAA are already available to us. For present purposes, then, much of the hard work has already been done. However, just as we had to supplement the formal vocabulary of PL in order to arrive at the language QL so we must go on to supplement the deductive apparatus of PL in order to characterise a formal system for QL. Most obviously, we must supplement the existing set of rules of inference with new rules which enable us to exploit formulas involving quantifiers in the course of proof-construction. But you need not be too daunted by that prospect. For the new rules governing the quantifiers in QL resemble the old rules governing the connectives in PL in at least one important respect. In the proof-theory of PL, each connective has both an *introduction-rule* which allows us to enter that connective onto a line of proof and an *elimination-rule* which allows us to eliminate a connective from a line of proof. Similarly, in QL each quantifier has both an introduction-rule and an elimination-rule which we can exploit precisely in order to introduce a quantified formula onto a line of proof or to eliminate a quantified formula from a line of proof. Therefore, we have only four new rules of inference to consider. When we have spelled these out we will have come very close to completing the formal system for QL and, shortly thereafter, we will be able to offer a precise definition of the notion of **proof-in-QL**.

Before we go on to consider each rule in detail, it is useful to consider first a brief illustration from science fiction. Moreover, although the illustration is quite light-hearted, it is none the less an accurate illustration of some crucial formal notions which we will consider again in the next chapter. Hence, I am not being misleading here and you should find that keeping the illustration in mind helps to clarify what is going on in this chapter. So, imagine a galaxy far, far away in which there is a fabulous world inhabited by a very small number of very lucky people. The name of this particular world is *Fabworld* just because everything in that world is absolutely fabulous. Moreover, everything in Fabworld is also fantastically groovy. So, Fabworld is far away indeed from Dullworld. At the last census, the population of Fabworld stood at a mere three people. But, of course, each of those people is both fabulous and groovy and so they are completely untroubled by the size of the population. We will use the proper names *a*, *b* and *c* from QL to denote each of the three inhabitants of Fabworld respectively. In fact, it turns out that the language QL is perfectly suited not just to naming everyone in Fabworld but also to describing everyone in Fabworld and, indeed, to describing just how things stand to one another at any given moment in Fabworld. In other words, Fabworld and its inhabitants constitute a perfectly legitimate, if highly restricted, domain for QL.

I
Reasoning with the Universal Quantifier 1: The Rule UE

With Fabworld clearly in mind, the easiest way to get a handle on the rules of inference for the quantifiers is to ask: what do quantified formulas imply in this context? Well, let's begin with the universal quantifier. For example, consider the simplest possible kind of universally quantified formula, say, $\forall x\ [Fx]$, and take as the domain of quantification Fabworld and its inhabitants. It is natural to interpret the predicate-letter F as standing for the property of being fabulous in this context. So, under the present interpretation, what does the QL-formula $\forall x\ [Fx]$ imply? In the previous chapter we noted that it is of the essence of any universally quantified formula that it implies each and every one of its instances, i.e. the sense of 'all' which the universal quantifier helps us to express is *distributive*. Hence, we can think of the universally quantified formula as predicating the property of being fabulous of each and every element of the domain. Under the present interpretation, then, the formula $\forall x\ [Fx]$ implies that every single inhabitant of Fabworld is fabulous. Now, there are only three individuals in Fabworld and, in QL, the proper names for those individuals are a, b and c respectively. So, spelling things out completely, what is implied by $\forall x\ [Fx]$ under the present interpretation is just that a is F and that b is F and that c is F. Therefore, we can think of the universally quantified formula as a sort of abbreviated conjunction. Waiving the rules about brackets for clarity, we can now cash out just what is meant by $\forall x\ [Fx]$ under the present interpretation in terms of the conjunction:

$$Fa \,\&\, Fb \,\&\, Fc$$

And that is a deep insight. For every universally quantified formula involving a predicate implies just such a conjunction, i.e. a conjunction each conjunct of which is a subject–predicate formula predicating the relevant property of an element of the domain and such that no element of the domain is left out.

Now, what is true of everything is true of any particular individual. Hence, it is undoubtedly valid to reason from any universally quantified formula such as $\forall x\ [Fx]$ to the conclusion that each and every individual instance of that formula is F, i.e. to infer that a is F and that b is F and, indeed, that c is F. Thus, we have identified one kind or pattern of valid reasoning with universally quantified formulas. A universal formula implies its instances. So, given a universally quantified formula, we can validly infer from that formula any of its instances. In the process, the universal quantifier is eliminated and so the variable in the matrix must be replaced with a proper name, e.g. we might instantiate $\forall x\,[Fx]$ to Fa. In QL this kind of reasoning exploits precisely the rule of inference known as **universal elimination**, or **UE**, for short. The rule UE takes only one line number, namely, the line number of the universally quantified formula as premise and takes as dependency-numbers all and only the dependency-numbers of that universally quantified formula. We can now represent each of our earlier inferences as formal proofs, e.g. we can prove formally that:

$$\forall x\,[Fx] \vdash Fa$$

{1}	1.	$\forall x\,[Fx]$	Premise
{1}	2.	Fa	1 UE

and that:

$$\forall x\,[Fx] \vdash Fb$$

{1}	1.	$\forall x\,[Fx]$	Premise
{1}	2.	Fb	1 UE

and again that:

$$\forall x\,[Fx] \vdash Fc$$

{1}	1.	$\forall x\,[Fx]$	Premise
{1}	2.	Fc	1 UE

Indeed, we can perfectly well make all three inferences in the course of a single proof and, given &-Introduction, we can formally prove that $\forall x\,[Fx]$ implies the conjunction: $((Fa \,\&\, Fb) \,\&\, Fc)$ as follows:

$$\forall x\,[Fx] \vdash ((Fa \,\&\, Fb) \,\&\, Fc)$$

{1}	1.	$\forall x\,[Fx]$	Premise
{1}	2.	Fa	1 UE
{1}	3.	Fb	1 UE
{1}	4.	Fc	1 UE
{1}	5.	$(Fa \,\&\, Fb)$	2,3 &I
{1}	6.	$((Fa \,\&\, Fb) \,\&\, Fc)$	4,5 &I

In terms of Fabworld, we have now exhausted the number of individuals to which we can attribute the property (remember: Fabworld has only three inhabitants). In general, however, this is an unrealistic feature of the example which sets an unnecessarily strict limit on the number and kind of thing we can talk about. We can use QL to talk about whatever we want and about as many such things as we like. Therefore, we can now set aside the severely restricted domain of Fabworld and consider a wholly unrestricted domain. In an unrestricted domain there is no limit to the possible size of the conjunction which the universally quantified formula abbreviates, i.e. no limit to the possible number of conjuncts. Therefore (omitting brackets for clarity), $\forall x\,[Fx]$ implies the possibly infinite conjunction:

$$Fa \mathbin{\&} Fb \mathbin{\&} Fc \mathbin{\&} \ldots$$

Unlike any finite conjunction, no such infinite conjunction could ever actually be asserted. None the less, it is of the essence of any universally quantified QL formula that it implies each and every single one of its instances. For precisely that reason, many older logic textbooks[2] describe universal elimination (quite properly) as 'universal instantiation', because the rule allows us to make explicit in proofs the fact that a universal formula implies all its instances. Here, we will retain the expression 'universal elimination' to exploit the familiar idea of an elimination-rule.

The validity of the pattern of reasoning identified here holds regardless of the kind of universal formula involved. So, for example, the universal conditional:

$$\forall x\,[Fx \rightarrow Gx]$$

implies each and every singular conditional (again, omitting brackets for clarity):

$$Fa \rightarrow Ga$$

and $Fb \rightarrow Gb$

and $Fc \rightarrow Gc$

and so on.

Equally, the universal conjunction:

$$\forall x\,[Fx \mathbin{\&} Gx]$$

implies each and every singular conjunction:

Fa & *Ga*

and *Fb* & *Gb*

and *Fc* & *Gc*

and so on.

Moreover, the same is true of each and every kind of universally quantified formula. You should now be able to make clear sense of the following informal rule-statement for UE:

> UE: Given any universal formula on any line of proof you may infer any particular instance of that formula on another line of proof. The new line should be annotated with the line number of the universal formula in question and 'UE'. The dependency-numbers of the new line are identical with those of the line of the original universal formula.

Further, with just a little new formal vocabulary, UE can be stated in fully formal terms, i.e. where ν is any variable, $\phi(\nu)$ is any expression in which ν may occur (i.e. predicate or relational expression), ρ is any proper name and $\phi(\rho)$ is the result of substituting ρ for ν, UE can be formally stated as follows:

$$\frac{\forall \nu\ [\phi(\nu)]}{\phi(\rho)}\ \text{UE}$$

The dependency-numbers of the formula inferred by UE are identical with those of the formula from which it was inferred.

Before we go on to consider some further examples, study the contents of Box 6.1 carefully.

Exercise 6.1 contains some simple examples for you to try on your own. When you do, note carefully that QL inherits not only all of the rules of inference belonging to PL but also all of the *strategies* for proof-construction in PL.

Unlike PL, QL involves quantified formulas and you will find that this has a bearing on proof-construction, particularly in contexts involving hypothetical reasoning. In such cases, it pays to consider carefully whether the best assumption to make is a quantified formula, such as $\forall x\ [Fx]$, or an instance of that formula, such as *Fa*. Hint: a little reflection on the conclusion you want to derive should help with this choice. Bear these points in mind when attempting proofs 7–10.

BOX 6.1

Universal Elimination: UE

♦ *UE informally:* Given any universal formula on any line of proof you may infer any particular instance of that formula on another line of proof. The new line should be annotated with the line number of the relevant universal formula and 'UE'. The dependency-numbers of the new line are identical with those of the line of the original universal formula.

♦ UE Formally: Where v is any variable, ϕ(v) is any expression in which v may occur, ρ is any proper name and ϕ(ρ) is the result of substituting ρ for v:

$$\frac{\forall v\ [\phi(v)]}{\phi(\rho)}\ \text{UE}$$

The dependency-numbers of the formula inferred by UE are identical with those of the formula from which it was inferred.

EXERCISE 6.1

1 Prove that the following are valid sequents of QL (the numbers in brackets beside each sequent indicate the number of lines in my proof of the sequent):

1. $\forall x\ [Fx] : Fa$ (2)
2. $\forall x\ [Fx] : (Fa\ \&\ Fb)\ \&\ (Fc\ \&\ Fd)$ (8)
3. $\forall x\ [Fx\ \&\ Gx] : (Ga\ \&\ Fa)$ (5)
4. $\forall x\ [Fx \rightarrow Gx], Fa : Ga$ (4)
5. $\forall x\ [Fx \rightarrow Gx], Fb : (Fb\ \&\ Gb)$ (5)
6. $\forall x\ [Fx \rightarrow Gx], \sim Gc : \sim Fc$ (4)
7. $\forall x\ [Fx \rightarrow Gx], \forall x\ [Gx \rightarrow Hx] : (Fa \rightarrow Ha)$ (8)
8. $\sim Fa : \sim\forall x\ [Fx]$ (5)
9. $\sim(Fa\ \&\ Fb) : \sim\forall x\ [Fx]$ (7)
10. $\forall x\ [Fx \rightarrow Gx] : \forall y\ [Fy] \rightarrow Gb$ (6)

II
Reasoning with the Universal Quantifier 2: The Rule UI

To date, the inferences we have considered have all involved inferring from a universal generalisation to one of its instances, i.e. an application of universal elimination. However, each quantifier has both an elimination-rule and an introduction-rule. In the present section, we must begin to consider the nature of the introduction-rule for the universal quantifier, i.e. **universal introduction**, or **UI**, for short. And here we should be wary. For while applications of UE are obviously and intuitively valid quite generally, there are some obviously invalid inferences which might be made using UI and we must take care to avoid them.

Many older texts refer to the introduction rule for the universal quantifier not as *universal introduction* but as 'universal generalisation'.[3] And that is entirely appropriate. For although we will stick with 'UI' here, we will use that rule precisely to generalise universally from a formula containing a name to a universally quantified formula. When we do, each occurrence of the name in the original formula must be replaced by one and the same variable in the new formula and, of course, a universal quantifier must be introduced to bind that variable. In this, then, UI is rather like UE in reverse. Now, we need have no real worries when applying UE. It is always of the essence of a universal generalisation that it should imply its instances. Hence, the rule UE was stated without any restriction. But there is a very real danger of going wrong when trying to generalise universally. For example, consider the following argument:

1. Joan is female.

Therefore,

2. Everyone is female.

Clearly, it certainly is not always valid to generalise universally on a formula involving a particular name! To ensure that UI is truth-preserving, then, we must restrict the application of that rule very carefully. And this we will certainly go on to do. However, a little reflection quickly reveals a class of inferences involving UI which are perfectly valid. In that context, applications of UI are all quite safe and, again, we need have no real worries. What is the relevant class of inferences? Well, consider Fabworld again. In Fabworld, everything is both fabulous and groovy and is only fabulous and groovy. Hence, in that restricted domain, if we infer from the premise 'Everything is fabulous' that the individual by the name of a is fabulous then we can surely validly infer from that consequence back to the original assertion 'Everything is fabulous.' In more formal terms, if we infer from $\forall x\ [Fx]$ that Fa then, again, it is perfectly valid to infer on that basis that $\forall x\ [Fx]$. After all, we know that everything in Fabworld is fabulous anyway and, in both cases, the conclusion that the individual a is fabulous was derived precisely from that universal premise. Therefore, given that an instance was derived from a universal generalisation in the first place, it is certainly valid to universalise generally from that instance.

Moreover, even in an unrestricted domain, generalising universally from a formula which was itself inferred from a universal generalisation in the first place is perfectly valid, i.e. in the real world as much as in Fabworld. Further, in practice, if we are at all anxious, we can readily identify when a formula has been inferred from a universal generalisation simply by checking the formula's dependency-numbers for numbers referring back to lines of proof containing universal generalisations.

Given this insight, the class of proofs involving obviously valid applications of the rule UI to which I alluded earlier can readily be identified. It is just that class of inferences which do nothing more than generalise universally from a formula which is itself an instance of a universal generalisation, i.e. straightforward inferences from universal generalisations to universal generalisations. In the context of any such proof, we know that applications of UI are safe. Therefore:

> Generalising universally on a formula which is itself an instance of a universal generalisation is perfectly valid. To apply UI in this way, write the line number of the formula in question together with 'UI' on the new line. The dependency-numbers of the new line are simply identical with the old line.

For example, consider the sequent $\forall x\ [Fx \ \& \ Gx] : \forall x\ [Fx] \ \& \ \forall x\ [Gx]$. The proof of this sequent proceeds by eliminating the universal quantifier involved in the premise and dismantling the resulting conjunction by &E. Further, because each conjunct has been inferred, and has only been inferred, from a premise which is itself a universal generalisation we can safely generalise universally on each instantiated conjunct. Finally, the proof is completed using &I. Here is the complete proof:

$$\forall x\ [Fx \ \& \ Gx] \vdash \forall x\ [Fx] \ \& \ \forall x\ [Gx]$$

{1}	1.	$\forall x\ [Fx \ \& \ Gx]$	Premise
{1}	2.	$Fa \ \& \ Ga$	1 UE
{1}	3.	Fa	2 &E
{1}	4.	$\forall x\ [Fx]$	3 UI

{1}	5.	Ga	2 &E
{1}	6.	$\forall x\,[Gx]$	5 UI
{1}	7.	$\forall x\,[Fx]\ \&\ \forall x\,[Gx]$	4,6 &I

Notice at once that each and every line of proof has an identical set of dependency-numbers, namely, the singleton set consisting only of 1. That number refers us back to line number one. The formula on that line is a universal generalisation. Therefore, each and every line of proof, including, crucially, lines 4 and 6 where we applied UI, depends on, and only depends on, that first universal generalisation. Therefore, applications of UI in this context are perfectly valid.

As a further example, consider the sequent $\forall x\,[Fx \rightarrow Gx], \forall x\,[Fx] : \forall x\,[Gx]$. This time, there are two premises to consider. But both are universal generalisations. Again, the proof proceeds by instantiating each premise and applying the rules of inference we inherit from PL; in this case MP. Finally, having derived the instantiated form of the conclusion, we complete the proof by applying UI to that formula, safe in the knowledge that the formula to which we apply UI depends only upon formulas which are themselves universal generalisations. Here is the complete proof:

$$\forall x\,[Fx \rightarrow Gx], \forall x\,[Fx] \vdash \forall x\,[Gx]$$

{1}	1.	$\forall x\,[Fx \rightarrow Gx]$	Premise
{2}	2.	$\forall x\,[Fx]$	Premise
{1}	3.	$Fa \rightarrow Ga$	1 UE
{2}	4.	Fa	2 UE
{1,2}	5.	Ga	3,4 MP
{1,2}	6.	$\forall x\,[Gx]$	5 UI

This time there is some variation in the sets of dependency-numbers belonging to each line just because we have two premises to work with. Thus, in the last two lines of proof both premises work together under an application of MP as we derive the instantiated form of the conclusion we desire. Finally, the proof is completed by applying UI to that formula. Again, note that its dependency-numbers refer back to formulas each of which is a universal generalisation. Hence, the application of UI is perfectly valid.

Confining the application of UI exclusively to proofs of this kind is less than satisfying and, in fact, such a restriction turns out to be far stronger than we actually require. Of course, we will have to impose some restriction upon the use of UI in order to prohibit unwarranted universal generalisations of the type considered above, e.g. inferring from the premise 'Joan is female' the conclusion 'Everyone is female.' To appreciate the nature of the required restriction, we have to think a little harder about the kind of inference involved in an application of UI.

UI enables us to generalise universally from saying something about some particular individual element of the domain to saying something about *every* element of the domain. In one sense, however, that is precisely the problem. When we generalise from the premise 'Joan is female' to the conclusion 'Everyone is female' we make a perfectly invalid inference in the process. Therefore, certain universal generalisations are invalid. It does not follow that every universal generalisation is invalid, e.g. universal generalisation from any instance which is itself derived from one or more universal generalisations is perfectly valid. Moreover, as we shall see, universal generalisation from an instance which does not itself depend upon a universal generalisation can also be perfectly valid. Indeed, in certain cases, it is perfectly valid to generalise universally from a single instance as itself a premise.

Here it may help to consider some examples. Suppose that we are in a mathematics class together and that the mathematics teacher wants to teach us something of the properties of lines. To that end, she marks two points on the board, calls them 'A' and 'B' and carefully draws a straight line between them. 'Note that the line on the board picks out the shortest possible path between the two marked points', the teacher says. 'Further', she continues, 'quite generally, a straight line is always the shortest possible path between two points.' But note carefully that what the teacher has said about the line is not only true of the line on the board. It is also true of every and any straight line. Hence, the teacher has quite legitimately generalised from a property of a particular instance of a straight line to a property of all straight lines.

Consider a further example. The western tradition of geometry has its roots in the work of a particular Ancient Greek geometer called 'Euclid'. In fact, Euclid's great text, the *Elements*, is the earliest example of a formal deductive system which we have in the western tradition. In the context of Euclidean geometry, reasoning of the kind we considered about the straight line is commonplace. For example, a geometer might say, 'Consider this particular triangle, call it "ABC".' He might then go on to make certain observations about the lines and angles constituting the triangle, perhaps demonstrating to us that this particular triangle has angles which sum to 180 degrees. But, now, the geometer might add, 'Look, there's nothing unique about this triangle. I could have picked any triangle, and could just as easily have demonstrated that its angles sum to 180 degrees. So, quite generally, a triangle is a three-sided, geometrical figure whose angles sum to 180 degrees.' Again, precisely what the geometer has done here is to generalise, perfectly legitimately, from a property of a particular instance of a triangle to a property of all triangles.

In these contexts, then, a universal conclusion can be validly inferred purely on the basis of a particular instance as premise. The question we have to ask is just: why is such reasoning valid in these cases? Consider carefully the particular line and the particular triangle in the examples given above. What was special about those particular instances as against the remaining members of their classes? The answer, surely, is absolutely nothing. In each case, don't we have the feeling that the particular instance employed was just employed as a perfectly *typical* example of its kind? Another way to put the point is to say that, in each case, there was no special reason for picking that particular instance rather than any other. Hence, we might say, the choice of each particular instance was quite *arbitrary*.

And now we have done enough to appreciate the kind of restriction we should impose upon UI. In short, we will restrict the application of UI to all and only those particular instances which are perfectly typical examples of their kind and which are genuinely arbitrarily chosen, in the sense that we know absolutely nothing special about that particular instance. In effect, all we should know about that individual is that it is an element of the domain and nothing else. For just that purpose, we could (and perhaps should) set aside a special set of symbols which are usually called *arbitrary names*. But if we are careful enough in our use of UI we can get away without any further complication in the vocabulary of QL. So, the remaining question is just how to apply the restriction we want to impose upon UI. In the practice of proof-construction, this can be perfectly straightforward. If we are tempted to generalise universally on a formula containing a particular name we can always look back at previous lines of proof, checking carefully that there is no line which contains a formula telling us something special about that named individual, e.g. that it has a particular property or properties. In other words, we must always take care to ensure that the particular name we use in such a context refers to an element of the domain which is genuinely typical in the context of that use. In effect, then, we try to give a kind of contextual definition, a definition in use, of arbitrariness for names.

The requisite restriction can now be set out more precisely. Remember that we want to prohibit the use of UI on names of individuals about which we know something special. Now, if the formula containing the name includes among its dependency-numbers any number referring back to a previous line which also has a formula containing that name then, at that earlier line, we do discover something about our named individual and so UI cannot be validly applied. And that is exactly the restriction we require:

> UI may not be applied to any formula containing a name which includes among its dependencies any formula which itself contains that name.

The informal rule-statement for UI can now be completed:

> UI: Given a formula containing a name on any line of proof you may replace each occurrence of that name with a variable, introduce the universal quantifier to that matrix and write the resulting formula on a new line provided that the original formula containing the name does not include among its dependencies any formula containing that name. Annotate the new line 'UI' together with the line number of the original line. The dependency-numbers of the new line are identical with those of the line of the original formula.

Given the new elements of formal vocabulary introduced in the previous section, UI can be stated in formal terms as follows. Where ρ is any proper name, $\phi(\rho)$ is any expression in which ρ may occur, ν is any variable and $\phi(\nu)$ is the result of substituting ν for ρ we can state UI formally as follows:

UI: $$\frac{\phi(\rho)}{\forall\nu\ [\phi(\nu)]}$$

Restriction: *provided that '$\phi(\rho)$' does not include among its dependencies any formula containing 'ρ'*

The dependency-numbers of the formula inferred by UI are identical with those of the formula from which it was inferred.

We can now proceed to test the restricted version of UI given here by trying to construct a proof of a sequent exemplifying the invalid inference we considered earlier, i.e. a sequent exemplifying the inference from the premise 'Joan is female' to the conclusion 'Everyone is female.' That inference is fairly represented by the following QL sequent:

$Fa : \forall x\ [Fx]$

Let's try to construct the proof:

{1}	1.	Fa	Premise
!!!!1!!!	2.	$\forall x\ [Fx]$	1 UI !!!!!!!!!!!

Clearly, UI cannot be applied at line 2 without breaking the restriction we have imposed. For the formula containing the name to which we want to apply UI has the dependency-number 1. That number refers us to line 1. But the formula on line 1 certainly does contain that very name. Therefore, UI cannot be applied and the restriction is seen to be effective. Moreover, the restriction is not only effective in just this instance. In fact, it will keep us right quite generally. For example, consider the following proof:

$$\forall x\,[Fx \rightarrow Gx] \vdash \forall x\,[Fx \rightarrow (Gx \text{ v } Hx)]$$

{1}	1.	$\forall x\,[Fx \rightarrow Gx]$	Premise
{2}	2.	Fa	Assumption
{1}	3.	$Fa \rightarrow Ga$	1 UE
{1,2}	4.	Ga	2,3 MP
{1,2}	5.	$Ga \text{ v } Ha$	4 vI
{1}	6.	$Fa \rightarrow (Ga \text{ v } Ha)$	2,5 CP
{1}	7.	$\forall x\,[Fx \rightarrow (Gx \text{ v } Hx)]$	6 UI

The important point is not so much that UI is validly applied at line 7 but rather that UI could not be validly applied any earlier than line 7. For example, suppose we tried to apply UI at line 3 to the formula on line 2. Just as in the previous case, UI cannot be applied here without breaking the restriction. Again, that formula depends upon itself. Hence, its dependencies certainly do include a formula which contains the name. Further, UI could not be applied either to the formula on line 4 or to the formula on line 5. Again, each of those formulas has a dependency-number which refers us to a line of proof with a formula containing the name in question, namely, line 2. Therefore, we must first apply CP at line 6 to discharge dependency-number 2 before we can legitimately apply UI, i.e. the application of CP enables the application of UI.

This brings us naturally to a final question about the use of UI. Can we legitimately apply UI to a formula whose set of dependency-numbers is wholly empty, i.e. to a formula which has no dependencies? The short answer is 'yes'. Any such formula is a theorem. The truth of a theorem is guaranteed by the nature and structure of the formula itself. As we put it in Chapter 2, the validity of a theorem follows from logic itself. Hence, we certainly do want to be able to universalise on these formulas, i.e. to assert that the theorem holds not just for some individual element of the domain but for every element of the domain. So, we can, indeed, legitimately apply UI to formulas whose sets of dependency-numbers are empty.

In Exercise 6.2, proof-construction generally begins with applications of UE in the early stages and closes with an application of UI at the last line of proof. This is a very common pattern of proof in QL. In between, if it's not immediately obvious how to proceed, strategy should be dictated by the golden rule. Hence, if the conclusion of the sequent is a universal conditional, assume the instantiated version of the antecedent and then try to derive the instantiated version of the consequent for CP before finally applying UI. If the conclusion of the sequent is not a universal conditional then ask: is any premise a universal disjunction or a disjunction of universal generalisations? If so, try to derive the conclusion you want by vE, i.e. assume the first disjunct and derive the conclusion from it. Next, repeat the process for the second disjunct and then apply vE. But take care at which stage you apply UI! As ever, when neither strategy applies, try RAA, i.e. assume the opposite of what you want and try to derive a contradiction from it before refuting it by reductio ad absurdum. Finally, while you may not need to have recourse to the golden rule in order to decide how to construct the first few proofs, you may well have to do so for later proofs.

Study the contents of Box 6.2 carefully before attempting Exercise 6.2.

BOX 6.2

Universal Introduction: UI

♦ *UI informally:* Given a formula containing a name on any line of proof you may replace each occurrence of that name with a variable, introduce the universal quantifier to that matrix and write the resulting formula on a new line *provided that the original formula containing the name does not include among its dependencies any formula containing that name.* Annotate the new line 'UI' together with the line number of the original line. The dependency-numbers of the new line are identical with those of the line of the original formula.

♦ *UI formally:* Where ρ is any proper name, ϕ(ρ) is any expression in which ρ may occur, ν is any variable and ϕ(ν) is the result of substituting ν for ρ:

UI: $$\frac{\phi(\rho)}{\forall \nu\ [\phi(\nu)]}$$

Restriction: *provided that* ϕ(ρ) *does not include among its dependencies any formula containing* ρ

The dependency-numbers of the formula inferred by UI are identical with those of the formula from which it was inferred.

EXERCISE 6.2

1 Prove that the following are valid sequents of QL (here and in subsequent exercises, the numbers in brackets beside each sequent indicate the number of lines in my proof of the sequent). Remember to check carefully each application you make of UI using the criterion provided in the preceding section!

1.	∀*x* [*Fx*] & ∀*y* [*Gy*] : ∀*z* [*Fz* & *Gz*]	(7)
2.	∀*x* [*Fx*] : ∀*x* [*Fx* v *Gx*]	(4)
3.	∀*x* [*Fx* → *Gx*] : ∀*x* [(*Fx* & *Hx*) → *Gx*]	(7)
4.	∀*x* [*Fx* → *Gx*], ∀*x* [*Gx* → *Hx*] : ∀*x* [*Fx* → *Hx*]	(9)
5.	∀*x* [*Fx* → *Gx*], ∀*x* [*Hx* → ~*Gx*] : ∀*x* [*Fx* → ~*Hx*]	(10)
6.	P → ∀*x* [*Fx*] : ∀*x* [P → *Fx*]	(6)
7.	∀*x* [*Fx*] v ∀*x* [*Gx*] : ∀*x* [*Fx* v *Gx*]	(10)
8.	∀*x* [*Fx* v *Gx*], ∀*x* [*Fx* → *Gx*] : ∀*x* [*Gx*]	(9)
9.	: ∀*x* [*Fx* → *Fx*]	(3)
10.	: ∀*x* [*Fx* v ~*Fx*]	(10)

III
Introducing the Existential Quantifier: The Rule EI

Having outlined and discussed both the introduction-rule and the elimination-rule for the universal quantifier in the proof-theory of QL we must now turn our attention to the rules of inference which govern the existential quantifier. To that end, we will first consider the introduction-rule for the existential quantifier; **existential introduction**, or **EI**, for short. EI is used to infer an existentially quantified formula from a formula which tells us something about some named individual. In general, when it is so used, each occurrence of the name in the original formula will be replaced with one and the same variable in the resulting formula.[4] Intuitively enough, the dependency-numbers of the new formula are identical with those of the old. Moreover, in sharp contrast to its universal counterpart, EI allows us to make inferences whose validity is perfectly obvious, quite generally. For just that reason, we will have no need to impose any sort of special restriction upon its use. For example, consider the simplest possible case of such an inference in natural language. Given 'Arlo Guthrie is a folk singer' I might go on to infer 'There exists at least one folk singer.' Such reasoning is naturally represented in QL by the following intuitively valid sequent:

$Fa : \exists x\ [Fx]$

Given EI, the validity of the sequent is quickly and easily demonstrated. The proof simply proceeds as follows:

		$Fa \vdash \exists x\ [Fx]$	
{1}	1.	Fa	Premise
{1}	2.	$\exists x\ [Fx]$	1 EI

Note that EI takes only one line number, the line number of the formula to which the rule was applied. Further, like UI, EI also exemplifies a process of generalisation. We start off with a formula about a particular named individual, i.e. a formula which tells us something about one element of the domain, and go on to infer a formula which tells us something about the domain itself, namely, that it contains some individual with that particular property. Hence, we use EI to derive a general conclusion from a singular premise. For just that reason certain older logic texts refer to EI as the rule of 'existential generalisation'. Again, that is entirely appropriate. We will stick with the name 'EI' here but, either way, it should be clear that this kind of generalisation is perfectly valid. In a sense, the conclusion we derive is always more modest than the premise it is derived from, i.e. the premise informs us that a given property belongs to a particular named individual while the conclusion tells us only that some element of the domain has that property. Thus, using EI, we move from a precise state of information to what Quine calls the 'studied ambiguity' of an existential generalisation. In this respect, existentially quantified formulas are almost secretive; they tell us something about some element of the domain without giving away which element is involved. Finally, you might also like to think of EI as being rather like UE in reverse, for while UE exemplifies the principle that what is true of everything is true of any particular thing, EI exemplifies the principle that what is true of some particular thing is, therefore, true of something. Informally then:

> EI: Given a formula containing a name on any line of proof you may replace one or all occurrences of that name with a variable. Introduce the existential quantifier to the resulting matrix and write the formula on a new line.

Annotate the new line 'EI' together with the line number of the original line. The dependency-numbers of the new line are identical with those of the line of the original formula.

Further, where ρ is any proper name, ϕ(ρ) is any expression in which ρ may occur, ν is any variable and ϕ(ν) is the result of substituting ν for ρ EI is formally stated as follows:

$$\text{EI} \qquad \frac{\phi(\rho)}{\exists\nu\,[\phi(\nu)]}$$

The dependency-numbers of the formula inferred by EI are identical with those of the formula from which it was inferred.

As you will discover, EI works well together with the existing stock of rules of inference. Moreover, its addition to the proof-theory of QL finally allows us to derive existentially quantified formulas as conclusions for the first time. Given that we have not yet considered the elimination-rule for the existential quantifier, we cannot yet derive an existentially quantified formula from an existentially quantified formula. None the less, the proofs of a great number of sequents do not require such inferences and EI can be very useful, for example, when we want to derive an existentially quantified conclusion from universally quantified premises, from unquantified premises or, indeed, from a mixture of both. For example, consider the sequent:

$\forall x\,[Fx \rightarrow Gx], Fa : \exists x\,[Gx]$

In order to derive the existentially quantified conclusion we must use EI. Given that rule, however, the proof is perfectly straightforward. Here is the completed proof:

$\forall x\,[Fx \rightarrow Gx], Fa \vdash \exists x\,[Gx]$

{1}	1.	$\forall x\,[Fx \rightarrow Gx]$	Premise
{2}	2.	Fa	Premise
{1}	3.	$Fa \rightarrow Ga$	1 UE
{1,2}	4.	Ga	2,3 MP
{1,2}	5.	$\exists x\,[Gx]$	4 EI

The usefulness of EI extends well beyond the context of proofs of this kind. For example, EI can be very useful in constructing proofs of certain of the quantifier equivalences which we noted in the previous chapter. Consider the following intuitively valid sequent:

$\sim\exists x\ [\sim Fx] : \forall x\ [Fx]$

Note carefully how EI is exploited in the construction of the proof of this sequent:

		$\sim\exists x\ [\sim Fx] \vdash \forall x\ [Fx]$	
{1}	1.	$\sim\exists x\ [\sim Fx]$	Premise
{2}	2.	$\sim Fa$	Assumption for RAA
{2}	3.	$\exists x\ [\sim Fx]$	2 EI
{1,2}	4.	$\sim\exists x\ [\sim Fx]\ \&\ \exists x\ [\sim Fx]$	1,3 &I
{1}	5.	$\sim\sim Fa$	2,4 RAA
{1}	6.	Fa	5 DNE
{1}	7.	$\forall x\ [Fx]$	6 UI

In the absence of any obvious clue to the overall strategy for proof-construction, we proceed by RAA. So, we must first assume the opposite of what we want. Now, it may seem natural simply to assume $\sim\forall x\ [Fx]$ here. But if we could derive an instantiated formula whose dependencies did not include any formula containing the relevant name then we could use UI to derive the universally quantified conclusion anyway. So, *Fa* might appear to be the obvious candidate. However, in order to apply RAA we must also derive a contradiction. If we assume $\sim Fa$ we can then exploit EI in order to derive a formula which contradicts the premise. Thus, we can go on to discharge the dependency-number of the assumption in a way which subsequently facilitates the desired application of UI, i.e. to complete the proof.

Finally, note that EI allows us to target particular occurrences of a name within a formula, i.e. we are not obliged to replace every occurrence of one and the same name with a variable. In effect, we can select which occurrence of the name we want to generalise upon. This should be fairly intuitive. After all, we are concerned with existential generalisation here, there are no special restrictions upon sets of dependencies, and so, while we certainly can generalise over each and every occurrence of a name in a formula, it is open to us simply to generalise on a single occurrence. The next two proofs illustrate the freedom of application which we enjoy around (and only around) EI:

1.	$\vdash \exists x\,[Fx \rightarrow Fx]$		
{1}	1.	Fa	Assumption for CP
---	2.	$Fa \rightarrow Fa$	1,1 CP
---	3.	$\exists x\,[Fx \rightarrow Fx]$	2 EI

1.′	$\vdash \exists x\,[Fx \rightarrow Fa]$		
{1}	1.	Fa	Assumption for CP
---	2.	$Fa \rightarrow Fa$	1,1 CP
---	3.	$\exists x\,[Fx \rightarrow Fa]$	2 EI

The proof of 1 is the familiar case. When we apply EI to the formula on line 2 we replace both occurrences of the name *a* with the variable *x*. However, the proof of 1′ reveals that we are not required to replace both occurrences of *a* and in constructing that proof we generalise only on the first occurrence of the name. Later, we will consider other contexts in which we might want to exploit this freedom. For the moment, it is sufficient to appreciate that such moves are legitimate. As ever, we let the requirements of proof-construction in any particular case determine when and whether to exploit that freedom.

Surprisingly perhaps, although EI exemplifies an obviously valid form of inference and, in itself, as it were, is wholly uncontroversial, the addition of the rule to the deductive apparatus of QL does have some implications which are very controversial indeed. In the next section I take some time to outline briefly and discuss the philosophical issues arising therefrom. Those readers who are concerned only to enable themselves to exploit the mechanics of QL can safely proceed by ignoring Section IV completely and jumping ahead to Section V after attempting Exercise 6.3.

Study the contents of Box 6.3 carefully before attempting Exercise 6.3.

Exercise 6.3 is intended both to provide you with some practice with the rule EI and to let you see how well that rule works together with our existing stock of rules of inference. You should find the first few proofs perfectly straightforward, but the examples are given in ascending order of difficulty and proof-construction for later cases may require a little thought. As ever, keep the applicability of the golden rule in mind.

BOX 6.3

Existential introduction: EI

- *EI informally:* Given a formula containing a name on any line of proof you may replace one or all occurrences of that name with a variable. Introduce the existential quantifier to the resulting matrix and write the formula on a new line. Annotate the new line 'EI' together with the line number of the original line. The dependency-numbers of the new line are identical with those of the line of the original formula.

- *EI formally:* Where ρ is any proper name, $\phi(\rho)$ is any expression in which ρ may occur, v is any variable and $\phi(v)$ is the result of substituting v for ρ:

$$\text{EI} \qquad \frac{\phi(\rho)}{\exists v\ [\phi(v)]}$$

The dependency-numbers of the formula inferred by EI are identical with those of the formula from which it was inferred.

EXERCISE 6.3

1 Prove that the following are valid sequents of QL:

1. $(Fa \ \&\ Ga) : \exists x\ [Fx \ \&\ Gx]$ (2)
2. $(Fa \ \&\ Ga) : \exists x\ [Fx] \ \&\ \exists x\ [Gx]$ (6)
3. $\forall x\ [Fx] : \exists x\ [Fx]$ (3)
4. $\forall x\ [Fx \ \&\ Gx] : \exists x\ [Fx] \ \&\ \exists x\ [Gx]$ (7)
5. $\forall x\ [Fx] : \exists x\ [Fx \text{ v } Gx]$ (4)
6. $\forall x\ [Fx \rightarrow (Gx \rightarrow Hx)], (Fa \ \&\ Ga) : \exists x\ [Hx]$ (8)
7. $\forall x\ [(Fx \ \&\ {\sim}Gx) \rightarrow Hx)], (Fa \ \&\ {\sim}Ga) : \exists x\ [Fx \ \&\ Hx]$ (7)
8. $(Fa \ \&\ Ga), \forall x\ [Hx \rightarrow {\sim}Gx] : \exists x\ [Fx \ \&\ {\sim}Hx]$ (9)
9. $\exists x\ [Fx] \rightarrow \text{P} : \forall x\ [Fx \rightarrow \text{P}]$ (6)
10. $\forall y\ [Gy \rightarrow Hy] : \exists x\ [Gx] \rightarrow \exists y\ [Hy]$ (7)

IV
A Brief Note on Free Logic

While EI exemplifies an obviously valid form of inference and is, as such, wholly uncontroversial, the addition of the rule to the deductive apparatus of QL has some implications which are very controversial indeed. These are nicely illustrated in the proof of the following sequent (which you should have proved yourself in Exercise 6.3), though they are by no means confined to that particular sequent. Consider the proof carefully:

$\forall x\ [Fx] \vdash \exists x\ [Fx]$

{1}	1.	$\forall x\ [Fx]$	Premise
{1}	2.	Fa	1 UE
{1}	3.	$\exists x\ [Fx]$	2 EI

The proof is quite straightforward and so, you may wonder, what exactly is the problem here? Recall our discussion of the nature and consequences of universal generalisations in the early sections of Chapter 5. There we agreed that sentences of the form: 'All *A*s are *B*s' are properly understood as universal conditionals. Moreover, we stipulated that sentences of that form should be translated into QL in that way precisely in order to avoid making ontological commitments. For example, recall that the sentence 'All trespassers will be shot' does not imply that there actually are any trespassers, i.e. that there exist any trespassers. Therefore, we resolved to make ontological commitments by using positive existentially quantified formulas and only by using positive existentially quantified formulas. But now consider the preceding proof. It seems to spoil the game completely here. For a universally quantified formula has turned out to imply a positive existentially quantified formula!

Worse still, the validity of this particular sequent is wholly sanctioned by the set of rules of inference for QL. In other words, logic itself seems to entail that something exists. But surely we can live with this rather odd consequence. After all, isn't it true that things exist? Certainly it is. However, it might well be the job of the natural sciences, for example, to inform us that things exist, but formal logic is not physics and it can hardly be the job of formal logic to tell us that things exist. Further, while natural science informs us both that things exist and, indeed, just what kinds of thing exist, formal logic seems only to tell us that something exists. To see the point more clearly, recall that we are currently considering the proof-theory of

QL. As such, we are considering syntax rather than semantics. Earlier, we agreed that syntax concerns the shape of formulas in QL considered as wholly uninterpreted. But that is the problem. The proof seems to inform us that there exists something which has the property *F*. But, strictly speaking, that predicate-letter is, so far, uninterpreted. So, we seem to know that something exists about which we know absolutely nothing other than that it exists! Curiouser and curiouser.

The problem we have identified here is not peculiar to the particular sequent considered above. Ontological commitments can also be generated in other ways. And, in fact, I generated just such a commitment above in this text when I proved that:

$\vdash \exists x\ [Fx \rightarrow Fx]$

In other words, I proved as a theorem that there exists something such that if it has the property *F* then it has the property *F*. Further, recall the last sequent you were invited to prove in Exercise 6.2 above, i.e. the sequent:

$: \forall x\ [Fx \text{ v} \sim Fx]$

The proof of this sequent is constructed in a way which should be familiar from PL, i.e. by means of the following double reductio:

		$\vdash \forall x\ [Fx \text{ v} \sim Fx]$	
{1}	1.	$\sim(Fa \text{ v} \sim Fa)$	Assumption
{2}	2.	Fa	Assumption
{2}	3.	$Fa \text{ v} \sim Fa$	2 vI
{1,2}	4.	$\sim(Fa \text{ v} \sim Fa)\ \&\ (Fa \text{ v} \sim Fa)$	1,3 &I
{1}	5.	$\sim Fa$	2,4 RAA
{1}	6.	$Fa \text{ v} \sim Fa$	5 vI
{1}	7.	$\sim(Fa \text{ v} \sim Fa)\ \&\ (Fa \text{ v} \sim Fa)$	1,6 &I
---	8.	$\sim\sim(Fa \text{ v} \sim Fa)$	1,7 RAA
---	9.	$Fa \text{ v} \sim Fa$	8 DNE
---	10.	$\forall x\ [Fx \text{ v} \sim Fx]$	9 UI

But look closely at lines 9 and 10. What is there to stop us applying EI rather than UI to the formula on line 9? Absolutely nothing. So, in a similar manner, we could just as easily have proved that

$\vdash \exists x\ [Fx \text{ v} \sim Fx]$

But, again, in so doing we would have generated another existential commitment; this time, to the existence of something which either has or lacks the property *F*. And again, in all this, we have at no point specified just which property the predicate-letter is supposed to stand for. Curiouser and curiouser, indeed. Moreover, certain critics of classical QL make another criticism at this point in the debate. This is not so much that QL itself generates ontological commitments but rather that QL provides no criterion for distinguishing between names which genuinely denote existent objects and so-called vacuous names, such as 'Santa Claus', which do not denote existent objects or, perhaps, denote non-existent objects. Hence, this last point brings us straight back to the problem of Plato's beard which we considered at the end of Chapter 5. How are we to respond to these, apparently quite grave, problems?

Well, these are properly philosophical problems which have provoked a wide range of responses from philosophers and logicians over the years. The present context precludes a full discussion of those responses and although I will state my own view of the matter here, that view is no doubt an idiosyncratic one. In the last analysis, we should all come to our own judgement on any such philosophical question. To that end, interested parties would be well advised to consult Chapter 5 of Stephen Read's *Thinking about Logic* for a sustained discussion of the complex issues involved here.[5] Read's discussion is both accessible and penetrating and is undoubtedly the place for interested parties to begin any serious investigation. Note also that the chapter ends with a list of suggested further reading on the topic. Having said all that, my own view is that the present problem is not, in the last analysis, a grave one and does not point up any great flaw in classical formal logic. But now I must try to make my case for that conclusion.

First, it is agreed on all hands that we should not look to formal logic itself in order to determine what our ontological commitments are. Second, such commitments vary from culture to culture and, indeed, from person to person. Third, we noted in the last chapter that truth and falsity in QL are relativised to particular interpretations of that language. Now, it is here, I argue, that questions about ontological commitments really arise.

Given an interpretation which commits us to the existence of a domain consisting of certain elements, say, the domain of human beings, what are we to make of the three proofs considered above? Well, suppose that, under that interpretation, *F* stands for ' . . . is a folk singer'. But now the proofs look harmless. First, if everyone is a folk singer (heaven forbid!) it follows that any given individual is a folk singer. But what of the two theorems? Again, given the interpretation, we can truly say that there is some individual such that if that person is a folk singer then that person is a folk singer. Equally, can't we truly say of any person that that person either is a folk singer or isn't a folk singer? Quite generally, then, when the domain is non-empty the proofs in question are not counterintuitive. Therefore, we should be careful not to put the cart before the horse here, i.e. given a choice of domain and an associated ontology we can certainly use formal logic to investigate what can validly be inferred, but it is surely a mistake to consider formal logic purely *in abstracto* and then to try to work out ontological commitments on that basis.

In order to get the proofs to bite, then, it looks as if we need to accept the special assumption that the domain is, in fact, empty. Given that assumption, the two theorems will seem to be invalid. After all, if there is nothing then it will not follow that there exists something with a given property or something to which the law of excluded middle applies. But, again, ontological commitments must be decided in advance, before we have any basis on which to worry about the existential import of the formal machinery. And, after all, as we agreed earlier, surely things do exist. So, why should we choose an empty domain for formal logical purposes?

However, simply taking care about the choice of domain will not solve all our problems here. For recall the final objection we considered above. Suppose that the domain is non-empty but that, under our interpretation, we have a set of names some of which refer to existent objects and some of which are vacuous. QL itself provides no criterion to distinguish between the two so how are we to avoid making invalid inferences, e.g. inferring from 'Everything is *F*' that '*a* is *F*' when *a* might turn out to be a vacuous name?

There are a number of things to be said here. But first note that the situation described again presupposes an interpretation. Further, it presupposes that under that interpretation some names refer to existent objects while some either fail to refer or refer to non-existent objects. But note that nothing, formal logic included, forces us to construct such an interpretation. Suppose, for example, that we want to construct an interpretation about horses. Further suppose that we want to formalise sentences involving the names: 'Shergar', 'Red Rum' and 'Pegasus'. Must we choose as names, say, *a*, *b* and *c* here? The short answer is 'no'. We are quite free to choose the names *a* and *b* and then to choose predicates such as ' . . . is a horse', ' . . . is winged' and ' . . . belongs to Bellerophon'. Following Russell and Quine, we could then treat sentences involving the apparent name 'Pegasus' as sentences which really involve a definite description, say, 'The winged horse belonging to Bellerophon'. Nothing in logic itself forbids us from Quining away any apparent proper name which we believe lacks a bearer. Hence, cultural and personal beliefs rather than formal logic do shape and should shape our ontological commitments. But suppose that we do want to recognise interpretations under which certain names are vacuous. Well, note that this could mean *either* that certain names do not refer to anything at all *or* that certain names somehow refer to things which are non-existent! Quite what is meant by the latter option is not at all clear to me, but note that either way all the proofs considered earlier really are worrying under either interpretation. After all, how are we to know for any given name whether or not it does refer to an actual existent?

If the name does not refer to an actual existent then the provably valid QL sequents outlined do look plainly invalid. Just these considerations have prompted a number of formal logicians to propose revising Classical QL at this point. The family of formal logical responses to this situation has come to be known as **free logic**, i.e. logic which is free of the kind of ontological implications which QL seems to force on us here. The key revision which the free logicians advocate is that we should build in to the rules of inference for each quantifier the further restriction that the rule can only be applied when, in addition to satisfaction of the usual requirements, we also have a formula on a line of proof which assures us of the actual existence of the referent of the name in question. Thus, quantification is restricted to the class of actually existent entities.

It follows that we now need a new expression to indicate the existence of the thing referred to by the name. Further, the rules EI and UI must now take two line numbers, as we require in addition to a formula of the familiar kind containing a name another formula which assures us of the existence of the thing referred to by that particular name. A little reflection reveals that all three of the sequents considered here cannot be proved in such a formal system unless and until those extra lines can be cited to enable applications of EI. And so our problems are solved. But are they?

I would argue that, in a deep sense, the introduction of a new expression which assures us of actual existence is really no help at all. After all, formal logic itself cannot tell us which particular names to apply that expression to. At the end of the day, then, ontological commitments must still be settled by personal and cultural beliefs and not formal logic. Further, if we consider any apparent name not to denote anything then, when we set up our interpretation, we are quite free to indulge in a spot of Quining and to make our ontological commitments clear in that way. The point is not that we require recourse to free logic only in a certain, very particular, class of cases but rather that we are never forced to have recourse to free logic at all. Moreover, the fact that free logicians tend to motivate their programme on syntactic grounds, i.e. precisely when the focus is not on how we might best construct a particular interpretation, leads me at least to suspect that the problem reflects something about certain logicians and the way in which they pursue their formal investigations rather than any great flaw in classical formal logic itself.

As noted, this view may well be idiosyncratic, and interested parties really would be well advised to consult Chapter 5 of Stephen Read's *Thinking about Logic* for what is certainly a much fuller discussion and, perhaps, a fairer hearing of the case. However, although you may well choose to disagree with my approach, note that it does have the merit of simplifying the rules of inference governing the quantifiers. In the last analysis, we will not complicate those rules via any added stipulations, and so we can proceed with the rule EI as defined in the previous section in all of its full unrestricted glory. Exercise 6.4 provides the opportunity for some further practice with our rather 'un-free' version of the rule EI. Note very carefully sequents 4–7 here which exploit negation to express *quantifier-equivalences*. Sequents 4 and 5 are particularly important for, as you will see, these are integral to the formal methods we exploit in the next and final chapter.

EXERCISE 6.4

1 Prove that the following are valid sequents of QL:

1.	$: \forall x\,[Fx] \rightarrow \exists x\,[Fx]$	(4)
2.	$\forall x\,[\mathrm{P}\ \&\ Fx] : \mathrm{P}\ \&\ \exists x\,[Fx]$	(6)
3.	$\sim\exists x\,[Fx] : \sim Fa$	(5)
4.	$\sim\exists x\,[Fx] : \forall x\,[\sim Fx]$	(6)
5.	$\sim\forall x\,[Fx] : \exists x\,[\sim Fx]$	(11)
6.	$\sim\exists x\,[\sim Fx] : \forall x\,[Fx]$	(7)
7.	$\sim\forall x\,[\sim Fx] : \exists x\,[Fx]$	(10)
8.	$: \exists x\,[Fx \vee \sim Fx]$	(10)
9.	$\exists x\,[Fx] \rightarrow \forall x\,[Gx]$, $\forall x\,[\sim Hx \rightarrow \sim Gx] : \exists x\,[Fx] \rightarrow \exists x\,[Hx]$	(11)
10.	$\forall x\,[Fx \leftrightarrow Gx] : \exists x\,[Fx] \leftrightarrow \exists x\,[Gx]$	(16)

V
Eliminating the Existential Quantifier: The Rule EE

We turn now to the last of the rules of inference governing the quantifiers in QL, namely, the elimination-rule for the existential quantifier, **existential elimination**, or **EE**, for short. Naturally enough, EE allows us to eliminate the existential quantifier from a formula on a line of proof and, in so doing, allows us to derive another formula from the original existentially quantified formula. Hence, the elimination-rule for the existential quantifier governs the nature of those inferences we can make from an existentially quantified formula as such. Again, we should be on our guard here. For, as we shall see, it is remarkably easy to make perfectly invalid inferences from existentially quantified formulas. Hence, EE must be formulated in a way which allows us to pick out just the class of valid inferences which can be made from existentially quantified formulas. The questions facing us are: when can we validly derive a conclusion from an existentially quantified formula? And, moreover, what kinds of conclusion can be inferred on that basis?

To answer these questions we need to think a little harder about the meaning of the existential quantifier. What does an existentially quantified formula imply? Recall Fabworld. Fabworld, we said, is close to Heaven because everything in Fabworld is absolutely fabulous. Further, we agreed, everything in Fabworld is also fantastically groovy. As you know, at the last census the population of Fabworld stood at a mere three people. Earlier, we used the proper names *a*, *b* and *c* from QL to stand for each of the three inhabitants of Fabworld respectively. For present purposes, then, we can again restrict the domain to the population of planet Fabworld. So, what do existentially quantified formulas imply in the context of Fabworld? Consider the simplest possible kind of existentially quantified formula, $\exists x\ [Fx]$, for example. The domain of quantification is Fabworld and so it is again natural to interpret the predicate-letter *F* as standing for the property of being fabulous. Under the present interpretation, then, what does $\exists x\ [Fx]$ imply? Intuitively, this formula surely implies 'Something in Fabworld is fabulous.' Now, that is quite correct. But, as ever, while we are told that something in Fabworld is fabulous we are not being told which particular thing that is (existentially quantified formulas can indeed be rather secretive). But Fabworld has only three inhabitants and we know their names. So, when we are told that some individual is *F* we surely know that either it is *a* that is *F* or that it is *b* that is *F* or that it is *c* that is *F*. Therefore, we can think of the existentially quantified formula as a sort of abbreviated disjunction. Waiving the rules about brackets for clarity, we can now cash out just what is meant by $\exists x\ [Fx]$ under the present interpretation in terms of the disjunction:

$$Fa \vee Fb \vee Fc$$

Again, this is a deep insight. For every existentially quantified formula involving a predicate does, indeed, imply such a disjunction, i.e. a disjunction each disjunct of which is a subject–predicate formula predicating the relevant property of an element of the domain and such that no element of the domain is left out.

Given that Fabworld has only three inhabitants, we have exhausted the number of individuals to which the property can be attributed. As ever, the size of the population of Fabworld is rather unrealistic compared to the real world. But we can use QL to talk about whatever world we want and about as many people in that world as we like. So, we can now consider a wholly unrestricted domain. Given an unrestricted domain, there is no limit to the possible size of the disjunction which the existentially quantified formula implies, i.e. no limit to the possible number of disjuncts. Therefore, (omitting brackets for clarity) $\exists x [Fx]$ implies the possibly infinite disjunction:

$$Fa \vee Fb \vee Fc \vee Fd \vee \ldots$$

Moreover, quite generally, any existentially quantified QL formula can be taken to imply just such a possibly infinite disjunction, even though no such infinite disjunction could actually be asserted by anyone!

Having unpacked the content of existentially quantified QL formulas in this way, we should go on to ask what follows about the nature of valid inference from such formulas. Well, we unpacked existentially quantified formulas precisely in terms of disjunction. And we have already considered the nature of valid inference from disjunctive formulas in the context of PL. There we noted that formal logicians are rather mean-spirited. For when we asked: 'Under which circumstances can we validly infer a conclusion from a disjunction?' the answer came back: 'If you can derive that conclusion from each disjunct then and only then can you legitimately infer the conclusion from the disjunction alone.' Just that principle of inference is exemplified in vE and, in a sense, the same principle is at the heart of EE. Here, however, the disjunctions we are considering are possibly infinite. So, we cannot impose exactly the same demand that the conclusion be derived from each and every disjunct. For although we could certainly start the job we certainly would never be able to finish it.

It follows that we must impose a requirement of a different kind here, and that we do. Imagine asking the formal logician: 'When can I validly infer a conclusion from a formula which implies a possibly infinite disjunction?' This time, the answer is: 'If you can pick, quite arbitrarily, an element of the domain which is a perfectly typical case just in the sense that you know nothing about that particular individual other than that it is an element of the domain, and you can derive the desired conclusion from a formula about that *typical disjunct* (TD), then you may infer that same conclusion from the original formula.' This gives us a very clear picture of the pattern of inference which any application of the rule EE encodes. And note that EE is simpler to use than vE. After all, vE involved five identifiable steps each of which had to be recorded whenever that rule was applied. In contrast, the pattern we discern when considering inferences from existentially quantified formulas has only three central elements.

First, we have the *original existential formula* (if we did not, we should not be tempted to apply EE) which, let's say, involves a predicate or predicates. Second, we must carefully choose a *typical disjunct* involving the same predicate, ensuring that, in the context in question, that disjunct really is genuinely typical. Finally, we can exploit the existing stock of rules of inference in order to derive the *desired conclusion* from that typical disjunct. Then and only then can EE be applied. Therefore, any line annotated 'EE' should also be annotated with three numbers:

1. The line number of the original existential formula in question.
2. The line number of the typical disjunct.
3. The line number of the conclusion derived from that typical disjunct.

Again, rather as with vE, at the line at which we apply the rule, EE enables us to discharge the dependency-number of the typical disjunct in favour of the dependency-number of the original existential formula. This reflects the fact that if the conclusion has been shown to follow from the typical disjunct then it does indeed follow from the original existential formula. Hence, like vE, EE is a discharge rule. Moreover, like vE, EE is not just a rule of inference. It is also a strategy for proof-construction. Further, and again as with vE, the real clue about when to apply the strategy for EE is the presence of an existentially quantified formula not in the conclusion but among the premises. Remember: we might well be able to construct an existentially quantified conclusion simply by using EI. However, if the set of premises you are faced with contains an existentially quantified formula it is much more likely that the strategy for proof-construction will at some point include an application of EE. Thus, EE does indeed govern just those inferences we make from existentially quantified formulas as such.

With all these points in mind, we can at last consider a case of EE in action. For example, consider the following sequent:

$\exists x\,[Fx\ \&\ Gx] : \exists x\,[Gx\ \&\ Fx]$

Here, the premise is indeed an existentially quantified formula. Moreover, the conclusion is also an existentially quantified formula. But that latter fact is a red herring just because existential generalisations can be inferred by EI (recall that Exercise 6.3 contained no fewer than seven such examples). The real clue then is the presence of the existentially quantified formula as premise. Thus, we must face up to an application of EE. So, our strategy will be first to assume a typical disjunct to reason from. Next, we will exploit the existing stock of rules in order to derive the desired conclusion. Finally, we will apply EE, cite the three line numbers and discharge the dependency-number of the typical disjunct in favour of the dependency-number of the original existentially quantified premise.

Let's construct the proof together carefully. First, set out the premise:

{1}	1.	$\exists x\ [Fx\ \&\ Gx]$	Premise

Next, carefully choose a typical disjunct. Here, the choice is quite straightforward as, so far, no names whatsoever have appeared in the proof. Hence, we can simply and safely pick *a* and attribute to it the same predicates that we find in the original existentially quantified premise, i.e. line 2 looks like this:

{2}	2.	$Fa\ \&\ Ga$	Assumption TD

So far so good. But now we must derive the desired conclusion from the typical disjunct. Here, that is perfectly straightforward. We simply use &E to break down the conjunction on line 2, use &I to get the conjuncts the right way round and, finally, apply EI to infer the desired conclusion. So, lines 3 to 6 look like this:

{2}	3.	Fa	2 &E
{2}	4.	Ga	2 &E
{2}	5.	$Ga\ \&\ Fa$	3,4 &I
{2}	6.	$\exists x\ [Gx\ \&\ Fx]$	5 EI

To complete the proof it only remains to apply EE. So, let's look carefully for the numbers we need to state the rule annotation properly. First, we need the line number of the original existential formula we are concerned with, i.e. the premise on line 1. So, the first number is just 1. Next, we need the line number of the typical disjunct. Again, this is obvious; we marked the typical disjunct 'TD' when we assumed it on line 2 (note that although this is not strictly required it is a useful practice for keeping track of what is where as the proof grows). Hence, the second number is 2. Finally, we need the line number of the conclusion derived from that typical disjunct. Well, we derived the conclusion on line 6 (note that the dependency-numbers belonging to the formula on that line assure us that we did, indeed, derive that formula from the typical disjunct). Thus, the final number is 6. So, on this occasion, the three numbers required for EE are 1,2 and 6. Therefore, we can now complete the proof as follows:

{1}	7.	$\exists x\ [Gx\ \&\ Fx]$	1,2,6 EE

To sum up, consider the whole as a unity:

$\exists x\ [Fx\ \&\ Gx] \vdash \exists x\ [Gx\ \&\ Fx]$

{1}	1.	$\exists x\ [Fx\ \&\ Gx]$	Premise
{2}	2.	$Fa\ \&\ Ga$	Assumption TD
{2}	3.	Fa	2 &E
{2}	4.	Ga	2 &E
{2}	5.	$Ga\ \&\ Fa$	3,4 &I
{2}	6.	$\exists x\ [Gx\ \&\ Fx]$	5 EI
{1}	7.	$\exists x\ [Gx\ \&\ Fx]$	1,2,6 EE

Note carefully that immediately preceding the application of EE we have an application of EI. This is perfectly natural. After all, the conclusion we want to derive is itself an existentially quantified formula and, as the principle of existential generalisation, EI allows us to infer existentially quantified formulas. Moreover, this pattern of proof-construction is a common one which exemplifies a useful rule of thumb: '*EI before EE*' (the prescription is supposed to be reminiscent of another rule of thumb, one for spelling, which prescribes 'I before E except after C'). Note that, rather like the spelling rule, 'EI before EE' has its share of exceptions. None the less, like its counterpart, it remains a rather useful rule of thumb, as we shall see.

We can now make an attempt at stating the rule EE. Later we shall have to tighten the rule-statement up in a way which will ensure that we will always apply EE validly. But, for present purposes, it is useful to have a summary of just what we've managed to carve out to date. So, let's call this proto-rule 'EE. Part One':

> EE. Part One: You may infer a conclusion from an existentially quantified formula if you first assume a *genuinely typical disjunct* and then derive the desired conclusion from that disjunct. Restate the conclusion on a new line of proof. Annotate the new line 'EE' together with three numbers: (i) the line number of the original existential formula; (ii) the line number of the typical disjunct; (iii) the line number of the conclusion derived from that typical disjunct. The dependency-numbers of the new line consist of all the dependencies belonging to the derived conclusion except that you may discharge the dependency-number of the typical disjunct and replace it with the dependency-number of the original existentially quantified formula.

While EE is obviously the most complex of the rules of inference governing the quantifiers, it certainly can be quite straightforward to apply and, with a little practice, you should quickly find yourself at home with it as a strategy for proof-construction (but do remember that the real clue as to when to apply EE is the presence of an existentially quantified formula as premise and not as conclusion). So, for example, when, in Section III, we considered the sequent:

$\forall x\,[Fx \rightarrow Gx], Fa : \exists x\,[Gx]$

we were perfectly able to construct a proof of the sequent without recourse to EE. Recall the proof:

$\forall x\,[Fx \rightarrow Gx], Fa \vdash \exists x\,[Gx]$

{1}	1.	$\forall x\,[Fx \rightarrow Gx]$	Premise
{2}	2.	Fa	Premise
{1}	3.	$Fa \rightarrow Ga$	1 UE
{1,2}	4.	Ga	2,3 MP
{1,2}	5.	$\exists x\,[Gx]$	4 EI

Having instantiated premise 1, *Ga* is derived by MP as a basis for EI. Thus, $\exists x\,[Gx]$ is derived and thereby the proof is completed. But now consider a rather similar sequent:

$\forall x\,[Fx \rightarrow Gx], \exists x\,[Fx] : \exists x\,[Gx]$

Both the first premise and the conclusion are identical with the sequent whose proof we have just considered. But there is a crucial difference. This time, the sequent contains among its premises the existentially quantified formula $\exists x\,[Fx]$ rather than the simple subject–predicate formula *Fa*. In order to construct a proof of the new sequent, then, we must have recourse to EE. Remember: only by EE can we infer a conclusion from an existentially quantified formula as such. Hence, a typical disjunct must be chosen and the desired conclusion derived from it. But note that this requirement need not complicate the proof to any great extent. For example, consider the following proof:

$\forall x\,[Fx \rightarrow Gx], \exists x\,[Fx] \vdash \exists x\,[Gx]$

{1}	1.	$\forall x\,[Fx \rightarrow Gx]$	Premise
{2}	2.	$\exists x\,[Fx]$	Premise
{3}	3.	Fa	Assumption TD

{1}	4.	$Fa \rightarrow Ga$	1 UE
{1,3}	5.	Ga	3,4 MP
{1,3}	6.	$\exists x\,[Gx]$	5 EI
{1,2}	7.	$\exists x\,[Gx]$	2,3,6 EE

This proof is only two lines longer than the previous proof. The extra lines are just those where we assumed the typical disjunct (line 3) and finally where we applied EE to complete the proof (line 7). And note how easy EE can be to use. This time we are not given *Fa* as a gift among the premises. But we can simply assume that formula as we did on line 3 and off we go. Easy as the rule may seem to use, however, never lose sight of the fact that for EE, as for UI, at the heart of the principle of inference is the notion of a member of the domain which is a perfectly typical element. Therefore, we must again impose the requirement that there is no special reason for picking the particular disjunct we choose rather than any other. In fact, this time, we must impose not one but *two* crucial requirements.

One of the requisite restrictions can be illustrated in terms of the proof we have just constructed. Imagine that in constructing that proof we ignored the rule of thumb which prescribes EI before EE. We might have been tempted to think: let's get EE out of the way and then apply EI to finish off. It is crucial to appreciate that this would be a fatal mistake. Why? Well, consider the variant purported proof. All is well up to line 5, i.e:

{1}	1.	$\forall x\,[Fx \rightarrow Gx]$	Premise
{2}	2.	$\exists x\,[Fx]$	Premise
{3}	3.	Fa	Assumption TD
{1}	4.	$Fa \rightarrow Ga$	1 UE
{1,3}	5.	Ga	3,4 MP

And so we might now try to apply EE on line 6. But recall that EE can be applied to a formula only if that formula was derived from a genuinely typical disjunct, i.e. one about which we know nothing special. For that purpose, *a* was selected on line 2. But now look at the line which is supposed to contain the conclusion for EE, i.e. line 5. Surely, we do learn something special about *a* here, namely, that it is *G*. Therefore, at line 5 the name *a* cannot belong to a genuinely typical disjunct. Hence, an application of EI at line 6 is required precisely in order to remove that name from the conclusion derived for EE. So, there is more than just charm to the prescription 'EI before EE'. In this instance, the application of EI on line 6 does the crucial job of removing the name *a* from the conclusion of the inference and so enables the application of EE. Again, this is a very common pattern of inference.

In general, then, let us indeed apply EI before EE. Finally, note carefully that the moral of this particular story is, again, that we must always take the utmost care to ensure that the typical disjunct is genuinely typical in the context of that use. As noted, that cannot be the case if the name we have chosen is itself contained in the conclusion we derive for EE. The first restriction exemplifies just that point:

> If the name in the typical disjunct is contained in the conclusion derived for EE then EE cannot be legitimately applied.

This may seem a rather stringent requirement, and perhaps it is. However, it is not yet stringent enough to ensure the validity of every application of the rule EE. Remember that we want to prohibit the use of EE on formulas containing names of individuals about which we know something special. So far, we have ensured that we cannot learn anything special about our named individual from the formula derived as conclusion. But there is another possibility which we have not yet ruled out. We can illustrate this particular worry by considering a purported proof of the following rather dubious-looking sequent:

$Fa, \exists x\, [Gx] : \exists x\, [Gx\ \&\ Fx]$

We might set out to try to prove the sequent as follows:

{1}	1.	Fa	Premise
{2}	2.	$\exists x\, [Gx]$	Premise
{3}	3.	Ga	Assumption TD?????
{1,3}	4.	$Fa\ \&\ Ga$	1,3 &I
{1,3}	5.	$\exists x\, [Gx\ \&\ Fx]$	4 EI
	6.	$\exists x\, [Gx\ \&\ Fx]$	1,2,5 EE !!!!!!!!!!!!

Again, an application of EI (at line 5) removes the relevant name from the formula which is (allegedly) the conclusion derived for EE. But we have still not ensured that the disjunct is genuinely typical. In a sense, the point is obvious: at line 1 we immediately learn something about *a*, namely that it has the property *F*. So, what we might have thought was a typical disjunct at line 3 was nothing of the sort. Hence, EE cannot be legitimately applied. In general, then, we must not only ensure that we don't learn anything about the individual named in the typical disjunct from the conclusion derived for

EE but we must also ensure that no premise or assumption which we use to derive the conclusion from the typical disjunct contains that name either (although, of course, the typical disjunct will itself contain the name). We can now generalise the point to formulate a second restriction:

> If the name in the typical disjunct is itself contained in any premise or assumption used to derive the conclusion for EE from the typical disjunct then EE cannot be legitimately applied to that disjunct.

Given these restrictions, EE can now be stated in full:

> EE: To infer a conclusion from an existentially quantified formula: first assume a genuinely typical disjunct and then derive the desired conclusion from that disjunct. Restate the conclusion on a new line of proof. Annotate the new line 'EE' together with three numbers: (i) the line number of the original existential formula; (ii) the line number of the typical disjunct; (iii) the line number of the conclusion derived from the typical disjunct. The dependency-numbers of the new line consist of all the dependencies belonging to the derived conclusion except that you may discharge the dependency-number of the typical disjunct and replace it with the dependency-number of the original existentially quantified formula.
>
> Note carefully that EE cannot be legitimately applied if (i) the name in the typical disjunct is itself contained in the conclusion derived for EE, or (ii) the name in the typical disjunct is itself contained in any premise or assumption used to derive the conclusion for EE from the typical disjunct.

Obviously, this is by far the longest informal rule-statement we have had to consider to date in the present chapter. It should be unsurprising, then, that the corresponding formal rule-statement is also a little more complex. None the less, where ν is any variable, $\phi(\nu)$ is any expression in which ν may occur (i.e. predicate or relational expression), ρ is any proper name, $\phi(\rho)$ is the result of substituting ρ for ν and C is any conclusion, we can state EE formally as follows:

$$\exists\nu\ [\phi(\nu)] \vdash C \text{ iff } \phi(\rho) \vdash C$$

and ρ does not occur in C or in any premise or assumption used to derive C from $\phi(\rho)$.

The dependency-numbers of the formula inferred by EE consist of all the dependencies belonging to C once the dependency-number of the typical disjunct is replaced by that of the original existentially quantified formula.

Exercise 6.5 provides you with ample opportunity to practise your use of EE. Study the contents of Box 6.4 carefully before attempting the exercise.

BOX 6.4

Existential elimination: EE

♦ *EE informally:* To infer a conclusion from an existentially quantified formula: first assume a genuinely typical disjunct and then derive the desired conclusion from that disjunct. Restate the conclusion on a new line of proof. Annotate the new line 'EE' together with three numbers: (i) the line number of the original existential formula; (ii) the line number of the typical disjunct; (iii) the line number of the conclusion derived from the typical disjunct. The dependency-numbers of the new line consist of all the dependencies belonging to the derived conclusion except that you may discharge the dependency-number of the typical disjunct and replace it with the dependency-number of the original existentially quantified formula.

♦ Note carefully that EE *cannot* be legitimately applied if (i) the name in the typical disjunct is itself contained in the conclusion derived for EE, or (ii) the name in the typical disjunct is itself contained in any premise or assumption used to derive the conclusion for EE from the typical disjunct.

♦ *EE formally:* Where v is any variable, $\phi(v)$ is any expression in which v may occur, ρ is any proper name, $\phi(\rho)$ is the result of substituting ρ for v and C is any conclusion, we can state EE formally as follows:

$$\exists v\ [\phi(v)] \vdash C \text{ iff } \phi(\rho) \vdash C$$

and ρ does not occur in C or in any premise or assumption used to derive C from $\phi(\rho)$.

The dependency-numbers of the formula inferred by EE consist of all the dependencies belonging to C once the dependency-number of the typical disjunct is replaced by that of the original existentially quantified formula.

EXERCISE 6.5

1 Prove that the following are valid sequents of QL. Remember to check carefully each application you make of UI and EE using the criteria provided in the relevant sections above.

1. $\exists x\ [Fx] : \exists y\ [Fy]$ (4)
2. $\exists x\ [Fx\ \&\ Gx] : \exists x\ [Fx]\ \&\ \exists x\ [Gx]$ (8)

3. $\forall x\,[Gx \rightarrow Hx], \exists x\,[Fx\ \&\ Gx] : \exists x\,[Fx\ \&\ Hx]$ (10)

4. $\forall x\,[Fx \rightarrow Gx], \exists y\,[Fy\ \&\ Hy] : \exists z\,[Gz\ \&\ Hz]$ (10)

5. $\exists x\,[Fx\ \&\ Gx], \forall x\,[Hx \rightarrow {\sim}Gx] : \exists x\,[Fx\ \&\ {\sim}Hx]$ (11)

6. $\forall y\,[Gy \rightarrow Hy] : \exists x\,[Gx] \rightarrow \exists y\,[Hy]$ (8)

7. $\exists x\,[Fx \vee Gx] : \exists x\,[Fx] \vee \exists x\,[Gx]$ (10)

8. $\exists x\,[Fx] \vee \exists x\,[Gx] : \exists x\,[Fx \vee Gx]$ (12)

9. $\exists x\,[Fx] : {\sim}\forall x\,[{\sim}Fx]$ (7)

10. $\exists x\,[Fx\ \&\ {\sim}Gx] : {\sim}\forall x\,[Fx \rightarrow Gx]$ (10)

VI Reasoning with Relations

So far, our study of the proof-theory of QL has been confined to sequents composed of subject–predicate formulas and/or formulas involving only one-place predicates, i.e. to sequents of *monadic QL*. In the present section, we extend the proof-theory to sequents involving relations, i.e. to sequents of *polyadic QL*. In fact, representing reasoning with relations in proofs requires no further formal machinery whatsoever, i.e. the four new rules of inference are already perfectly adequate for that purpose. However, the move to polyadic QL does introduce some new complexity to proof-construction and some new opportunities to go wrong in applying the rules. So, I will spell out a few rules of thumb here which should be observed when constructing proofs with relations. In truth, the pitfalls are really very obvious and, with a little care, you will avoid them completely. In what follows then, I am undoubtedly stating the obvious. But, in the last analysis, it is perhaps better to err on the side of caution. So, what are the points to note here?

First, note carefully that although handling relations in proofs does not require any new formal machinery, we will be dealing for the first time with two-place expressions rather than one-place expressions. So, where we had only one variable to consider before we will now have two variables to consider. Further, each variable must be bound by a quantifier and so where we had only one quantifier to consider before we may now have two quantifiers to consider. Finally, although exactly the same quantifier rules apply, each rule can only be applied to one quantifier at a time. In effect, then, handling relational expressions in proofs will slow down proof-construction to some extent. Moreover, as we move to consider the construction of proofs with relations it is equally important to maintain the policy of uniformly substituting the same name with the same variable when applying the rules for quantifiers. So, the second point to note is just that we must carefully maintain the practice of *uniform substitution*.

Finally, we can bring both points together and draw a rather obvious consequence. Every quantified two-place relational expression has two variables and two quantifiers. But the quantifier rules only allow us to handle one quantifier at a time. Further, when we introduce or eliminate a quantifier we must uniformly substitute the same variable with the same name. It follows that, in the process of proof-construction, we will at some points have rather odd-looking formulas which are a mixed bag of names and variables. While such formulas might seem unfamiliar at first they can certainly be perfectly well formed if arrived at in the right way (e.g. provided no variable occurs free) and so they need not worry us. In the last analysis, the presence of such formulas simply reflects the fact that although we are working with formulas governed by iterated quantifiers the rules only allow us to deal with one quantifier at a time.

With these points in mind, it is instructive to compare and contrast three proofs which illustrate each of the points I have just made. Please note that the point here is *just* to illustrate the mechanics of uniform substitution. So, do not expect the proofs to be terribly informative. Remember: we're not really concerned with content here. So, first, consider the familiar case of a proof involving only one-place predicate-letters and quantifiers which are not iterated. In fact, the following proof should be quite familiar to you; you should already have constructed the proof in Exercise 6.2.

$$\forall x\,[Fx \rightarrow Gx] \vdash \forall x\,[(Fx \,\&\, Hx) \rightarrow Gx]$$

{1}	1.	$\forall x\,[Fx \rightarrow Gx]$	Premise
{2}	2.	$Fa \,\&\, Ha$	Assumption for CP
{2}	3.	Fa	2 &E
{1}	4.	$Fa \rightarrow Ga$	1 UE
{1,2}	5.	Ga	3,4 MP
{1}	6.	$(Fa \,\&\, Ha) \rightarrow Ga$	2,5 CP
{1}	7.	$\forall x\,[(Fx \,\&\, Hx) \rightarrow Gx]$	6 UI

In this case, we never have more than one quantifier and one variable to worry about. None the less, note carefully the practice of uniform substitution both in the application of the elimination-rule for the universal quantifier at line 4 and in the application of the introduction-rule at line 7. Further, as ever, each application of UI should be checked in terms of the criteria given earlier. For, in what follows, exactly the same rules of inference apply and, therefore, exactly the same restrictions also apply. In the present case, then, we must take care to apply CP before UI so as to eliminate any dependency containing the name on which we want to universalise. Having done so, we subsequently apply UI at line 7 to the formula on line 6 whose only dependency refers to a formula which is itself a universal generalisation, i.e. premise 1.

Next, consider a sequent whose conclusion contains not one but two variables and, therefore, two quantifiers. Further, although no relational expressions are involved (yet!) the quantifiers are iterated this time and so we must take care that in the proof of this sequent we apply the quantifier rules to one quantifier at a time. As a result, in the later stages of the proof, we will generate formulas which have a mixture of names and variables. Consider the following proof carefully:

$\forall x\ [Fx \rightarrow Gx] \vdash \forall x\ [\forall y\ [(Fx\ \&\ {\sim}Gy) \rightarrow (Gx\ \&\ {\sim}Fy)]]$

{1}	1.	$\forall x\ [Fx \rightarrow Gx]$	Premise
{1}	2.	$Fa \rightarrow Ga$	1 UE
{3}	3.	$Fa\ \&\ {\sim}Gb$	Assumption for CP
{3}	4.	Fa	3 &E
{3}	5.	${\sim}Gb$	3 &E
{1,3}	6.	Ga	2,4 MP
{1}	7.	$Fb \rightarrow Gb$	1 UE
{1,3}	8.	${\sim}Fb$	5,7 MT
{1,3}	9.	$Ga\ \&\ {\sim}Fb$	6,8 &I
{1}	10.	$(Fa\ \&\ {\sim}Gb) \rightarrow (Ga\ \&\ {\sim}Fb)$	3,9 CP
{1}	11.	$\forall y\ [(Fa\ \&\ {\sim}Gy) \rightarrow (Ga\ \&\ {\sim}Fy)]$	10 UI
{1}	12.	$\forall x\ [\forall y\ [(Fx\ \&\ {\sim}Gy) \rightarrow (Gx\ \&\ {\sim}Fy)]]$	11 UI

The conclusion of this particular sequent involves not one but two universal quantifiers, i.e. for the first time the quantifiers are iterated here. Further, the quantifiers in question are iterated around a conditional matrix. Hence, it is natural to assume the antecedent and then try to derive the consequent for CP. For simplicity, we assume the antecedent in unquantified form and then try to derive the consequent in unquantified form. But note very carefully the form of the antecedent which is assumed on line 3. For the first time, the assumption we require involves two distinct names. This precisely reflects the fact that we want to derive a conclusion which involves two quantifiers each of which looks after a distinct variable. Hence, we require two distinct names, for, given the practice of uniform substitution, if we assume a formula such as (*Fa* & ~*Ga*) or (*Fb* & ~*Gb*) we will be forced to replace each and every name as soon as we introduce the very first quantifier, i.e. it would then be impossible to derive any conclusion involving more than one quantifier. Had we simply wanted to derive a conclusion involving only one quantifier and one variable such assumptions would be perfectly appropriate and quite in order in terms of the rules. But because we desire a conclusion with two distinct quantifiers and two distinct variables it is very much in our interest to assume the antecedent as we have, i.e. with two distinct names. As ever, we let the pragmatics of proof-construction dictate our strategy for making assumptions here.

Having assumed the appropriate antecedent we should consider the form of the corresponding consequent we require. Clearly, the consequent in question is itself a conjunction and so we must derive each conjunct first. Keeping a watchful eye on the conclusion, however, note that we again want to derive the consequent in a form which involves the same pair of distinct names which we chose when we assumed the antecedent. The derivation of the first conjunct is straightforward, i.e. it is derived by MP at line 6. Now, because the second conjunct involves a distinct name, deriving that conjunct may not seem so obvious. However, what is true of everything is not just true of any one element of the domain but of each and every such element. Hence, to derive the second conjunct we can simply perform a second universal elimination to the name *b* before finally inferring the conjunct we want by MT at line 8.

Next, we apply &I and, on line 10, we can at last apply CP. To complete the proof, we must apply UI to the result. Now, we can certainly get what we want here but only one step at a time. First, at line 11, we apply UI to the name *b*, i.e. to all and only the occurrences of that name. Hence, we uniformly substitute the variable *y* in place of the name *b* throughout the formula on line 10. (Note also that this application of UI is perfectly valid. Again, we apply UI to a formula whose only dependency is itself a universal generalisation, i.e. premise 1.) This generates the anticipated 'mixed-bag' formula on line 11. Finally, we can now go on to apply UI to the name *a* and uniformly substituting *x* for *a* gives us the desired conclusion at line 12 (again the application of UI is perfectly valid – the only dependency involved is itself a universal generalisation).

The slowed-down pattern of inference which generates mixed-bag type formulas of the kind that we found at the end of the preceding proof (as we start to apply the introduction-rules for quantifiers) is very similar to the pattern of inference we find when reasoning with relations in proofs. Moreover, when reasoning with relations, a similar pattern is often found around the application of elimination-rules in the early stages of proof-construction.

In both cases, the key point is to treat one quantifier at a time. The same pattern can be seen in the closing stages of a similar proof which, for the first time, does actually involve two-place relational expressions. The proof is intended only to illustrate the mechanics of proof-construction; nothing more. In fact, this particular sequent is wholly trivial. (Can you see why? Think hard about the content of the conclusion and how that stands to the content of the premise.) However, the proof of the sequent does usefully illustrate important elements of the mechanics of reasoning with relations:

$\forall x\,[Fx \rightarrow Gx] \vdash \forall x\,[\forall y\,[(Fx \,\&\, Rxy) \rightarrow (Gx \,\&\, Rxy)]]$

{1}	1.	$\forall x\,[Fx \rightarrow Gx]$	Premise
{2}	2.	$Fa \,\&\, Rab$	Assumption for CP
{1}	3.	$Fa \rightarrow Ga$	1 UE
{2}	4.	Fa	2 &E
{1,2}	5.	Ga	3,4 MP
{2}	6.	Rab	2 &E
{1,2}	7.	$Ga \,\&\, Rab$	5,6 &I
{1}	8.	$(Fa \,\&\, Rab) \rightarrow (Ga \,\&\, Rab)$	2,7 CP
{1}	9.	$\forall y\,[(Fa \,\&\, Ray) \rightarrow (Ga \,\&\, Ray)]$	8 UI
{1}	10.	$\forall x\,[\forall y\,[(Fx \,\&\, Rxy) \rightarrow (Gx \,\&\, Rxy)]]$	9 UI

Again, the conclusion is a universally quantified conditional involving iterated universal quantifiers and so we assume the antecedent in unquantified form before trying to derive the consequent in unquantified form. Equally, we must take care to choose two distinct names here just because the conclusion we require involves two distinct variables. Once more, the consequent is a conjunction and so we must derive each conjunct. This time, the derivations are very straightforward and only involve &E, MP and, lastly, &I. Next, we conditionalise and, on line 9, we apply UI for the first time and arrive at a mixed-bag formula (a brief glance at the relevant dependencies assures us that we have observed the relevant restriction). To complete the proof, we apply UI once more, again, quite properly, this time to the name *a*. Uniform substitution of *x* for *a* generates exactly the conclusion we desire.

Before we proceed to an exercise, it's worth making two further points explicit. The first is simply the one made above that when reasoning from formulas involving relations as premises and not merely to a formula involving relations as conclusion, we must be just as careful in our application of elimination-rules as we have been when applying introduction-rules. We can see both patterns of inference at work in the following proof:

$\forall x\,[\forall y\,[Rxy]] \vdash \forall y\,[\forall x\,[Rxy]]$

{1}	1.	$\forall x\,[\forall y\,[Rxy]]$	Premise
{1}	2.	$\forall y\,[Ray]$	1 UE
{1}	3.	Rab	2 UE
{1}	4.	$\forall x\,[Rxb]$	3 UI
{1}	5.	$\forall y\,[\forall x\,[Rxy]]$	4 UI

This time, we need have no anxiety about satisfying the restrictions on UI; throughout the proof, the only dependency involved is that a universal generalisation. But note carefully that when we first apply UE we do, indeed, find the same type of formula involving both names and variables that we found earlier when applying UI. In fact, this particular proof requires nothing more than repeated applications of just these rules. This brings me to a final point. To date, we have only considered reasoning with relations involving universally quantified formulas. But we are by no means confined to the universal quantifier when it comes to reasoning with relations. In any given case, the quantifiers which are iterated may well include an existential quantifier or they may be exclusively existential. The familiar quantifier rules allow us to handle such formulas with ease but, again, the same restrictions and, indeed, the same strategic considerations all apply.

So, let's consider together an example involving the iteration of different kinds of quantifier. For example, consider the sequent:

$\exists x\,[\forall y\,[Rxy]] : \forall y\,[\exists x\,[Rxy]]$

In this case, the premise is an existentially quantified formula (note that the quantifier in initial position, i.e. literally the first quantifier, dictates the kind of formula involved). To derive a conclusion from that formula, then, we must use EE. Therefore, we must first choose a typical disjunct. As yet, no names occur in the proof and so we can safely pick *a* for that purpose. But note carefully that the resulting typical disjunct is the formula:

$\forall y\,[Ray]$

This is a universally quantified formula (remember: the quantifier in initial position dictates the kind of formula involved). Hence, UE can be applied to that formula and the final variable *y* can be eliminated in favour of the name *b*, i.e. we can validly infer from $\forall y\,[Ray]$ that *Rab*. To derive the desired conclusion and complete the proof, the quantifiers must be reintroduced in reverse order. Again, each time we apply UI and EE we must check that we do not thereby violate the relevant restrictions. Provided we do not, we have complete freedom about the order in which we reintroduce the quantifiers, i.e. about which quantifier we use to bind which variable. Once more, we let the requirements of proof-construction dictate the order in which the quantifiers are introduced. Here is the proof in full:

$$\exists x\ [\forall y\ [Rxy]] \vdash \forall y\ [\exists x\ [Rxy]]$$

{1}	1.	$\exists x\ [\forall y\ [Rxy]]$	Premise
{2}	2.	$\forall y\ [Ray]$	Assumption TD
{2}	3.	Rab	2 UE
{2}	4.	$\exists x\ [Rxb]$	3 EI
{2}	5.	$\forall y\ [\exists x\ [Rxy]]$	4 UI
{1}	6.	$\forall y\ [\exists x\ [Rxy]]$	1,2,5 EE

The applications of UI at line 5 and EE at line 6 are perfectly valid (it is a very useful exercise to check for yourself that each restriction really has been met in each case).

Finally, one last word of warning. Earlier I noted that applying an elimination-rule to a formula with iterated quantifiers can result in inferring a different kind of quantified formula. So, for example, when we first considered premise 1 of the preceding proof, i.e. the existentially quantified formula $\exists x\ [\forall y\ [Rxy]]$, we realised that to reason from that formula we would have to choose a typical disjunct for it. That typical disjunct turned out to be a universally quantified formula, namely, the formula $\forall y\ [Ray]$. But now imagine the case where premise 1 was any formula which began $\forall y\ [\exists x\ [\ldots]]$. Obviously, we can apply UI to any such formula. But consider carefully the kind of formula we would thereby infer and its implications for proof-construction. Exercise 6.6 contains a number of proofs for you to try yourself.

EXERCISE 6.6

1 Prove that the following are valid sequents of QL (as ever, remember to check carefully each application you make of UI and EE using the criteria provided in the relevant sections above):

1. $\forall y\ [\forall x\ [Rxy]] : \forall x\ [\forall y\ [Rxy]]$ (5)
2. $\forall x\ [\forall y\ [\forall z\ [Rxyz]]] : \forall z\ [\forall y\ [\forall x\ [Rxyz]]]$ (7)

3. $\exists x\, [\forall y\, [Rxy]] : \exists x\, [Rxa]$ (5)

4. $\exists x\, [\exists y\, [Rxy]] : \exists y\, [\exists x\, [Rxy]]$ (7)

5. $\forall x\, [\forall y\, [Rxy \rightarrow Ryx]] : \forall x\, [\exists y\, [Rxy] \rightarrow \exists y\, [Ryx]]$ (10)

6. $\forall x\, [Fx \rightarrow Gx] : \forall x\, [\exists y\, [Fy\, \&\, Rxy] \rightarrow \exists y\, [Gy\, \&\, Rxy]]$ (12)

7. $\forall x\, [\exists y\, [\forall z\, [Rxyz]]] : \forall x\, [\forall z\, [\exists y\, [Rxyz]]]$ (8)

8. $\exists x\, [Fx]\ \&\ \forall y\, [Gy \rightarrow Rxy]$,
$\forall x\, [Fx] \rightarrow \forall y\, [Hy \rightarrow {\sim}Rxy] : \forall x\, [Gx \rightarrow {\sim}Hx]$ (16)

2 Recall the discussion of different kinds of relation in Section VIII of the previous chapter. There we adopted a standard vocabulary for expressing the formal properties of relations; *R* for relation, etc. Using that standard vocabulary construct proofs which demonstrate:

(i) that any relation which is *asymmetrical* is *irreflexive*

and:

(ii) that any relation which is *intransitive* is *irreflexive*.[6]

3 Consider the following sequent carefully:

$$\forall y\, [\exists x\, [Rxy]] : \exists x\, [\forall y\, [Rxy]]$$

Earlier, in Section IX of Chapter 5, we noted that this sequent exemplifies the quantifier switch fallacy and is, as such, invalid. Therefore, the sequent in question cannot be proved in QL. None the less, a sufficiently misguided person might try to prove it. Explain to the misguided individual exactly why such an attempt is doomed to fail just in terms of the rules of inference governing the quantifiers.

VII
Proof-Theory for Identity: The Rules =I and =E

The set of rules of inference for QL is completed by adding in two further rules, namely, the introduction- and elimination-rules for identity. The introduction-rule for identity, **identity-introduction**, or **=I**, for short, is both extremely intuitive and particularly easy to use. The intuition which identity-introduction exploits is just the one noted in the discussion of identity in Section XIII of Chapter 5, namely, that everything is identical with itself, i.e. the *law of identity*. If everything is identical with itself, then, certainly, *a* is identical with *a* and, indeed, *b* is identical with *b* and *c* is identical with *c*, and so on, ad nauseam. Quite generally, then, where ρ is any proper name, any formula of the form:

$(\rho = \rho)$

may be entered on any line of proof. Any such line should be annotated '=I'. Further, because the very form of any such formula guarantees the truth of that formula and because all such formulas follow immediately from the law of identity, from logic itself, as it were, any line of proof annotated '=I' has no dependency-numbers whatsoever, i.e. any such formula is introduced gratis. Hence, we can add to the list of strategic possibilities for proof-construction in QL the possibility of introducing an identity statement for any particular name at any line of proof without thereby adding a single dependency-number to the existing proof.

Here is the rule-statement in all its glorious simplicity:

=I: Any formula of the form $(\rho = \rho)$ may be entered on any line of any proof. Any such line should be annotated '=I' and should have no dependency-numbers.

Given the existing rules of inference for QL a proof of the law of identity in QL follows almost immediately. First, we may use '=I' to enter any formula of the form $(\rho = \rho)$ on any line of proof, free of all dependencies. Therefore, we can do exactly that on line 1 using the name *a*, as follows:

---	1.	$a = a$	=I

Next consider the possibility of applying the rule UI to the formula on line 1. Remember: the only restriction on UI concerns the set of dependency-numbers belonging to the formula to which that rule is applied. But the formula on line 1 has no dependency-numbers. Therefore, UI can be safely applied. Thus, in QL, we can demonstrate the validity of one of the three traditional laws of logic recognised throughout the history of western philosophy in a mere two lines:

$\vdash \forall x\, [x = x]$

---	1.	$a = a$	=I
---	2.	$\forall x\, [x = x]$	1 UI

Moreover, the proof of the law of identity in QL is a perfect illustration of the very idea which motivated the extension of UI beyond sequents involving only universally quantified formulas in Section II of this chapter. There we considered the straight line and the triangle and noted that we could use any particular instance to illustrate properties belonging to all. Here, in the proof of the law of identity, we have another prime example of that kind of reasoning: all we know about the element of the domain denoted by a is that it is identical with itself. But, of course, that is not just true of that element of the domain. It is true of every element of the domain.

It only remains to consider the elimination-rule for identity, **identity-elimination**, or **=E**, for short. Again, this rule is also highly intuitive and, again, it is by no means difficult to use. To appreciate the kind of reasoning involved this time recall our earlier discussion of the nature of identity in Section XIV of the previous chapter. One of the first points we made there was that although 'is' is involved in identity statements that sense of 'is' is not the same as the 'is' of predication. When we form an identity statement we do not simply attribute a property to a subject. Rather, we explicitly identify two things, i.e. we construct an identity statement precisely in order to assert either that one name refers to something which is the same as itself or that two different names refer to one and the same element of the domain. Consider the latter case carefully. Suppose we know that two distinct names refer to one and the same object, 'Bob Dylan' and 'Robert Zimmerman', for example. Further suppose that we also know something about the individual denoted by the name. Say, for example, we know that Bob Dylan is bald. But if Bob Dylan is bald and Bob Dylan is the very same person as Robert Zimmerman it must follow that Robert Zimmerman is bald.

Equally, if we know that Cicero was an eminent Roman philosopher whose work greatly influenced David Hume and we know that Cicero and Tully were one and the same individual then, again, it must follow that Tully was an eminent Roman philosopher whose work greatly influenced David Hume. Quite generally, then, if two names denote one and the same element of the domain and we know some particular property of one of those elements it follows that the same property must also hold of the other. It is precisely this pattern of reasoning which identity elimination allows us to exploit in the course of a proof. Given both an identity statement and a formula which tells us something about one of the elements denoted by a term in that identity we may validly infer that the same thing also holds of the element denoted by the other term. Suppose, then, that a is identical to b. And further suppose that a is F. It follows that b is F. Omitting brackets for clarity, this argument is fairly represented in QL by the sequent:

$$a = b, Fa : Fb$$

This sequent is obviously valid. Moreover, identity-elimination makes its proof perfectly straightforward. Consider the proof carefully:

$a = b, Fa \vdash Fb$

{1}	1.	$a = b$	Premise
{2}	2.	Fa	Premise
{1,2}	3.	Fb	1,2 =E

Note carefully that the annotation for =E includes two line numbers. This simply reflects the fact that we use =E to draw an inference from an identity statement together with a formula which tells us something about the element of the domain denoted by a term in the identity. So, to apply =E we must have, on two separate lines of proof, an identity statement and a formula which tells us something about the element of the domain denoted by a term in that identity statement. Hence, the annotation for =E requires two numbers. Note also the way in which the application of =E to the two formulas brings the dependency-numbers of both together at the line annotated =E. Finally, although we very often exploit =E when we discover that a predicate applies to a term in an identity statement, we are not confined in our use of =E to predicates. So, for example, if I know that *a* is identical with *b* and I find out that, in fact, *b* is identical with *c*, I can certainly go on to infer validly that, therefore, *a* is identical with *c*. Consider the corresponding sequent:

$a = b, b = c : a = c$

Again, this should certainly be intuitive. After all, the sequent exemplifies the *transitivity of identity*. And again, we can easily use =E to demonstrate the validity of the sequent:

$a = b, b = c \vdash a = c$

{1}	1.	$a = b$	Premise
{2}	2.	$b = c$	Premise
{1,2}	3.	$a = c$	1,2 =E

With these illustrations clearly in mind, consider the rule-statement for = E:

> =E: Given an identity statement on a line of proof and a formula on another line of proof which asserts a predicate or relation of a term in the identity you may infer that the same predicate or relation applies to the other term in the identity. The new line should be annotated '=E' together with the line number of both formulas used. The dependency-numbers of the new line consist of all of the dependency-numbers belonging to the two formulas used.

BOX 6.6

Proof-theory for identity: the rules = I and =E

- =*I*: Any formula of the form (ρ = ρ) may be entered on any line of any proof. That line should be annotated '=I' and should have no dependency-numbers.

- =*E*: Given an identity statement on a line of proof and a formula on another line of proof which ascribes a predicate or relation to a term of the identity you may infer that the same predicate or relation applies to the other term of the identity. The new line should be annotated '=E' together with the line number of both formulas used. The dependency-numbers of the new line consist of all of the dependency-numbers belonging to the two formulas used.

Finally, note that we can use any negation-introducing rule of inference to infer a *non-identity*, i.e. a negated identity statement, such as:

$\sim(a = b)$

Quite generally, you will find that the rules of inference for identity work extremely well with the existing stock. Exercise 6.7 contains a number of proofs which both illustrate that point and allow you to practise with the extremely intuitive rules of inference governing identity. As ever, take care to check each application of UI and EE in terms of the restrictions outlined earlier. Study the contents of Box 6.6 before attempting the exercise.

EXERCISE 6.7

1 Prove that the following are valid sequents of QL.

Hint: when constructing proofs of certain of the sequents below, it is helpful to keep in mind the kind of freedom we enjoy with the Rule EI. Not to put too fine a point on it, recall that EI does not necessarily require us to replace every name occurring in a formula with a variable.

1. $: \exists x\,[x = x]$ (2)
2. $(Fa \,\&\, Ga), (a = b) : (Fb \,\&\, Gb)$ (7)

3. $\forall x\ [Fx \rightarrow (x = a)] : \exists x\ [Fx \rightarrow Fa]$ (7)

4. $: ((a = b)\ \&\ (b = c)) \rightarrow (a = c)$ (5)

5. $: \forall x\ [\exists y\ [x = y]]$ (3)

6. $Fa : \exists x\ [(x = a)\ \&\ Fx]$ (4)

7. $\exists x\ [(x = a)\ \&\ Fx] : Fa$ (6)

8. $\forall x\ [Fx \rightarrow [Gx \vee (x = a)] : Ga \rightarrow \forall x\ [Fx \rightarrow Gx]$ (12)

9. $\forall x\ [Fx \rightarrow Gx] : \forall x\ [\forall y\ [(Fx\ \&\ {\sim}Gy) \rightarrow {\sim}(x = y)]]$ (13)

10. $\exists x\ [\forall y\ [Fy \rightarrow (x = y)]] : \forall x\ [\forall y\ [(Fx\ \&\ Fy) \rightarrow (x = y)]]$ (14)

VIII
Strategies for Proof-Construction in QL #1

In Chapter 3 of this text we took some time to discuss questions relating to strategies for proof-construction in PL, i.e. we addressed the question of how, seated in the examination hall facing a number of daunting proofs some Monday morning, we might go about trying to crack those proofs. It is equally important to consider just this question in the context of QL and, having considered each of the new rules of inference, it is now time to try to answer it. QL introduces a new level of complexity as regards proof-construction just because it requires us to take into account sequents composed of quantified formulas and, indeed, subject–predicate formulas. And that is the fundamental problem, i.e. unlike PL, QL requires us to consider quantified formulas and/or subject–predicate formulas. Further, as regards quantified formulas in particular, we have two different possibilities to consider: universally quantified formulas and existentially quantified formulas. So, the first point to make about strategy for proof-construction in QL is simply this:

> Consider very carefully just which kind or kinds of quantified formula are involved in the particular sequent and whether any subject–predicate formulas are involved.

This may well state the obvious but, in fact, we can give some useful rules of thumb for strategies for proof-construction of QL sequents simply on the basis of the different kinds of formula which compose those sequents. Again at the risk of stating the obvious, the proof-theory of QL also provides us with both introduction-rules and elimination-rules for the quantifiers and, as I'm sure you've noticed, these rules can be quite straightforward to use. As you will have realised from the early exercises in this chapter, the proof of certain (extremely straightforward) sequents involving, for example, universally quantified formulas as premises and an unquantified or subject–predicate formula as conclusion requires nothing more than a simple application of the relevant elimination-rule, e.g.:

$\forall x\ [Fx] \vdash Fa$

Obvious as this point may seem, it allows us to pick out a particular class of QL sequents, namely, those sequents whose premises consist solely of universally quantified formulas (or of a mixture of universally quantified formulas and unquantified formulas) and whose conclusion is an unquantified formula. Hence, we can now formulate the first general rule of proof-construction in QL for just this class of sequents:

1 For any sequent whose premises consist solely of universal formulas (or of a mixture of universal and unquantified formulas) and whose conclusion is an unquantified formula:

(i) Apply UE to each universal premise so as to infer formulas about the individual(s) referred to in the unquantified formula(s), i.e. to infer formulas involving the same name(s) as the unquantified formulas.

(ii) Exploit the PL rules of inference to derive the desired conclusion about the named individual(s).

Further, we can easily go on to identify another class of QL sequents whose premises also involve only universally quantified formulas but whose conclusions are existentially quantified formulas, e.g.:

$\forall x\ [Fx] \vdash \exists x\ [Fx]$

We can now formulate a second general rule of proof-construction in QL for just this class of sequents:

2 For any sequent whose premises consist solely of universal formulas (or of a mixture of universal and unquantified formulas) and whose conclusion is an existentially quantified formula:

(i) Apply UE to each universal premise to infer formulas about the individual(s) referred to in the unquantified formula(s), i.e. to infer formulas involving the same name(s) as the unquantified formulas, or any name if none occurs.

(ii) Exploit the PL rules of inference to derive the desired conclusion about the named individual(s).

(iii) Apply EI to the conclusion to complete the proof.

So far we have ignored another easily identifiable set of QL sequents whose premises also involve only universally quantified formulas, i.e. that class of sequents whose premises and conclusions are all universally quantified formulas. The proof of many sequents in this class can be extremely straightforward. Frequently, constructing a proof of a sequent involving only universally quantified formulas simply requires an application of the relevant elimination-rule followed by a subsequent application of the appropriate introduction-rule, e.g.:

$\forall x\,[Fx] \vdash \forall y\,[Fy]$

More commonly, however, we will have to do a little work in between the application of the elimination-rule and the closing application of the relevant introduction-rule. For example, consider the proof of the sequent:

$\forall x\,[Fx\ \&\ Gx] \vdash \forall y\,[Gy\ \&\ Fy]$

{1}	1.	$\forall x\,[Fx\ \&\ Gx]$	Premise
{1}	2.	$Fa\ \&\ Ga$	1 UE
{1}	3.	Fa	2 &E
{1}	4.	Ga	2 &E
{1}	5.	$Ga\ \&\ Fa$	3,4 &I
{1}	6.	$\forall y\,[Gy\ \&\ Fy]$	5 UI

In this case, the work between the application of the elimination-rule and the introduction-rule is simply carried out by applying &E and &I. These rules, of course, belong as much to PL as to QL. But now think of the vast number of sequents involving universally quantified formulas whose proof-construction follows that pattern of eliminating the quantifiers and then exploiting PL rules of inference in order to derive a conclusion to which we can then apply the introduction-rule.

Again, we can easily generalise this procedure to characterise a third rule of thumb for proof-construction in QL:

3. For any sequent whose premises consist solely of universally quantified formulas (or of a mixture of universal and unquantified formulas) and whose conclusion is also a universally quantified formula:

(i) Apply UE to each premise to infer formulas about one particular individual, i.e. to infer formulas involving an appropriate name.

(ii) Exploit the PL rules of inference to derive the desired conclusion about that specific individual.

(iii) Make sure all assumptions (if any) are discharged in the sub-proof and apply UI to complete the proof.

So far, we have ignored those classes of sequent which involve existentially quantified premises. Here again, however, we can readily identify classes of sequent whose proofs all involve a similar strategy. For example, consider that class of sequents whose premises are all existential and whose conclusions are either existential or unquantified. In order to derive a conclusion from an existentially quantified formula as such we must apply EE. Hence, we must show that the conclusion in question can be derived from a typical disjunct and so we must first assume such a disjunct and take care that the disjunct we assume is, indeed, typical, i.e. that we have no prior knowledge about the named individual we pick. In other words, we must carefully observe the restrictions on EE in the process of constructing any such proof. For recall that EE cannot be legitimately applied if the name in the typical disjunct is contained in the conclusion derived for EE or if the name in the typical disjunct is itself contained in any premise or assumption used to derive the conclusion for EE from the typical disjunct. None the less, constructing a proof by EE can be perfectly straightforward and we can always check the dependencies of the relevant lines of proof before applying EE. Exactly this strategy is illustrated in the proof of the following sequent:

$$\exists x\,[Fx\ \&\ Gx] \vdash \exists x\,[Fx]\ \&\ \exists x\,[Gx]$$

{1}	1.	$\exists x\,[Fx\ \&\ Gx]$	Premise
{2}	2.	$Fa\ \&\ Ga$	Assumption TD
{2}	3.	Fa	2 &E
{2}	4.	$\exists x\,[Fx]$	3 EI
{2}	5.	Ga	2 &E
{2}	6.	$\exists x\,[Gx]$	5 EI
{2}	7.	$\exists x\,[Fx]\ \&\ \exists x\,[Gx]$	4,6 &I
{1}	8.	$\exists x\,[Fx]\ \&\ \exists x\,[Gx]$	1,2,7 EE

Hence, we can formulate a fourth rule of thumb for proof-construction in QL:

4. For any sequent whose premises are all existentially quantified formulas and whose conclusions are either unquantified, existential or a combination of existentially quantified formulas:

   (i) Construct an argument from a typical disjunct, i.e. carefully select a disjunct which is genuinely typical in the context of proof and assume a formula of the relevant form which features that disjunct.

   (ii) Exploit the PL rules of inference to derive the desired conclusion from the formula featuring the typical disjunct.

   (iii) Apply EI before EE and, if necessary, re-apply PL rules.

   (iv) Ensure that the restrictions on EE are met, apply EE and, if necessary, re-apply PL rules.

Finally, we can identify a last set of QL sequents and a further general rule of strategy, namely, the class of QL sequents whose premises are a mixed bag of universal and existential formulas and whose conclusion is an existentially quantified formula. In such cases, we must again take a little care in the process of proof-construction. Here, the key tip is always to select a typical disjunct first, i.e. look to the application of EE before any other rule of inference for quantifiers. In other words, assume the typical disjunct before applying UE to any formula. Having done so, apply UE to the remaining premises to infer formulas about the same individual(s). Next, exploit the PL rules of inference to derive the desired conclusion from the typical disjunct. Take care to apply EI before EE and check relevant dependencies before applying EE to complete the proof. All of these points are illustrated in the following proof:

$$\forall x\,[Gx \rightarrow Hx], \exists x\,[Fx \,\&\, Gx] \vdash \exists x\,[Fx \,\&\, Hx]$$

{1}	1.	$\forall x\,[Gx \rightarrow Hx]$	Premise
{2}	2.	$\exists x\,[Fx \,\&\, Gx]$	Premise
{3}	3.	$Fa \,\&\, Ga$	Assumption TD
{1}	4.	$Ga \rightarrow Ha$	1 UE
{3}	5.	Fa	3 &E
{3}	6.	Ga	3 &E
{1,3}	7.	Ha	4,6 MP
{1,3}	8.	$Fa \,\&\, Ha$	5,7 &I

{1,3}	9.	$\exists x\,[Fx \mathbin{\&} Hx]$	8 EI
{1,2}	10.	$\exists x\,[Fx \mathbin{\&} Hx]$	2,3,9 EE

Note carefully that we assume the typical disjunct first, at line 3, before applying UE to infer formulas about the same individual. The point to note here is just that if we had applied UE to instantiate to the name *a* first we could not then have assumed a typical disjunct involving that name. Hence, it is a good procedural rule to assume the typical disjunct(s) before applying UE.

Once again, we can generalise this approach as follows:

5. For any sequent whose premises are a mixture of universal and existential formulas and whose conclusion is an existentially quantified formula:
   (i) *First*: assume genuinely typical disjunct(s) as required. *Then*: apply UE to the remaining premises to infer formulas about the *same* individual(s).
   (ii) Exploit the PL rules of inference to derive the desired conclusion about the relevant individual.
   (iii) Apply EI before EE.
   (iv) Check the relevant dependencies before applying EE to complete the proof.

Hopefully, with these general strategic rules clearly in mind, the process of proof-construction in QL should not seem so intimidating. As ever, there are exceptions to our rules of thumb, but you should find that a vast number of sequents yield to just these strategic considerations. Before we proceed to an exercise, however, it is useful and important to dig a little deeper as regards questions of strategy for proof-construction in QL. Hence, I postpone presenting a full summary and setting out any exercises until the end of the following section.

IX
Strategies for Proof-Construction in QL #2

Digging Deeper Again: Proof and Sub-Proof in QL

Recall the 'golden rule' for proof-construction in PL. Although we set that rule up as a useful rule of thumb for proof-construction specifically in PL, you may well have realised by now that the rule can also be helpful in QL. As ever, proof-construction in QL requires us to take account of quantified formulas and, indeed, subject–predicate formulas and you may feel that the golden rule doesn't really address these kinds of formula. In fact, that rule can usefully be applied to questions of strategy for sequents composed of quantified formulas and/or subject–predicate formulas. For example, consider once more the following proof:

$\forall x\,[Fx\,\&\,Gx] \vdash \forall y\,[Gy\,\&\,Fy]$

{1}	1.	$\forall x\,[Fx\,\&\,Gx]$	Premise
{1}	2.	$Fa\,\&\,Ga$	1 UE
{1}	3.	Fa	2 &E
{1}	4.	Ga	2 &E
{1}	5.	$Ga\,\&\,Fa$	3,4 &I
{1}	6.	$\forall y\,[Gy\,\&\,Fy]$	5 UI

The proof follows a perfectly straightforward pattern which can be generalised as follows:

1. First, apply the relevant elimination-rules.
2. Next, exploit the PL rules of inference to derive the desired conclusion in unquantified form.
3. Complete the proof by application of the appropriate introduction-rule.

The golden rule is most obviously useful at Stage 2, e.g. in the preceding proof, the work between the application of the elimination-rule and the introduction-rule is simply carried out by applying &E and &I. More commonly, however, we will have to do a little more work in between the application of the elimination-rule and the closing application of the relevant introduction-rule! None the less, for a large number of sequents involving universally quantified formulas, proof-construction follows this pattern of eliminating the quantifiers and then exploiting PL rules of inference in order to derive a conclusion to which the relevant introduction-rule can then be applied. For all these proofs, we have, in effect, a PL sub-proof within a QL overall proof. Hence, in all such cases, the golden rule can usefully be applied.

Moreover, we can often usefully apply the golden rule in our earliest reflections about proof-construction. To apply the golden rule in the context of QL, we must always look extremely closely not just at the quantified formulas themselves but at the internal structure of those formulas, i.e. ask yourself: which connectives are involved? And that is another extremely useful rule of thumb for proof-construction:

> Having identified the kind or kinds of quantifier involved look very closely at the internal structure of the formulas involved and identify the relevant connectives.

For an impressive number of sequents, simply answering these two questions and then applying the golden rule will allow you to crack strategic questions immediately. For example, consider the following sequent:

$\forall x\, [Fx \vee Gx], \forall x\, [Fx \rightarrow Gx] : \forall x\, [Gx]$

Let's apply the two points spelled out above. First, identify the kinds of quantified formula involved. In this case, only universally quantified formulas are involved. Hence, we will apply UE to infer formulas involving one named individual before trying to derive the unquantified form of the conclusion we desire about that given individual, e.g. *Ga*. If we succeed, we can then complete the proof by applying UI. In this case, because only universally quantified formulas are involved, we know at once that the application of UI will be safe. Thus, we know how to begin the proof and, indeed, we also know what the last line of proof should look like. So, let's spell things out as we did in our earlier discussion, i.e. let's set out the premises and the conclusion and try to bridge the gap:

{1}	1.	$\forall x\, [Fx \vee Gx]$	Premise
{2}	2.	$\forall x\, [Fx \rightarrow Gx]$	Premise
	.		
	.		
{1,2}	n.	$\forall x\, [Gx]$	??

Further progress can now be made by adding in the relevant UE and UI steps, i.e. we can flesh out the skeleton of the proof more fully, as follows:

{1}	1.	$\forall x\, [Fx \vee Gx]$	Premise
{2}	2.	$\forall x\, [Fx \rightarrow Gx]$	Premise
{1}	3.	$Fa \vee Ga$	1 UE

{2}	4.	$Fa \rightarrow Ga$	2 UE
	.		
	.		
	n – 1	Ga	????
{1,2}	n	$\forall x\,[Gx]$	8 UI

Having identified the kind of quantifier involved and drawn the inferences we can, let's apply our second point, i.e. carefully consider the internal structure of the formulas involved and identify the relevant connectives. Notice at once that the instantiated form of Premise 1 is a disjunction while the instantiated form of Premise 2 is a conditional. To bridge the gap, then, we must derive *Ga* from both these formulas. Again, here is where the golden rule can help. Let's apply that rule. So, first, ask yourself: *is the main connective in the conclusion we want to derive a conditional?* Well, the conclusion is the simple subject–predicate formula *Ga* which is not a conditional. Therefore, the answer to the first question is 'no'. Of course, if the answer to that question had been 'yes' then, according to the golden rule, the overall strategy would simply be conditional proof. Given that the answer is 'no', we proceed to the second part of the Rule and ask: *is the main connective in any or all of the formulas we want to derive the conclusion from a disjunction?* This time, the answer is 'yes', the instantiated form of Premise 1 is, indeed, a disjunction. Therefore, the correct strategy for deriving *Ga* must involve vE. And now this is familiar territory. The strategy for vE is just: assume the first disjunct and try to derive the conclusion, then, assume the second disjunct and try to derive the conclusion. If we succeed in those derivations, we can discharge the 'extra' dependency-numbers of the assumptions when we apply vE, leaving the conclusion depending only on its premises. In the process, we will also enable a subsequent application of UI. Simply applying the strategy for vE allows us to bridge the gap to *Ga* as follows:

$\forall x\,[Fx \vee Gx], \forall x\,[Fx \rightarrow Gx] \vdash \forall x\,[Gx]$

{1}	1.	$\forall x\,[Fx \vee Gx]$	Premise
{2}	2.	$\forall x\,[Fx \rightarrow Gx]$	Premise
{1}	3.	$Fa \vee Ga$	1 UE
{2}	4.	$Fa \rightarrow Ga$	2 UE
{5}	5.	Fa	Assumption (1st disjunct)
{2,5}	6.	Ga	4,5 MP (1st conclusion)
{7}	7.	Ga	A (2nd disjunct *and* 2nd conclusion)

{1,2}	8.	Ga	3,4,6,7,7 vE
{1,2}	9.	$\forall x\ [Gx]$	8 UI

Note that vE must be applied before UI in order to discharge the remaining assumptions about *a*, i.e. to discharge all dependencies involving the name *a* from the line to which we want to apply UI. Remember: that line must not depend upon any formula containing the name *a*. Hence, the application of vE enables the subsequent safe application of UI. Therefore, the golden rule certainly can play a very helpful role in strategic thinking in QL.

A little reflection on the kind of quantified formula involved in a sequent will usually reveal the *overall strategy* for proving that sequent. But it will generally be the identification of the kinds of PL connective involved in those formulas, together with the golden rule, which will reveal the appropriate kind of *sub-proof*. Moreover, given the golden rule, you should find that strategic questions can be answered completely and immediately simply by reflecting on (i) the kinds of quantified formula involved and (ii) the kinds of PL connective involved in those formulas. Indeed, in the proof of the last sequent, given the answers to these questions and an application of the golden rule, all we had to do was make one application of MP.

Further, remember that applications of the golden rule can be iterated, i.e. having identified the right basic strategy for sub-proof, if you are still puzzled, simply ask again: is the conclusion I want to derive a conditional . . . ? Remember: some or even all of the strategies spelled out by the golden rule can work together within a single proof, one strategy providing a sub-proof of something useful for another. So, in QL as in PL, not only can we apply the golden rule to sub-proofs within an overall proof but we can also go on to apply the golden rule to sub-proofs within sub-proofs, digging deeper and deeper as we go. Hence, questions about strategy in QL need not be so daunting. For example, consider the following sequent:

$\exists x\ [Fx \vee Gx] : \exists x\ [Fx] \vee \exists x\ [Gx]$

Here, we are faced with a sequent involving an existentially quantified formula as sole premise. But proof-construction need not be daunting. After all, we certainly know how the proof should begin and how the proof should end. Again, let's spell things out:

{1}	1.	$\exists x\ [Fx \vee Gx]$	Premise
	.		
	.		.
{1}	n.	$\exists x\ [Fx] \vee \exists x\ [Gx]$	???

In fact, the strategic considerations involved in this case are really very straightforward. As we noted earlier, to derive any conclusion from an existentially quantified formula as such we must apply EE. But EE is as much a strategy for proof-construction as it is a rule of inference. According to that strategy, we must first carefully select a typical disjunct and then attempt to derive the desired conclusion from it before EE can be applied. So far, no name is involved in the proof. Hence, we can simply take (*Fa* v *Ga*) as the typical disjunct in this context and so a further step can be added as follows:

{1}	1.	$\exists x\ [Fx \text{ v } Gx]$	Premise
{2}	2.	$Fa \text{ v } Ga$	Assumption
	.		
	.		
{1}	n.	$\exists x\ [Fx] \text{ v } \exists x\ [Gx]$	???EE

The question facing us now is how to bridge the gap. Again, here is where the golden rule can help. In fact, we need only ask the first two questions before we get a decisive answer to the question about strategy, first ask: *is the main connective in the conclusion we want to derive a conditional?* Here, again, the answer is 'no'. Next, ask: *is the main connective in any or all of the formulas we are deriving the conclusion from a disjunction?* Again, the answer to that question is 'yes', the typical disjunct is, indeed, a disjunction. Therefore, the strategy for the sub-proof is vE. Moreover, that sub-proof will deliver a conclusion to which EE can finally be applied. Thus, the proof is completed as follows:

$$\exists x\ [Fx \text{ v } Gx] \vdash \exists x\ [Fx] \text{ v } \exists x\ [Gx]$$

{1}	1.	$\exists x\ [Fx \text{ v } Gx]$	Premise
{2}	2.	$Fa \text{ v } Ga$	Assumption TD
{3}	3.	Fa	Assumption
{3}	4.	$\exists x\ [Fx]$	3 EI
{3}	5.	$\exists x\ [Fx] \text{ v } \exists x\ [Gx]$	4 vI
{6}	6.	Ga	Assumption
{6}	7.	$\exists x\ [Gx]$	6 EI
{6}	8.	$\exists x\ [Fx] \text{ v } \exists x\ [Gx]$	7 vI
{2}	9.	$\exists x\ [Fx] \text{ v } \exists x\ [Gx]$	2,3,5,6,8 vE
{1}	10.	$\exists x\ [Fx] \text{ v } \exists x\ [Gx]$	1,2,9 EE

And, again, the extent of the usefulness of the golden rule in the context of QL should be quite clear, even when existentially quantified formulas are involved.

So far, the examples we have considered have not required us to have recourse to the third part of the golden rule, namely, the simple prescription that, failing all else, we should try RAA. In fact, that strategy is at least as useful in QL as it is in PL. Again, in QL as in PL, when the strategy for proof-construction just is not obvious RAA is well worth a try. But remember: generally, the trick with this particular strategy is to assume the opposite of what you want then try to derive a contradiction from that assumption together with any other formula or formulas already involved in the proof. The double negation rules will allow you to finish things off to suit your purposes. Hence, we can add a final strategic prescription:

> Whenever the desired conclusion is the negation of a quantified formula assume the unnegated version of the formula, apply UE or EE as appropriate and derive a contradiction. Apply RAA to complete the proof.

The golden rule is not absolutely fail-safe. And QL contains a number of valid sequents for which proof-construction is genuinely rather difficult. In such cases, although you are well advised to try RAA you may find that you just have to keep bashing away, trying first one line of attack then another.

In all honesty, the true golden rule of proof-construction is simply that practice makes perfect. To that end, I present three sets of revision exercises for proof-construction. This time the sequents belong to QL rather than to PL. Moreover, within each exercise the level of difficulty is generally on the increase from one proof to the next and, again, each exercise involves proofs of a higher level of difficulty than its predecessor. Further, also note that although the first two revision exercises are simply designed to recap on the proof-theory of QL as we have considered it in the present chapter, Revision Exercise III is composed of ten sequents which we have not considered together before. None the less, proofs of each of the sequents composing Revision Exercise III certainly can be constructed just in terms of the existing stock of rules of inference with which you are already familiar. Practice at proof-construction in Revision Exercises I and II will develop your intuitions about strategies for proof and sub-proof in QL in a way which should fully prepare you for Revision Exercise III. Once you have mastered every proof, you will have mastered the proof-theory of quantificational logic. Box 6.7 offers a brief recap on strategies for proof-construction in QL. So, again, study the contents of the box carefully before attempting the subsequent exercises and then Examination 4.

BOX 6.7

Strategies for proof-construction in QL #1

	Premises:	*Conclusion:*
♦ 1.	Universal and unquantified	Unquantified

Procedure: Apply UE to each universal premise to infer formulas about the individual(s) referred to in the unquantified formula(s), i.e. to infer formulas involving the same name(s) as the unquantified formulas. Apply the golden rule to determine the strategy which will yield the desired conclusion about the named individual(s).

	Premises:	*Conclusion:*
♦ 2.	Universal or universal and unquantified	Existential

Procedure: Again, apply UE to each universal premise to infer formulas about the individual(s) referred to in the unquantified formula(s), i.e. to infer formulas involving the same name(s) as the unquantified formulas, any name if none occurs. Apply the golden rule to determine the strategy which will yield the desired conclusion about the named individual(s). Apply EI to the conclusion to complete the proof.

	Premises:	*Conclusion:*
♦ 3.	All universal	Universal

Procedure: Again, apply UE to each premise to infer formulas about one particular individual, i.e. to infer formulas involving a name. Apply the golden rule to determine the strategy for sub-proof which will yield the desired conclusion about that individual. Make sure all assumptions are discharged in the sub-proof. Apply UI to complete the proof.

Strategies for proof-construction in QL #2

	Premises:	*Conclusion:*
♦ 4.	All existential	Existential or unquantified

Procedure:

(i) Construct an *argument from a typical disjunct*, i.e. carefully select and assume a disjunct which is genuinely typical in the context of proof.

(ii) Apply the golden rule to determine the correct strategy for deriving the desired conclusion from the formula assumed as typical disjunct.

(iii) Apply EI before EE and, if necessary, re-apply PL rules.

(iv) Ensure that the restrictions on EE are met, apply EE and, if necessary, re-apply PL rules.

Premises:	*Conclusion:*
♦ 5. Universal and existential	Existential

Procedure:

(i) *First:* carefully select and assume genuinely typical disjunct(s). *Then:* apply UE to the remaining premises to infer formulas about the *same* individual(s).

(ii) Apply the golden rule to determine the correct strategy for sub-proof.

(iii) Apply EI before EE.

(iv) Check the relevant dependencies before applying EE to complete the proof.

♦ 6. Finally, whenever the desired conclusion is the negation of a quantified formula assume the unnegated version of the formula and derive a contradiction. Apply RAA to complete the proof.

REVISION EXERCISE I

1 Prove that the following are valid sequents of QL:

1. $\sim Fa : \sim\forall x\,[Fx]$ (5)
2. $\forall x\,[Fx \rightarrow Gx] : \forall y\,[Fy] \rightarrow Gb$ (6)
3. $\forall x\,[Fx \rightarrow (Gx \rightarrow Hx)], (Fa\ \&\ Ga) : \exists x\,[Hx]$ (8)
4. $\exists x\,[Fx] \rightarrow \mathrm{P} : \forall x\,[Fx \rightarrow \mathrm{P}]$ (6)
5. $\forall x\,[Fx]\ \mathrm{v}\ \forall x\,[Gx] : \forall x\,[Fx\ \mathrm{v}\ Gx]$ (10)
6. $\forall x\,[Gx \rightarrow Hx], \exists x\,[Fx\ \&\ Gx] : \exists x\,[Fx\ \&\ Hx]$ (10)
7. $\forall y\,[Gy \rightarrow Hy] : \exists x\,[Gx] \rightarrow \exists y\,[Hy]$ (8)
8. $\forall y\,[\forall x\,[Rxy]] : \forall x\,[\forall y\,[Rxy]]$ (5)
9. $\forall x\,[\forall y\,[Rxy \rightarrow \sim Ryx]] : \forall x\,[\sim Rxx]$ (8)
10. $\forall x\,[Fx \rightarrow (x = a)] : \exists x\,[Fx \rightarrow Fa]$ (7)

REVISION EXERCISE II

1 Prove that the following are valid sequents of QL:

1. $\forall x\ [Fx \rightarrow Gx], \forall x\ [Hx \rightarrow \sim Gx] : \forall x\ [Fx \rightarrow \sim Hx]$ (10)
2. $\sim\exists x\ [Fx] : \sim Fa$ (5)
3. $\forall x\ [Fx \rightarrow Gx], \exists y\ [Fy\ \&\ Hy] : \exists z\ [Gz\ \&\ Hz]$ (10)
4. $\exists x\ [Fx]\ \mathrm{v}\ \exists x\ [Gx] : \exists x\ [Fx\ \mathrm{v}\ Gx]$ (12)
5. $\exists x\ [Fx\ \&\ \sim Gx] : \sim\forall x\ [Fx \rightarrow Gx]$ (10)
6. $\exists x\ [\exists y\ [Rxy]] : \exists y\ [\exists x\ [Rxy]]$ (7)
7. $\forall x\ [\forall y\ [Rxy \rightarrow Ryx]] : \forall x\ [\exists y\ [Rxy] \rightarrow \exists y\ [Ryx]]$ (10)
8. $\forall x\ [Fx \rightarrow (Gx\ \mathrm{v}\ (x = a))] : Ga \rightarrow \forall x\ [Fx \rightarrow Gx]$ (12)
9. $: \forall x\ [\exists y\ [x = y]]$ (3)
10. $: \forall x\ [Fx\ \mathrm{v}\ \sim Fx]$ (10)

REVISION EXERCISE III

1 Prove that the following are valid sequents of QL:

1. $: \exists x\ [x = a]$ (2)
2. $\forall x\ [Fx] : \sim\exists x\ [\sim Fx]$ (10)
3. $\exists x\ [\exists y\ [\forall z\ [Rxyz]]] : \forall z\ [\exists y\ [\exists x\ [Rxyz]]]$ (9)
4. $\exists x\ [\sim Fx] : \exists x\ [Fx \rightarrow \mathrm{P}]$ (13)
5. $: \exists x\ [Fx]\ \mathrm{v}\ \forall y\ [\sim Fy]$ (12)
6. $: \forall x\ [Fx]\ \mathrm{v}\ \exists x\ [\sim Fx]$ (15)
7. $\forall x\ [Fx \rightarrow ((x = a)\ \mathrm{v}\ (x = b))], \exists x\ [Fx\ \&\ Gx] : Ga\ \mathrm{v}\ Gb$ (15)
8. $\exists x\ [Fx\ \&\ Gx], \exists x\ [Fx\ \&\ \sim Gx] : \sim\forall x\ [\forall y\ [(Fx\ \&\ Fy) \rightarrow (x = y)]]$ (18)
9. $\exists x\ [Fx\ \&\ Gx], \exists x\ [Fx\ \&\ \forall y\ [Gy \rightarrow \sim Rxy]] : \exists x\ [Fx\ \&\ \sim\forall y\ [Fy \rightarrow Ryx]]$ (19)
10. $\forall x\ [\forall y\ [(Fx\ \&\ Fy) \rightarrow (x = y)]] : \exists x\ [\forall y\ [Fy \rightarrow (x = y)]]$ (26)

Examination 4 in Formal Logic

Answer every question.

1. Prove that the following are valid sequents of QL:

   (i) $: \forall x\, [Fx \lor Gx] \rightarrow (\forall x\, [Fx] \rightarrow \forall x\, [Gx])$

   (ii) $: \exists x\, [Fx \,\&\, Gx] \rightarrow (\exists x\, [Fx] \,\&\, \exists x\, [Gx])$

   (iii) $: \forall x\, [Fx \,\&\, Gx] \leftrightarrow (\forall x\, [Fx] \,\&\, \forall x\, [Gx])$

   (iv) $: \exists x\, [\exists y\, [Rxy]] \leftrightarrow \exists y\, [\exists x\, [Rxy]]$

2. Complete the following proof:

{1}	1.	$\exists x\, [Fx \rightarrow (a = x)]$	Premise
	2.		Assumption TD
	3.		Assumption
{4}	4.	Fb	Assumption
	5.		2,4 MP
	6.		=I
{2,4}	7.	$(b = a)$	5,6 =E
{2,3,4}	8.	$(b = a) \,\&\, {\sim}(b = a)$	3,7 &I
	9.	${\sim}Fb$	
{2}	10.		3,9 CP
{2}	11.		10 EI
{1}	12.		1,2,11 EE

3. In Examination 3 you should have formalised the following argument as a sequent of QL:

   1. All horses are animals

   Therefore,

   2. All horse's heads are animal's heads.

   This argument may be represented in QL by the following sequent:

$\forall x\ [Fx \rightarrow Gx] : \forall x\ [\exists y\ [Fy\ \&\ Rxy] \rightarrow \exists y\ [Gy\ \&\ Rxy]]$

(i) Construct a QL-interpretation which shows that this formalisation fairly represents the original natural language argument.

(ii) Construct a proof of the sequent in question.

4. Prove that the following is a valid sequent of QL:

$\exists x\ [Fx\ \&\ \forall y\ [Gy \rightarrow Rxy]],$

$\forall x\ [Fx \rightarrow \forall y\ [Hy \rightarrow {\sim}Rxy]] : \forall x\ [Fx \rightarrow {\sim}Hx]$

Notes

1 The system of predicate calculus introduced in this chapter and many of the most insightful illustrations of that system are due to Lemmon, E.J., [1965], *Beginning Logic*, London, Thomas Nelson and Sons. In adopting a Lemmon-style system, I do not mean to imply that this approach is more pleasing formally or philosophically than systems which exploit free and bound variables, for example, or proper and arbitrary names. However, Lemmon's approach does seem to me to have pedagogical advantages both in allowing the rules of inference to be explained intuitively (e.g. without introducing various further symbols for variables) and in providing students with simple but effective checks for misuse of the rules. Any mistakes in the current presentation are mine alone.

2 See, for example, Faris, J.A., [1964], *Quantification Theory*: Monographs in Modern Logic, London, Routledge & Kegan Paul.

3 Again, see, for example, Faris, [1964].

4 As we will see below, certain applications of existential introduction constitute exceptions to this rule.

5 See Read, Stephen, [1995], *Thinking about Logic: An Introduction to the Philosophy of Logic*, Oxford, Oxford University Press, Chapter 5: 'Plato's Beard: On What There Is and What There Isn't'.

6 The idea for this (very useful and appropriate) question is due to Newton-Smith, W.H., [1985], *Logic: An Introductory Course*, London, Routledge, Chapter 7.

7
Formal Logic and Formal Semantics #2

7
Formal Logic and Formal Semantics #2

I
Truth-Trees Revisited

In Chapter 5, we considered QL as a formal language into which we could translate fairly complex natural language arguments in a way which clearly showed up both the logical form of those arguments and the internal, logical structure of the sentences which compose them. Subsequently, in Chapter 6, we went on to develop a deductive apparatus for QL and so established a formal system for QL in terms of which we were able to construct formal proofs of even the most subtle and complex sequents. In contrast, our discussion of the semantics of QL has been limited to informal methods. Hence, in the present chapter we turn to a more detailed examination of the formal semantics for QL and to one particularly useful method for testing QL sequents for **semantic validity**.

As you will see, this new focus on the formal semantics for QL and the associated method of testing for validity allows us to bring together a number of familiar points from both of those earlier chapters and, hopefully, will help to clarify those points still further. To that end, recall the discussion of the *truth-tree method* for PL sequents given in Chapter 4. In that earlier discussion I rather gave the game away when I pointed out that this particular semantic method has an applicability which stretches beyond PL into QL. Therefore, it should be no surprise to hear that the formal semantic method which we will exploit for QL is, indeed, the truth-tree method. Further, you will no doubt be pleased to hear that, in one sense, we have already done much of the work required to appreciate the nature of the formal method for sequents of QL. So, let's quickly recap on that earlier discussion of the nature of the truth-tree method in order to refresh our memories.

First, recall that the truth-tree method provides a way of testing formulas for *consistency*. As you know, the idea of consistency between sentences in a language is fundamental to formal logic. Now although the term 'consistency' is used in a number of importantly different senses in Logicspeak, for our purposes, the consistency of a set of well-formed formulas simply implies that each and every well-formed formula in that set can be true at one and the same time. Hence:

> A set of well-formed formulas is consistent if and only if every member of that set can be true simultaneously.

It follows that a proof of consistency in this sense is a proof of the existence of a true interpretation of all the members of that set. Earlier, we defined a true interpretation as a *model*. So, in the present context, we can equally well say that a proof of consistency is a proof of the existence of a model. As regards PL, we said that the truth-tree method was precisely a test for consistency in that sense, i.e. it provides an answer to the question: could the formulas constituting a given set of formulas all be true together? Exactly the same is true of the method at the level of QL, except, of course, that any or all of the formulas involved this time may be quantified formulas.

In other words, just as we could exploit the truth-tree method to answer questions about consistency by constructing consistency-trees for any given set of formulas of PL so we can exploit the same method to construct consistency-trees for QL. Again, just as in PL, when we construct a consistency-tree in QL we first list each formula on a separate line, one underneath the other. Again, this procedure reflects the most obvious way in which the formulas might all be true together. Having listed the complex formulas we are interested in we then begin to break down or develop each complex formula in terms of development rules. Just as the proof-theory of QL inherited all the rules of inference for PL, so the **truth-tree method for QL** inherits all of the development rules belonging to that method in PL. There is undoubtedly no harm in reminding ourselves of that basic stock of rules here. So, consider the PL rules once more. They are given in Box 7.1.

As we can see, the PL development rules simply exploit the type of formula involved, i.e. conjunction, disjunction and so on. However, we also exploited the distinction between branching and non-branching formulas to formulate the following procedural rule: *always develop every non-branching formula before any branching formula*. Adherence to that procedural rule ensures that we preserve the trunk-like effect for as long as possible, i.e. that all the branching occurs at the top of the tree. Finally, having constructed the tree, we must carefully study the formulas in each branch, reading up from the tip of the branch back to the very beginning of the trunk to identify any contradictions among the formulas lying on that branch. Remember: if a branch does contain a contradiction then that branch is dead and we record that fact by writing an 'X' under it. If every branch dies then the tree is dead.

BOX 7.1

PL development rules

```
1.  A & B     2.   A v B      3.   A → B     4.  ~(A → B)
      |            /   \           /   \             |
      A           A     B        ~A     B            A
      B                                             ~B

5.  ~(A & B)   6. ~(A v B)   7.  A ↔ B     8.  ~(A ↔ B)     9.  ~~A
     /   \          |            /   \          /   \            |
   ~A    ~B        ~A           A    ~A        A    ~A           A
                   ~B           B    ~B       ~B     B
```

In QL as in PL, each branch again represents a different possible way in which all the formulas involved might be true together. So, for example, in the case of disjunction, our splitting the branch still represents two possible ways in which the disjunction might be true, i.e. if either disjunct is true. Again, because there is only one way in which a conjunction can be true, namely, when both conjuncts are true, we do not split the branch and that practice reflects the fact that there is only one possibility involved in any such case. Thus, each branch represents an attempt to assign truth-values to the formulas involved in such a way as to bring those formulas out as true simultaneously, i.e. each branch represents an interpretation.

Further, because each branch represents an interpretation of the components of each formula, we know that when a branch dies that particular interpretation results in a contradiction, i.e. an *inconsistency*. If every branch dies then there is no interpretation which does not result in inconsistency. But, if that is so, then there is no way in which all the formulas being tested could all be true together. Therefore, that set of formulas must be strictly inconsistent. Conversely, if there is even one live branch then there is one way in which all the formulas being tested could be true together, i.e. a true interpretation, a model. So, that set is a consistent set.

Now, we can easily go on to exploit this insight about consistency in order to construct a test for semantic validity. After all, any well-formed formula is *semantically valid* if and only if the negation of that formula is inconsistent. Further, a sequent is valid if and only if the set consisting of the conjunction of the premises of that sequent together with the negation of the conclusion is inconsistent. It follows at once that, for any sequent, if the set consisting of the premises of that sequent together with the negation of the conclusion of the sequent is consistent then there is an interpretation under which the premises of the original sequent are true and the conclusion is false. If that is so, then the original sequent must be invalid. Conversely, if that same set is inconsistent then there is literally no way in which the premises can be true and the conclusion false and, therefore, the sequent must be valid.

So much for the truth-tree method[1] for PL sequents. But how are we to extend the method's applicability to sequents of QL? Well, just as the rules of inference for PL had to be supplemented to cope with quantified formulas when we established our formal system for QL, so the set of development rules and procedural considerations for PL **consistency-trees** must also be supplemented at the level of QL. After all, none of the existing rules is any help when it comes to developing any universally or existentially quantified formula. Hence, we must go on to introduce development rules for precisely these kinds of formula if we are ever to get the method off the ground for QL. But this is relatively straightforward. When constructing truth-trees in QL, we are only ever concerned with the **instantiation** of quantified formulas. For that purpose, we can simply adopt two rules of instantiation which are both very elegant and easy to use. Moreover, these particular rules of instantiation are fundamental to the version of the truth-tree method which is to be found in what is currently the most important textbook on the metatheory of quantificational logic. So, the reader who is familiar with this version of the method will be the one who is best prepared for a course of further study in formal logic and, therefore, it is that version of the truth-tree method which I will outline here. To that end, note the following terminological point carefully: the new rules allow us to instantiate quantified formulas to what I will now refer to as **terms**.

In effect, terms are simply proper names. Thus, quite generally, quantified formulas are instantiated to the same generic kinds of name which are represented by lower-case letters from the beginning of the alphabet, i.e. *a*, *b*, *c* . . . With that point clearly in mind, we can now go on to consider each of the development rules for quantified formulas.

First, consider the case of universally quantified formulas. In Chapter 6 we noted that it is of the essence of any such formula that it should imply its instances. So, for example, $\forall x\ [Fx]$ implies *Fa* and *Fb* and *Fc* and so on. Hence, it should be no surprise to learn that in constructing a QL consistency-tree any such inference from a universally quantified formula to one of its instances is perfectly legitimate. In Chapter Six, we went on to make explicit one particular rule, the rule UE, as the rule governing any such inference from a quantified formula to any particular instance, i.e. to any individual referred to by a proper name. In constructing a QL consistency-tree any such inference from a universally quantified formula to one of its named instances is still perfectly legitimate. Hence, in essence, we can legitimately handle universally quantified formulas in tree-construction in exactly the same way, i.e. simply by inferring from them one or more of their instances. For present purposes, however, we are not concerned to contrast elimination-rules with introduction-rules. Rather, we are solely concerned with instantiating quantified formulas. To keep that point in focus, then, from now on we will describe any such inference made in the process of tree-construction as an instance of **universal instantiation**, or **UIN**, for short. Where ν is any variable, $\phi(\nu)$ is any expression in which ν may occur, ρ is any proper name and $\phi(\rho)$ is the result of substituting ρ for ν, we can define universal instantiation as follows:

$$\text{UIN:} \quad \frac{\forall \nu\ [\phi(\nu)]}{\phi(\rho)}$$

Having added this new rule to the existing stock of PL rules we can immediately go on to demonstrate the validity of certain QL sequents, i.e. those which involve only positive universally quantified formulas and subject–predicate formulas. For example, consider the following QL sequent:

$: \forall x\ [Fx] \rightarrow Fa$

In this case, no premises are involved and so we simply write down the negation of the whole formula on line 1. Further, because the formula in question is a conditional, its negation has the form of a negated conditional. To develop any such formula, however, we can exploit the relevant PL rule. Having done so, it only remains to apply UIN to the resulting universally quantified formula in order to complete the truth-tree test. In the process, we do generate a contradiction and, therefore, prove that the original sequent is valid. Consider the completed tree carefully:

1.	$\sim(\forall x\ [Fx] \rightarrow Fa)$	Negated conclusion
2.	$\forall x\ [Fx\ ]$	From 1 by $\sim(A \rightarrow B)$
3.	$\sim Fa$	From 1 by $\sim(A \rightarrow B)$
4.	Fa	From 2 by UIN
	X 3,4	

Clearly, although this particular tree does not branch, the trunk does contain an obvious contradiction and, therefore, the original sequent is, indeed, shown to be valid.

Next, we must consider the case of existentially quantified formulas in the context of tree-construction. In the previous chapter we discussed at some length the ways in which we can go wrong in reasoning from existentially quantified formulas as such. However, we also noted that, in general, we could reason quite safely from an existentially quantified formula provided that our reasoning always concerned an element of the domain about which we had absolutely no special knowledge. Hence, we took care always to assume a formula of the relevant kind first, i.e. a *typical disjunct*. It should be no surprise, then, that we endeavour to preserve and maintain that restriction on inferences from existentially quantified formulas in the process of tree-construction. In fact, the new rule of instantiation we require here ensures that this is the case in a very similar way, i.e. we avoid legitimating the kind of invalid inferences we were concerned about in the previous chapter by rigorously imposing the following simple restriction:

> Whenever you make an inference from an existentially quantified formula the term to which you instantiate must always be a term which does not already occur in any formula in the branch, including the formulas in the trunk.

For example, in the simplest possible kind of case we can select *Fa* as an instance of $\exists x\ [Fx]$ provided only that the term *a* does not already occur anywhere in the branch. If there are no terms already occurring in the branch we can safely begin with *a* or any other term we care to choose. Hence, we can formally state the new rule of **existential instantiation**, or **EIN** for short as I shall now refer to it, and the new restriction on that rule as follows:

EIN:

$$\frac{\exists \nu[\phi(\nu)]}{\phi(\rho)}$$

Restriction: Provided ρ does not occur in any formula in the branch or trunk.

When constructing a truth-tree we are always seeking to identify and make explicit contradictions among the formulas involved. Now, precisely in order to maximise the possibility of generating contradictions, we will go on to impose a restriction on inferences from universally quantified formulas here. Quite simply, although we can certainly validly infer from a universally quantified formula any of its instances, we will impose the requirement that:

> Any universal instantiation is to a term which already occurs in some formula on the branch or any term if none occurs.

Hence, we can formally state the new rule of *universal instantiation*, or *UIN* as I shall now refer to it, and the new restriction on that rule as follows:

$$\text{UIN:} \quad \frac{\forall v[\phi(v)]}{\phi(\rho)}$$

Restriction: Provided ρ does occur in some formula on the branch or in the trunk (if no terms occur, instantiate to any term).

Note carefully how well these two simple restrictions can work together to maximise the possibility of generating contradictions in the process of tree-construction. If we always develop every existentially quantified formula before any universally quantified formula then we will first introduce a stock of terms into the tree. Having done so, we can then go on to develop any universally quantified formula to just those existing terms. In the process of so doing, we maximise the possibility of generating contradictions among formulas by covering each and every term introduced into the tree by any existentially quantified formula(s). We can now embody both of these restrictions in a single procedural rule which is both perfectly straightforward to understand and easy to use. Because the new rule governs the order in which we introduce terms in the process of tree-construction, I will call it the **rule of term-introduction**. The heart of the rule of term-introduction can now be expressed by the following simple prescription:

Always develop each and every existentially quantified formula before you develop any universally quantified formula.

With the new rules governing the development of quantified formulas and the crucial procedural rule of term-introduction clearly in mind we can now consider the truth-tree method for QL as applied to sequents involving different kinds of quantified formula for the first time. For example, consider the following sequent:

$\forall x\,[Fx] : {\sim}\exists x\,[{\sim}Fx]$

In this case, the premise is a universally quantified formula. But note carefully that the conclusion is in fact a negated existentially quantified formula. In order to get any truth-tree test for validity off the ground, we must assume not the conclusion itself but the negation of the conclusion. Here, then, we assume the double-negative: ${\sim}{\sim}\exists x\,[{\sim}Fx]$. Given DNE, that formula reduces to the simple existentially quantified formula $\exists x\,[{\sim}Fx]$. Now, the rule of term-introduction requires us to develop that formula before we develop the universally quantified premise. When we do instantiate the relevant existentially quantified formula, EIN requires that we instantiate to a term not already occurring in the tree. Finally, when we instantiate the premise, UIN requires that we instantiate to a term which does already occur, i.e. precisely to the existing term.

Consider the tree in full:

1.	$\forall x\ [Fx]$	Premise
2.	$\sim\sim\exists x\ [\sim Fx]$	Negated conclusion
3.	$\exists x\ [\sim Fx]$	2 DNE
4.	$\sim Fa$	3 EIN (term not already occurring)
5.	Fa	1 UIN (term already occurring)
	X 4,5	

Again, this tree does not branch, but again the trunk ends in a contradiction. Therefore, the original sequent is valid. And note very carefully how well the rules governing the quantifiers work together with the rule of term-introduction. Because we are required to develop every existentially quantified formula before we develop any universally quantified formula we are forced to develop the existentially quantified formula $\exists x\ [\sim Fx]$ first, at line 3. Further, the restriction we imposed on the rule governing the development of existentially quantified formulas requires us to introduce a term not already occurring in the tree. At this stage, however, no terms occur and so we simply pick *a* for the purpose. Finally, when we develop the last remaining formula $\forall x\ [Fx]$ we are required precisely to instantiate that formula to a term which does already occur. Thereby, we maximise the possibility of generating a contradiction and, indeed, we very quickly arrive at the contradiction (Fa & $\sim Fa$).

The preceding tree may already have made you aware of a final point which we have not yet considered. And that is just this: we know how to develop universally and existentially quantified formulas in the process of tree-construction, but how are we to develop negated such cases? That is, how are we to develop formulas of the form $\sim\forall x\ [Fx]$ and $\sim\exists x\ [Fx]$? The answer, as you may already have guessed, is just that we solve this problem simply by exploiting two of the *quantifier-equivalences* we established in Chapter 6, i.e. in Exercise 6.4. Hence, in the course of tree-construction:

1. Replace any formula of the form $\sim\forall x\ [Fx]$ with a formula of the form $\exists x\ [\sim Fx]$.
2. Replace any formula of the form $\sim\exists x\ [Fx]$ with a formula of the form $\forall x\ [\sim Fx]$.

Note carefully that when we do exploit the first equivalence for $\sim\forall x\ [Fx]$ what we infer is an existentially quantified formula. As ever, we want such formulas to have priority over universally quantified formulas in terms of development. So, we must expand our procedural rule. Hence, we can now state the rule of term-introduction in full as follows:

> Always develop each and every existentially quantified formula and every negated universally quantified formula (i.e. any formula whose development generates an existentially quantified formula or a negated universally quantified formula) before you develop any universally quantified formula or negated existentially quantified formula.

Finally, when you are faced with developing a number of quantified formulas of the same kind, i.e. when all the formulas involved are universally quantified or all are existentially quantified, the particular order in which you choose to develop the formulas internally really does not matter. As a result there will be some individual variation from tree to tree, as it were, but, in the last analysis, if properly constructed, there is no real basis for preference here. Equally, when faced with formulas of the form $\sim\forall x\ [Fx]$ and formulas of the form $\exists x\ [\sim Fx]$ in a single tree, again, the particular order in which you choose to develop the formulas involved really does not matter. However always remember to take care to follow the fully expanded version of the rule of term-introduction in the process of tree-construction.

Before we proceed to an exercise, let's consider together a tree which illustrates a number of the points made so far in this chapter, i.e. the tree for the following sequent:

$: (\forall x\ [Fx] \rightarrow \mathrm{P}) \rightarrow \exists x\ [Fx \rightarrow \mathrm{P}]$

In this instance, there are no premises to consider and so we begin the tree by writing the negation of the entire sequent on a line as follows:

1.	$\sim ((\forall x\ [Fx] \rightarrow \mathrm{P}) \rightarrow \exists x\ [Fx \rightarrow \mathrm{P}])$	Negated conclusion

Clearly, this formula is a negated conditional. Therefore, we apply the appropriate development rule (ultimately, from PL) and rewrite that formula on lines 2 and 3 as follows:

2.	$\forall x\ [Fx] \rightarrow \mathrm{P}$	From line 1
3.	$\sim\exists x\ [Fx \rightarrow \mathrm{P}]$	From line 1

The two resulting formulas are a conditional and a negated existentially quantified formula respectively. We cannot develop the conditional without branching, of course. However, we can exploit a quantifier equivalence to rewrite the negated existential formula without branching. Therefore, we rewrite that formula on a new line as follows:

4.	$\forall x\, [{\sim}(Fx \rightarrow \mathrm{P})]$	From line 3

At this point, we could instantiate the formula on line 4. However, we are seeking to maximise the possibility of generating contradictions and, as yet, there are no terms in the tree. A little reflection quickly reveals that if we now develop the formula on line 2 we will generate a formula which in turn converts into an existential formula, i.e. a formula we could use to introduce a term not already occurring. Having developed that existentially quantified formula, we could then develop the universally quantified formula on line 4 in a way which might well generate a contradiction. Hence, we should develop the formula on line 2 first before developing the formula on line 4. Thus, we can sum up and complete the tree as follows:

1.	$\sim((\forall x\, [Fx] \rightarrow \mathrm{P}) \rightarrow \exists x\, [Fx \rightarrow \mathrm{P}])$		Negated conclusion
2.	$\forall x\, [Fx] \rightarrow \mathrm{P}$		From line 1
3.	$\sim\exists x\, [Fx \rightarrow \mathrm{P}]$		From line 1
4.	$\forall x\, [{\sim}(Fx \rightarrow \mathrm{P})]$		From line 3
	/	\	
5.	$\sim\forall x\, [Fx]$	P	From line 2
6.	$\exists x\, [{\sim}Fx]$	\	5 Quantifier-equivalence
7.	$\sim Fa$	\	6 EIN (term *not* occurring)
8.	$\sim(Fa \rightarrow \mathrm{P})$	$\sim(Fa \rightarrow \mathrm{P})$	From line 4
9.	Fa	Fa	From line 8
10.	$\sim\mathrm{P}$	$\sim\mathrm{P}$	From line 8
	× 7,9	× 5,10	

In this instance, both branches end in contradiction, the tree is dead and, therefore, the original sequent is shown to be valid. Again, note carefully how well the PL rules (both procedural and developmental) can work together with their QL counterparts in the process of tree-construction.

In all honesty, I should admit that because each of the new rules we have introduced in this section is ultimately a development rule which exploits the shape of QL formulas those rules are in fact syntactical rather than semantic. Hence, we have in fact supplemented QL with an alternative syntax. It follows that, so far, we have only really characterised another method of proof in QL. In the next section, we will in turn supplement the existing set of development rules with genuinely semantic rules. None the less, the new syntactical rules remain essential to the method even when supplemented so as to become a genuinely semantic method. So, practise with the existing stock of rules is crucially important. Exercise 7.1 is designed to give you the opportunity to practise all that you have learned about the method to date, but first study the contents of Box 7.2 carefully. Further, I again include a Jeffrey-style flow chart for tree-construction. Consider the flow chart (Figure 7.1) carefully.

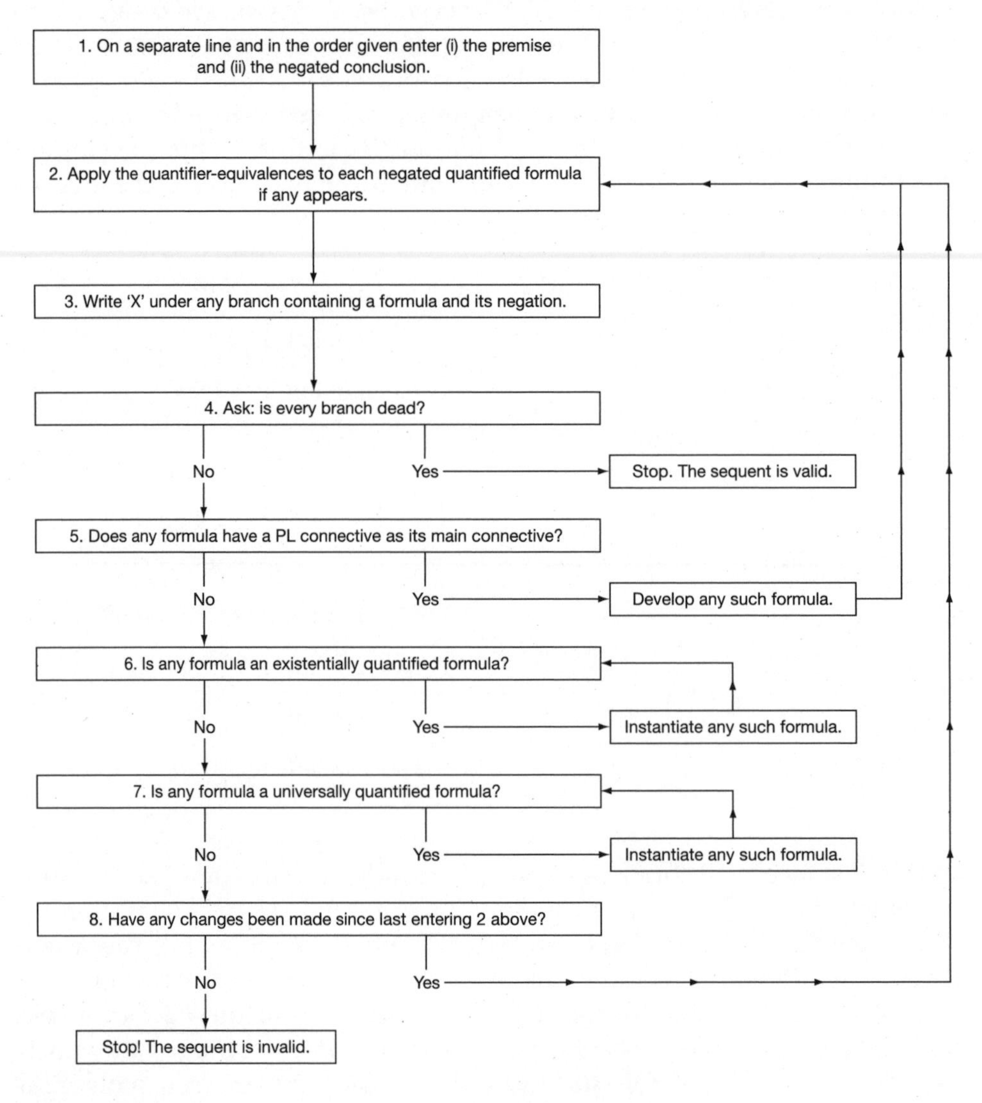

Figure 7.1 A Jeffrey-style flow chart for QL truth-trees

BOX 7.2

The truth-tree method for QL

♦ *Procedural rules:*

1. List the premises (if any) and the negation of the conclusion on separate lines one underneath the other to form the trunk of the truth-tree. Consider the kinds of formula involved carefully and:

2. Replace any formula of the form $\sim\forall x\,[Fx]$ with a formula of the form $\exists x\,[\sim Fx]$.

3. Replace any formula of the form $\sim\exists x\,[Fx]$ with a formula of the form $\forall x\,[\sim Fx]$.

4. Apply the *rule of term-introduction,* i.e.: Develop each and every existentially quantified formula and every negated universally quantified formula (i.e. any formula whose development generates an existentially quantified formula or a negated universally quantified formula) before you develop any universally quantified formula or negated existentially quantified formula.

♦ *Development rules:*

1. UIN – universal instantiation:

Informally: Instantiate every universally quantified formula to any term {t1 . . . tn} where {t1 . . . tn} are individual terms *already occurring* in some formula on the branch or in the trunk (any such term if none occurs).

Formally:

$$\text{UIN:}\quad \frac{\forall v\,[\phi(v)]}{\phi(\rho)}$$

Restriction: Provided ρ does occur in some formula on the branch or in the trunk (if no terms occur, instantiate to any term).

2. EIN – existential instantiation:

Informally: Instantiate every existentially quantified formula to a specific term {t1 . . . tn} provided that the specific term does not already occur in any formula in the branch or trunk.

Formally:

$$\text{EIN:} \quad \frac{\exists v\ [\phi(v)]}{\phi(\rho)}$$

Restriction: Provided ρ does not occur in any formula in the branch or trunk.

EXERCISE 7.1

1. Test the following sequents for validity using the truth-tree method for QL:

    1. $: \exists x\ [Fx \vee {\sim}Fx]$
    2. $\forall x\ [Fx \rightarrow Gx], \forall x\ [Fx] : \forall x\ [Gx]$
    3. $\forall x\ [Fx \rightarrow Gx], \exists x\ [Fx] : \exists x\ [Gx]$
    4. $\forall x\ [Fx \rightarrow Gx], \forall x\ [{\sim}Gx]: \forall x\ [{\sim}Fx]$
    5. $: \forall x\ [Fx] \rightarrow \exists x\ [Fx]$
    6. $\forall x\ [Fx]\ \&\ \forall x\ [Gx] : \forall x\ [Fx\ \&\ Gx]$
    7. $: \forall x\ [Fx \rightarrow Gx] \rightarrow (\forall x\ [Fx] \rightarrow \forall x\ [Gx])$
    8. $: \forall x\ [Fx\ \&\ Gx] \rightarrow (\forall x\ [Fx]\ \&\ \forall x\ [Gx])$
    9. $\exists x\ [(Fx\ \&\ Gx) \rightarrow Hx] : \forall x\ [(Fx\ \&\ Gx) \rightarrow Ha]$
    10. $: \exists x\ [Fx\ \&\ Gx] \rightarrow (\exists x\ [Fx]\ \&\ \exists x\ [Gx])$

II
More on QL Truth-Trees

So far, you should have found that each of the QL sequents we have tested using the truth-tree method is, in fact, a valid sequent. As you know, however, it is certainly not the case that every QL sequent is valid! Before we go on to consider some examples of invalid QL sequents let's reflect on the definition of validity for QL which we considered earlier in Chapter 5, i.e. recall that:

> A QL sequent is valid if and only if there is no possible interpretation under which all the premises of that sequent are true and the conclusion is false.

It follows at once that to identify and articulate a possible interpretation under which a given sequent is shown to have true premises and a false conclusion is to demonstrate the invalidity of the sequent in question. I will describe any such interpretation as an **invalidating QL interpretation**, or an **IQLI**, for short. Every IQLI is a *counterexample* to a QL sequent. But how are we to locate IQLIs in any given case? Again, here is where the truth-tree method can help.

Recall that much of the value and insight of the truth-tree method for PL derived precisely from that fact that we could exploit that method in order to identify invalid sequents. Moreover, we were also able to use the method to demonstrate the invalidity of sequents by identifying an IPLI of that sequent and, subsequently, by constructing an actual counterexample on that basis. For example, recall (from Section XI of Chapter 4) the tree for the following obviously invalid PL sequent:

$$P \rightarrow Q, Q : P$$

1.	$P \rightarrow Q$	Premise
2.	Q	Premise
3.	$\sim P$	Negated conclusion
	/ \	
4.	$\sim P$ Q	From line 1
	✓ ✓	

As the ticks indicate, and just as we would expect, we are faced with two live branches here. Therefore, the sequent is invalid. Each live branch represents a particular interpretation under which that sequent is invalid, i.e. a particular way in which we might assign truth-values to the formulas on that branch such that, under that assignment, the premises will be true while the conclusion is false. To spell out such an interpretation is to make clear an IPLI for that sequent which indicates the general character of a counterexample to the sequent. Further, as we also noted earlier, for the purpose of constructing an actual counterexample all the requisite work has already been done here: all we need do is read up any live branch assigning T to any unnegated atom and F to any negated atom (in this case, note that no matter which branch we choose we arrive at the same interpretation and, therefore, the counterexample has the same general character in both cases). So, for example, reading up the left-hand branch, we arrive at the following IPLI:

IPLI P: {F} Q: {T}

Finally, in order to construct an actual counterexample, we simply substitute any obviously false sentence for P, any obviously true sentence for Q, and then go on to mimic the form of the sequent accordingly. For clarity and simplicity, we agreed to use basic arithmetical sentences for these purposes. So, for example, we might pick: '0 = 1' for P and '2 + 2 = 4' for Q and then go on to mimic the form of this particular sequent. In so doing, we clearly demonstrate the invalidity of the sequent in question and, hence, the invalidity of the traditional fallacy of affirming the consequent.

In practice, much of what we have just said about PL truth-trees carries over and applies to QL truth-trees. Again, if in the process of tree-construction in the present fragment of QL we succeed in identifying a live branch we can rest assured that the sequent in question is invalid.[2] However, in terms of constructing an invalidating interpretation there is a little more work to be done at the level of QL. And that should be unsurprising. After all, a mere assignment of truth-values is not yet a QL interpretation. At this level, truth and falsity are no longer absolute but are relativised to a choice of domain together with a specification of how the elements of formal vocabulary involved hook up with elements of that domain. In other words, truth-values are not simply assigned analogously to PL cases. Rather, the QL interpretation should *generate* the set of truth-values we require.

None the less, the live branch does indeed exemplify what we might call our **target set of truth-values**, or **TST** for short, which an IQLI should generate, i.e. a set of truth-values for the formulas involved which, if such values resulted from a QL interpretation, would make that interpretation an IQLI which would thus invalidate the sequent. As a first step towards constructing an IQLI, we can still proceed by reading up the live branch to determine the target set we require. Moreover, we can consider T to be the target value for any unnegated formula and F the target value for any negated formula. Hence, a live branch still tells us a considerable amount about the invalidating interpretation we require. In a sense, it shows us what the final goal is, even if it does not tell us how to get there! When considering a QL sequent our basic statements are not in general, *atomic* formulas, i.e. sentence-letters, but rather *molecular* formulas, i.e. subject–predicate formulas. Hence, for invalid QL sequents, we need to determine target values not for simple sentence-letters but rather for subject–predicate formulas occurring on the live branch. This idea may seem unfamiliar but, again, in practice, it is very easy to use:

> To construct an IQLI for a QL sequent:
>
> First: determine the target set of truth-values for the formulas on the live branch by:
>
> (i) taking T to be the target value for any unnegated subject–predicate formula and
> (ii) taking F to be the target value for any negated subject–predicate formula.

Before we go any further, let's consider as an example the following obviously invalid QL sequent:

$\sim Fa \;\&\; \sim Fb : \exists x\, [Fx]$

As ever, we begin by writing down the premise followed by the negation of the conclusion, i.e.:

1.	$\sim Fa \;\&\; \sim Fb$	Premise
2.	$\sim\exists x\, [Fx]$	Negated conclusion

Now, in order to develop the formula on line 2 we must exploit a quantifier-equivalence as follows:

3.	$\forall x\, [\sim Fx]$	2 Quantifier-equivalence

We can now develop either the formula on line 1 or the formula on line 3 without branching. Hence, we may perfectly well develop the formula on line 1 first as follows:

4.	$\sim Fa$	From line 1
5.	$\sim Fb$	From line 1

Finally, to complete the tree, we apply UIN twice, i.e. we instantiate to both terms which already occur in the tree:

6.	*~Fa*	3 UIN
7.	*~Fb*	3 UIN

Again, the tree does not branch. But this time there is no contradiction among the formulas in the trunk. Therefore, the sequent is invalid. So, let's consider how we might arrive at an IQLI. Well, here we have no choice about which branch to examine. We simply read up the trunk. Now, there are only two subject–predicate formulas in the trunk: *~Fa* and *~Fb*, respectively. In order to arrive at an IQLI, then, we follow the procedure outlined above and determine the target set of truth-values we require. In this case, both formulas are negated. F is our target for any negated subject–predicate formula. Hence, we arrive at the following TST:

TST: *Fa*: {F} *Fb*: {F}

Hence, the IQLI we require is one under which both subject–predicate formulas *Fa* and *Fb* are false. To understand how to construct such an interpretation we need to appreciate how such values might be generated by a QL interpretation. To that end, we must think a little harder about the nature of the notions of truth and falsity in QL.

First, recall that in our earlier discussion of these notions in Chapter 5 we agreed that no sentence of QL is ever actually true or false just in itself, absolutely, as it were. Rather, we agreed, whether a given statement of QL is true or false depends upon what we take that statement to be about. It follows logically that truth and falsity in QL are now relativised to whatever set of things we understand ourselves to be making statements about, i.e. relativised to the universe of discourse, the domain. Further, we also noted that the truth-value of any given QL formula depends not just on the choice of domain but also on how the names, predicates and relations hook up with the elements of that domain, i.e. what stands for what in any given case. Finally, we noted that both to specify a domain and to spell out exactly how names, predicates and relations hook up with the elements of that domain is to give an interpretation for QL, i.e. a QL-interpretation. Earlier I also promised that such a QL-interpretation would, in fact, provide all the truth-conditions we require in order to decide the truth-value of each and every sentence of QL in a given context. It follows at once that our TST for *Fa* and *Fb* presupposes such an interpretation. So, we must now spell out a possible QL interpretation relative to which or, as we shall say, *under* which *Fa* and *Fb* are both false. But that is perfectly straightforward. For example, suppose that the domain consists of animals and that the predicate-letter *F* stands for the predicate ' . . . is a dog'. As you know, I am the proud owner of two cats, namely, Tiffin and Zebedee. Unsurprisingly, neither of my cats is a dog. Now, consider the following interpretation:

$\mathfrak{I}$ **D**: {**animals**}

F: . . . is a dog

a: Tiffin

b: Zebedee

Keeping things purely intuitive, it should be quite clear that neither Tiffin nor Zebedee belongs to the class of animals which are dogs. Rather, they both belong to the class of animals which are cats. Therefore, under this interpretation both *Fa* and *Fb* are indeed obviously false. And that is exactly the target set of truth-values we require to construct an IQLI here.

All this may seem rather light-hearted but some important points have been illustrated none the less. In fact, in what follows it is particularly useful to keep the intuitive notion of *belonging* clearly in mind. Unsurprisingly, formal logicians tend not to construct their IQLIs in terms of Tiffins and Zebedees. Rather, formal logicians tend to construct their IQLIs in terms of numbers, i.e. they simply let the domain consist of a small set of numbers. A very important logical result underpins this practice: while there is no guarantee that IQLIs can always be constructed in terms of Tiffins and Zebedees, if an IQLI can be constructed, there is a guarantee that it can always be constructed using the natural numbers. The guarantee in question is a profound metatheorem known as the Lowenheim–Skolem Theorem, after Leopold Lowenheim (who first proved the result) and Thoralf Skolem. Given this guarantee, we will adopt the practice of using numbers and will generally let the domain consist of as small a set of numbers as we can manage. So, for example, consider the following domain:

D: {**1,2,3**}

Following the same guidelines as we did for Tiffin and Zebedee, we construct our QL-interpretation by assigning some particular element of D to each of our names and some subset of the elements of D as the extension of our lone predicate. So, for example, we might construct the following, terribly simple, interpretation:

$\mathfrak{I}$ **D**: {**1**,**2**,**3**}

F: {1}

a: 2

b: 3

Simple as this interpretation undoubtedly is, note immediately that, again, we have already done everything we need to spell out just the kind of interpretation we need for the example considered above. Further, note carefully what we might call the *structural* similarity of this interpretation to the previous one, i.e. in both cases no element of the domain assigned to either name also belongs to the set of elements assigned to the predicate. Now, note the following formal terminology carefully. First, when we assign an individual element of the domain to a name we assign that name a **referent**, i.e. literally that element of the domain which the name stands for, denotes or *refers* to. Further, the set of elements of the domain assigned to the predicate is the **extension** of that predicate.

So far, we have considered how to assign elements of the domain to proper names and predicates so as to falsify subject–predicate formulas. But what of truth here? When is a subject–predicate formula true under an interpretation? Given what you already know, the answer to this question is perfectly straightforward. Again, it is useful to consider an example. So, consider the tree for the following QL sequent:

$\exists x\ [Fx] : {\sim}\exists x\ [{\sim}Fx]$

1.	$\exists x\ [Fx]$	Premise
2.	${\sim}{\sim}\exists x\ [{\sim}Fx]$	Negated conclusion
3.	$\exists x\ [{\sim}Fx]$	From 2 DNE
4.	${\sim}Fa$	3 (term *not* occurring)
5.	Fb	1 (term *not* occurring)
	✓	

This time, the tree does not branch and there is no contradiction in the trunk. Therefore, the original sequent is invalid. Hence, we must consider how we might arrive at an IQLI for the sequent. First, we read up the trunk to identify the TST we require. Again, there are only two subject–predicate formulas in the trunk: *Fa* and *~Fb*, respectively. The target truth-value for any unnegated such formula is, of course, T while the target value of any negated such formula is F. Thus, the TST we require is:

TST: *Fa*: {T} *Fb*: {F}

We have already considered how to falsify a subject–predicate formula and so it only remains to consider how we construct interpretations which generate true subject–predicate formulas. Consider the previous tree once more. Again, suppose that the domain consists of animals. As you know, I have a cat called Zebedee. However, you may not know that I also have a dog named Rover. Consider the following interpretation.

$\mathfrak{I}$ **D**: {**animals**}

F: . . . is a cat

a: Rover

b: Zebedee

It should be quite clear that while Zebedee as a cat does indeed belong to the class of animals which are cats Rover, as a dog, does not belong to the class of animals which are cats! Quite intuitively, then, under this interpretation *Fa* is obviously false and, equally, *Fb* is obviously true. In other words, just as a subject–predicate formula is false under any interpretation which assigns a referent to the subject which does not belong to the set of elements assigned to the predicate, so a subject–predicate formula is true under any interpretation which assigns a referent to the subject which does belong to the set of elements assigned to the predicate. (Again, it is useful to keep the intuitive notion of *belonging* clearly in mind or, equally, you might think of a predicate and a name *sharing* an element of the domain.)

Now let's consider the formal counterpart to our intuitive interpretation, i.e. let's consider an interpretation using numbers rather than cats and dogs. For example:

$\mathfrak{I}$ **D**: {**1,2**}

F: {2}

a: 1

b: 2

The previous point is perhaps even clearer under this interpretation. Compare the extension of the predicate with the element denoted by *b*. In fact, they are identical. Hence, we can certainly say that the element denoted by *b* already belongs to the extension of the predicate *F* and, equally, that the predicate and the name share that common element. And now compare the element denoted by *a*. It does not belong to the extension of the predicate *F*.

In formal terms, the intuitive notion of sharing or belonging which we have been working with is known as **satisfaction**. Thus, in this case, we can say that while *b* satisfies *F*, *a* does not satisfy *F*. Moreover, we can then say on that basis that while *Fb* is **true under I** here *Fa* is **false under I**. Now, satisfaction is not the same thing as truth, but we can exploit the notion of satisfaction to define the notion *true under I* which we need here. In effect, the definition of satisfaction allows us to bridge the gap between the assignment of elements of the domain which any QL-interpretation makes to the formal vocabulary of QL and the subsequent evaluation of QL formulas composed from that vocabulary as *true under I* or *false under I*. To enable us to grasp the truth-tree method as quickly as possible we will not pause to give a complete definition here. Relevant definitions can be found in full in a text to which I refer in the next section.[3] The definition in question is a complex one but, fortunately, we need only the simplest and most intuitive part of it here. Intuitively, then, we will say that a simple subject–predicate formula is *true under I* if and only if the element assigned to the name involved in that formula also belongs to the extension of the predicate, i.e. if the referent of that name satisfies the predicate. Otherwise, such a formula is false. Hence, for simple subject–predicate formulas, we define the notions *true under I* and *false under I* as follows:

> Under any given QL-interpretation, a subject–predicate formula is *true under I* if and only if the referent of the name in that formula satisfies its predicate. Otherwise, the subject–predicate formula is *false under I*.

Returning to our example, we can now ask: is the subject–predicate formula *Fa true under I* in this case? What we are asking is just: does the referent of *a* satisfy *F* under this interpretation? Obviously not. It follows at once that *Fa* is false under this interpretation. Next we ask: is the simple subject–predicate formula *Fb true under I* in the present case? This time, we are asking if the referent of *b* satisfies *F* and, clearly, it does. In other words, under this interpretation:

Fa: {F} *Fb*: {T}

And that is exactly the IQLI which we required for the example above.

Unsurprisingly, we will not always be able to construct IQLIs on the basis of quite such a restricted domain. None the less, when constructing any IQLI it is a good rule of method to try first to construct the interpretation you require on the basis of a very small domain and only to increase the number of elements involved as necessary. For example, consider the tree for the following sequent:

$\exists x\ [Fx \vee Gx], \exists x\ [\sim Fx] : \exists x\ [Gx]$

1.	$\exists x\ [Fx \vee Gx]$	Premise
2.	$\exists x\ [\sim Fx]$	Premise
3.	$\sim\exists x\ [Gx]$	Negated conclusion
4.	$Fa \vee Ga$	1 EIN term-introduction
5.	$\forall x\ [\sim Gx]$	3 Quantifier-equivalence
6.	$\sim Fb$	2 EIN term-introduction
7.	$\sim Ga$	5 UIN term-introduction
8.	$\sim Gb$	5 UIN term-introduction
	/ \	
9.	*Fa* *Ga*	From line 4
	✓ × 7,9	

Clearly, the left-hand branch is live. Therefore, the original sequent has been shown to be invalid. Reading up the live branch, we can list the following target set of truth-values:

TST:

Fa: {T}

Fb: {F}

Ga: {F}

Gb: {F}

Now, we might again try to construct an IQLI just on the basis of a domain consisting of the numbers 1 and 2. In this instance, however, that domain is too restricted. Obviously, because we want *Fa* to be true and *Ga* to be false, we could, for example, construct the following interpretation:

$\mathfrak{I}$ **D**: {**1,2**}

F: {1}

G: {2}

a: 1

b: 2

Under this interpretation, it is indeed the case that *Fa* is true and *Ga* false. But notice at once that *Gb* is also true! We can easily remedy the situation, however, merely by increasing the number of elements in the domain by one, i.e. D: {1,2,3}. Given the new domain, we can construct the IQLI we require, for example, as follows:

$\mathfrak{I}$ **D**: {**1**,**2**,**3**}

F: {1}

G: {3}

a: 1

b: 2

(Check for yourself that this is indeed an IQLI.)

BOX 7.3

To use the truth-tree method to test sequents of QL for validity and invalidity:

♦ Proceed by constructing a truth-tree and exploiting the new development rules to *instantiate* each quantified formula involved and thus to arrive at particular instances of the sequent in question, i.e. at quantifier-free exemplifications of the sequent.

♦ Next, consider each branch in order to determine whether any branch is live. If any branch is live make an IQLI explicit.

To make an IQLI explicit:

♦ (i) Carefully read up the live branch from the tip.

♦ (ii) Identify the *target set of truth-values*, the *TST*, required.
NB The target value of any unnegated subject–predicate formula is T. The target value of any negated subject–predicate formula is F.

♦ (iii) Keep clearly in mind that, under any given QL-interpretation, a subject–predicate formula is *true under I* if and only if the referent of the name in that formula satisfies its predicate. Otherwise, the subject–predicate formula is *false under I.*

♦ (iv) Finally, make explicit a QL-interpretation under which individual elements of the domain are assigned to names and subsets of the domain are assigned to predicates in such a way as to generate exactly the TST as listed in (ii).

Quite generally, you will find that it pays to begin with a domain consisting of a very small number of elements and then to expand the domain only as required. Exercise 7.2 gives you the opportunity to practise this minimalist strategy for constructing IQLIs. Study the contents of Box 7.3 carefully before attempting the exercise.

EXERCISE 7.2

1 Test the following sequents for validity using the truth-tree method for QL. If you find any to be invalid make an invalidating QL-interpretation explicit:

1. $\exists x\, [Fx] : \forall x\, [Fx]$
2. $\forall x\, [\sim Fx] : \sim\exists x\, [Fx]$
3. $\exists x\, [Fx \vee Gx], \exists x\, [\sim Fx] : \exists x\, [Gx]$
4. $\forall x\, [Fx \rightarrow Gx], \forall x\, [Gx \rightarrow Hx] : \forall x\, [Fx \rightarrow Hx]$
5. $\forall x\, [Fx \rightarrow Gx], \exists x\, [\sim Fx] : \exists x\, [\sim Gx]$
6. $\exists x\, [Fx \rightarrow Gx] : \exists x\, [Fx] \rightarrow \exists x\, [Gx]$
7. $: (\exists x\, [Fx]\ \&\ \exists x\, [Gx]) \rightarrow (\exists x\, [Fx\ \&\ Gx])$
8. $\forall x\, [Fx \rightarrow Gx], \exists x\, [Hx\ \&\ \sim Fx] : \exists x\, [Hx\ \&\ \sim Gx]$
9. $\exists x\, [Fx\ \&\ Gx], \exists x\, [Gx\ \&\ Hx] : \exists x\, [Fx\ \&\ Hx]$
10. $\forall x\, [(Fx\ \&\ Gx) \rightarrow Hx], \exists x\, [Fx], \exists x\, [Gx] : \exists x\, [Hx]$

III
Relations Revisited: The Undecidability of First Order Logic

As you may have noticed, so far in the present chapter we have studiously avoided applying the truth-tree method to any sequent involving any relation and, thereby, we have completely avoided sequents involving multiple generality. Hence, each and every QL sequent we have tested together (including all those in Exercises 7.1 and 7.2) has only ever involved predicate-letters, i.e. formal expressions which have one and only one place to be filled. As noted in Chapter 5, predicate-letters such as these are *monadic* and the fragment of QL which concerns itself only with monadic predicates is *monadic QL*. Ultimately, monadic QL is only a part of the larger whole that is quantificational logic for, as you know, quantificational logic also includes two-place relational expressions, three-place relational expressions and so on. None the less, monadic QL is a particularly interesting and, we might say, particularly well-behaved fragment of QL just because that monadic fragment has all of the formal properties which so commended PL to us. In short, monadic QL is Sound and Complete in just the senses we considered earlier as regards PL.

Hence, we can again rest assured that every provable sequent of the monadic fragment of QL is also a semantically valid sequent of QL and that every semantically valid sequent of that fragment of QL is also a provable sequent of that fragment of QL. So, monadic QL is, indeed, Sound and Complete.

Given the enormous gains in expressive power we make as we move on from PL to monadic QL it is surely extremely satisfying and even, perhaps, rather startling to realise that all of the tremendous formal integrity of PL is fully preserved in that fragment of quantificational logic. The first proofs of these formal results about the monadic fragment of QL are due to Leopold Lowenheim [1878–1940] and interested parties can find proofs of them in Part Three of Geoffrey Hunter's *Metalogic*.[4]

For all the increased expressive power of monadic QL as against PL, and its undoubted formal integrity, the monadic fragment of QL none the less confines our formal explorations of validity to formulas and sequents involving only monadic predicates. As you know, the language of quantificational logic can easily be extended beyond the monadic to the polyadic, i.e. to include relations involving, in principle, any number of places. We took the first step away from merely monadic concerns when considering relations in Section VIII of Chapter 5, but there we were mainly concerned with questions of translation and were content to consider questions of validity and invalidity purely informally. However, in Chapter 6, we supplemented the formal language QL with a formal method of proof and, thus, we were able to prove the validity of a number of sequents involving two-place relational expressions. Moreover, the exercises for Chapter 6 gave you the opportunity to prove the validity of a number of sequents involving three-place relational expressions.

Hence, we do at least know that a number of valid sequents of polyadic QL are provable formally. In fact, this is a hint (though nothing more) at a very important formal property of polyadic QL, namely, that polyadic QL is, again, Complete, precisely in the sense that every semantically valid sequent of polyadic QL is also a provable sequent of polyadic QL. Again, it is surely extremely satisfying to realise that this crucially important formal property of our system of formal logic is preserved at the highest level.

As we move beyond monadic QL, however, something very important happens to our system of formal logic and we should pause to take stock of the fact. This important fact is one which is often glossed over in introductory texts, but it is extremely important none the less and we will make it as explicit as the present context allows. The truth of the matter is that to take that first step away from monadic QL and into polyadic QL is to sacrifice a very valuable property of our system of formal logic, and, further, one which should be close to every logic student's heart. Obviously, given what I have said above, the property that is lost in the course of the transition is neither Soundness nor Completeness. These crucial properties are preserved. Rather, the property in question is known as **decidability**. But what is decidability? And what are the implications of undecidability? Keeping things intuitive, we can fairly say that, in a given system of formal logic, a given property is *decidable* if and only if for any and every sequent of that system there exists an *effective decision-procedure* in terms of which we can conclusively establish that the sequent has or lacks the property in question. In other words, as Hunter puts the point:

> In logic and mathematics, an *effective method* for solving a problem is a method for *computing* the answer that, if followed correctly and as far as may be necessary, is logically bound to give the right answer (and no wrong answers) in a finite number of steps.[5]

In the present context, then, to possess an effective method for the property we are most interested in, namely, validity, is to possess an effective decision-procedure, i.e. an effective way of deciding, for any and every sequent, whether that sequent is valid or not. In PL, for example, the comparative truth-table test for validity constitutes an effective method which, therefore, constitutes a decision-procedure for validity in PL. And this is a very clear example: for any PL sequent, given the truth-tables, we (or a machine which we have programmed) could literally compute a decision about the validity of the sequent, i.e. in a finite number of steps, we or the machine could determine whether any given PL sequent is valid or invalid. Equally, the truth-tree method is also an effective method for PL and, again, it too gives us a decision-procedure for the property of validity in that context, i.e. it too allows us to determine whether or not any given sequent has or lacks the property of validity.

Hence, still keeping things intuitive, we can fairly say that a system of formal logic is decidable as regards validity if, given a good, effective, formal method such as comparative truth-tables or truth-trees, then, for any sequent of the system which you give to me, I can use that method to give you back a decision about the validity of that sequent. The comparative truth-table test may not obviously seem to carry over into QL but, in this chapter, we have explicitly extended the truth-tree method into monadic QL. Further, you yourself will already have tested twenty QL sequents using that method and so you may well be tempted to think that the truth-tree method is equally a decision-procedure for validity in monadic QL. And that is certainly the case. I cannot prove that this is so in the present context but, in fact, both PL and monadic QL have, indeed, been proved to be decidable by Emil Post in 1920 and Leopold Lowenheim in 1915, respectively. However, not only has no formal logician proved the decidability of polyadic QL, but, in 1936, an extremely important formal logician, Alonzo Church, proved (on a certain assumption) that quantificational logic beyond the monadic fragment is actually *undecidable*.[6] Indeed, the undecidability result has come to be known as *Church's Theorem* (and the assumption underpinning it as *Church's Thesis*).

As noted above, a system of formal logic is decidable as regards validity if, given an effective formal method, then, for any sequent of the system which you give to me, I can use that method to give you back a decision about the validity or invalidity of that sequent. Obviously, decidability is an enormously attractive formal property, so what of the implications of losing it?

As regards polyadic QL, the situation is quite a complex one in this respect. In fact, for any given valid QL sequent, a formal method such as the truth-tree method will, in a finite number of steps, arrive at a proof that the given sequent is, indeed, valid. For just this reason, Polyadic QL is *semi-decidable*, i.e. our formal method is effective as regards establishing that the valid sequents are valid.[7] Therefore, if we program the hypothetical machine with our favourite formal method, namely, the truth-tree method, we can rely on the machine to produce the relevant trees and, quite mechanically, to give us the answer: 'Yes. This one is a valid one' in each valid case.

However, earlier we noted that a given property is decidable if and only if for each and every sequent of the system there exists an effective method in terms of which we can conclusively establish that the sequent either has or lacks the property in question. In the present context, then, determining that the valid sequents of QL are, indeed, valid is only half the story, i.e. we also want the method to determine that invalid QL sequents are, indeed, invalid. In other words, an effective method should not only be able to tell us: 'Yes. This one is valid' but should also be able to tell us: 'No. This one is not valid.' And it is here that the problem sets in. For there is no effective mechanical procedure for determining the invalidity of invalid sequents in polyadic QL. Hence, it is precisely as regards invalidity that polyadic QL is undecidable.

To date, each of the sequents I have tested using the truth-tree method in this text and each of the sequents you have tested in the relevant exercises has invariably ended in one of two ways, i.e. either, after a finite number of lines, each branch has ended in contradiction, the tree has died and, therefore, the original sequent has been shown to be valid or, in a finite number of steps, we have fully developed every formula and one or more branches remain open, proving the sequent to be invalid. Hence, so far, we have seen no indication of any undecidability whatsoever and you may find it hard to imagine what such an indication might look like. However, in polyadic QL, there is in fact a third kind of result which a truth-tree test may deliver, namely, it may not stop at all. In other words, the tree itself or one of its branches may keep developing ad infinitum. We can see this possibility vividly illustrated when we attempt a truth-tree test on the following sequent:[8]

$: {\sim}\forall x\ [\exists y\ [Rxy]]$

Now, there are no premises in this case. The formula in question is masquerading as a theorem, a logical truth. A little intuitive reflection should quickly reveal that this formula is certainly not logically true but the question is: can we show purely mechanically that the sequent is invalid? As ever, we will try to apply the truth-tree test for validity here. Hence, we write the negation of the sequent on a line and proceed to apply the rules, as follows:

1.	${\sim}{\sim}\forall x\ [\exists y\ [Rxy]]$	

Obviously, we can simply eliminate the double negation and proceed to develop the resulting formula using the quantifier rules as below:

2.	$\forall x\ [\exists y\ [Rxy]]$	1 DNE
3.	$\exists y\ [Ray]$	from line 2 by UIN
4.	Rab	from line 3 by EIN

Note very carefully that applying EIN to line 3 results in a new term being introduced at line 4. Now, we want to maximise our chances of generating a contradiction here, UIN allows us to develop the formula on line 2 to any existing term, and so it is natural to reinstantiate 2 again. But note what happens when we do:

5.	$\exists y\ [Rby]$	from line 2 by UIN
6.	Rbc	from line 5 by EIN

Again, we are confronted with a new term as soon as we apply EIN to line 5. We still want to maximise our chances of generating a contradiction here, of course, and UIN still allows us to develop the formula on line 2 to any existing term, so we might try to reinstantiate 2 once again. But note carefully what happens when we do:

7.	$\exists y\ [Rcy]$	from line 2 by UIN
8.	Rcd	from line 7 by EIN

To cut a long story short, if we continue to reinstantiate we could carry on reinstantiating forever. A close look at what is happening in the tree as a whole should assure you that in this case we will never be able to stop developing the formulas involved in the tree:

	$: \sim\forall x\ [\exists y\ [Rxy]]$	
1.	$\sim\sim\forall x\ [\exists y\ [Rxy]]$	Negated conclusion
2.	$\forall x\ [\exists y\ [Rxy]]$	from line 1 DNE
3.	$\exists y\ [Ray]$	from line 2 by UIN
4.	Rab	from line 3 by EIN
5.	$\exists y\ [Rby]$	from line 2 by UIN
6.	Rbc	from line 5 by EIN
7.	$\exists y\ [Rcy]$	from line 2 by UIN
8.	Rcd	from line 7 by EIN

.

.

.

.

Trees of this type, which are potentially infinitely extensible, are known as **infinite trees**. The existence of infinite trees vividly illustrates the possible third result I mentioned above, i.e. that the tree may not stop at all. It follows that we may never be able to establish mechanically that an invalid sequent such as this is actually invalid. And this give us a very clear manifestation of Church's discovery, i.e. that polyadic QL is undecidable precisely as regards invalidity. At just this point, then, we have lost the formal property of decidability but, we should ask: what exactly are the implications of that loss?

First, it is crucial to appreciate that the problem is not peculiar to our method. Rather, we are confronted here with the kind of problem which is at the heart of undecidability. For although it is in fact true that we will discover the validity of any valid QL-sequent in a finite number of steps, it is true quite generally, i.e. for any method, that we could find ourselves running mechanical tests for invalidity which simply take us on and on with no guarantee that the test will ever terminate. Hence, Church's point is not that there is a problem with the truth-tree method. Rather, his point is that there is no effective method for determining invalidity in polyadic QL.

However, we should carefully consider what follows from this fact. Keeping things intuitive, we can fairly say that Church has pointed up an important limit to what can be done purely mechanically about deciding the logical status of sequents in any formal system of quantificational logic beyond the monadic fragment. And that is exactly why I said earlier that the problem of undecidability should be close to every logic student's heart. For, in the good old days of monadic QL, given any sequent, we could quite mechanically run a truth-tree test knowing that, in a finite number of steps, we would get the desired result. Now, however, things are rather different and there is no such guarantee. But consider the sequent whose tree turned out to be infinite very carefully:

$: {\sim}\forall x\ [\exists y\ [Rxy]]$

As I noted above, a little reflection surely reveals that this particular sequent is not a logical truth. After all, we know that this sequent is invalid, don't we? Moreover, we can certainly prove that it is invalid. Remember: if the sequent is logically true then there can be no possible interpretation under which that sequent is false. But consider the domain which simply consists of human beings. If it is a trifle too optimistic to hope that 'Everyone loves someone' (even if only himself or herself!) it is certainly not too optimistic to hope that 'Everyone knows someone', i.e. it is surely true that everyone is at least acquainted with someone (again, even if only himself or herself). Equally, it is also true that everyone has a biological mother, i.e. that everyone is someone's offspring, as we might put it. Alternatively, consider the domain which consists of the natural numbers and the relation: ' . . . is divisible without remainder by ---'. Of course, every natural number is divisible without remainder at least by itself and 1. Under all these interpretations, then, $\forall x\ [\exists y\ [Rxy]]$ is obviously true and, equally obviously, its negation ${\sim}\forall x\ [\exists y\ [Rxy]]$ is false. There certainly exist interpretations under which ${\sim}\forall x\ [\exists y\ [Rxy]]$ is false, then, and, therefore, $: {\sim}\forall x\ [\exists y\ [Rxy]]$ is invalid, i.e. each of the above interpretations is an IQLI of that sequent.

The point to remember, then, is not that we cannot determine the invalidity of polyadic QL sequents. Clearly, we can. Church's point is rather that no mechanical procedure is effective as regards invalidity in polyadic QL. And now the loss of decidability need not fill us with horror. Church has pointed up a limit to what can be done purely mechanically; he has not pointed up a limit to human thinking and reasoning or to logic itself, as we might put it.

In other words, Church has shown that formal logic cannot be completely mechanised and that even at the highest level there is a crucial role for human thinking and natural reasoning, for art and imagination. At least some readers will undoubtedly find this particular implication refreshing, and some may even see the potential for a way of distinguishing between the nature of human thinking and the nature of mechanical thinking on that basis. Once again, however, the pursuit of such properly philosophical questions lies beyond the scope of the present text.

For formal logical purposes it is important to realise that although we now have no guarantee that we won't end up with an infinite tree in any given case we certainly should not give up on the truth-tree method at this point. For it absolutely does not follow that truth-tree tests for validity or, indeed, for invalidity at the level of polyadic QL are not possible. Remember: we certainly can show valid QL sequents to be valid and the procedure involved can be perfectly mechanical. For example, consider the tree for the following sequent which involves two-place relational expressions and multiple quantification:

$$\exists x\ [\forall y\ [Rxy]] \vDash \forall y\ [\exists x\ [Rxy]]$$

1.	$\exists x\ [\forall y\ [Rxy]]$	Premise
2.	$\sim\forall y\ [\exists x\ [Rxy]]$	Negated conclusion
3.	$\exists y\ [\sim\exists x\ [Rxy]]$	2 Quantifier-equivalence
4.	$\forall y\ [Ray]$	1 EIN (term not occurring)
5.	$\sim\exists x\ [Rxb]$	3 EIN (term not occurring)
6.	$\forall x\ [\sim Rxb]$	5 Quantifier-equivalence
7.	$\sim Rab$	6 UIN (term occurring)
8.	Rab	4 UIN (term occurring)
	× 7,8	

Hence, a contradiction is very quickly generated in the trunk and, therefore, we have shown quite mechanically that the original sequent is valid.

Moreover, the truth-tree method can also deliver decisions as to the invalidity of certain sequents involving two-place relational expressions and multiple quantification, again, quite mechanically. In other words, although the lack of a generally effective procedure means that we no longer have an absolute guarantee of success in any such case, that does not entirely rule out the possibility of demonstrating the invalidity of a polyadic QL sequent mechanically.

For example, consider the tree for the following sequent:

	$: \forall x\ [\forall y\ [Rxy \rightarrow Ryx]]$	
1.	$\sim\forall x\ [\forall y\ [Rxy \rightarrow Ryx]]$	Negated conclusion
2.	$\exists x\ [\sim\forall y\ [Rxy \rightarrow Ryx]]$	1 Quantifier-equivalence
3.	$\exists x\ [\exists y\ [\sim(Rxy \rightarrow Ryx)]]$	2 Quantifier-equivalence
4.	$\exists y\ [\sim(Ray \rightarrow Rya)]]$	3 EIN (term not occurring)
5.	$\sim(Rab \rightarrow Rba)$	4 EIN (term not occurring)
6.	Rab	From line 5
7.	$\sim Rba$	From line 5
	✓	

Simply applying the usual procedures quite mechanically to this particular sequent very quickly demonstrates the invalidity of the sequent. Therefore, even at this level, there are indeed cases where truth-tree tests for invalidity do succeed. To complete the demonstration of the invalidity of the sequent, however, we ought to construct an IQLI. Initially, this may seem to pose a new problem, i.e. how are we to treat relational expressions in this context?

In sharp contrast to one-place predicates, it will not do to assign to a relational expression any single individual element of the domain under an interpretation. Intuitively, no such expression ever holds of any individual element of the domain. Hence, a QL-interpretation must assign to any two-place relational expression *R* pairs of elements of the domain. Moreover, as we saw in Chapter 5, the sense of a relational QL formula can be sensitive to the order in which names are used to fill its places. Thus, an interpretation must assign to two-place relational expressions a set of *ordered pairs* of elements of the domain. So much for that part of a QL-interpretation which constitutes an assignment. But we also require an evaluation of this kind of QL formula, i.e. a definition of *true under I* and *false under I* for formulas of this kind. Again we bridge the gap between assignment and evaluation by defining the notion of *satisfaction* for formulas of this kind. And this is quite straightforward. As noted above, a QL-interpretation must assign to any two-place relational expression a set of ordered pairs of elements of the domain. It follows that a two-place relational expression is satisfied under a QL-interpretation by just the set of ordered pairs of elements of the domain assigned to it by that interpretation. Intuitively, then, any formula composed of names and a two-place relational expression is *true under I* if and only if the pair of elements of the domain denoted by the names taken in order is identical with one of the ordered pairs assigned to the relational expression, i.e. if the ordered pair in question satisfies the relational expression. Hence:

> Under any given QL-interpretation, any formula involving only terms and a two-place relational expression is *true under I* if and only if the referents of those names taken in order satisfy the relational expression. Otherwise, such a formula is *false under I*.

Given this insight, we can easily identify the target set of truth-values we require in the present case simply by reading up the live branch as usual, i.e.:

Rab: {T}

Rba: {F}

Therefore, when constructing the IQLI, we must specify a relation under which, while *a* bears *R* to *b*, *b* does not bear *R* to *a*. In fact, the example given in Section VII of Chapter 5 is adequate here, i.e. for the relation 'loves' and the domain of cats and their owners, where *a* denotes Paul and *b* denotes Zebedee, the ordered pair ⟨Paul, Zebedee⟩ does indeed satisfy *R* while the ordered pair ⟨Zebedee, Paul⟩ no doubt does not! Again, we can simplify matters by using small sets of numbers here, taking care always to assign ordered pairs of elements from the domain to any relational expressions involved. To define a relation in this way we should specify all the ordered pairs for which the relation holds and those for which it does not. To simplify things we can simply consider the set of ordered pairs assigned to *R* to be exhaustive. In so doing, we specify what is often called the **solution set** for *R*. Check for yourself that given the specified solution set for *R* the following interpretation really is an IQLI:

$\mathfrak{I}$ **D**: {**1**,**2**}

R: {⟨1,2⟩}

a: 1

b: 2

To sum up then, while Church's result shows that we can have no fail-safe, purely mechanical method for deciding the invalidity of sequents involving relational expressions, the truth-tree method remains a useful tool at this level. For not only can we use it to establish the validity of sequents of QL beyond the monadic fragment but we can frequently also use that method to demonstrate the invalidity of such sequents. Moreover, even in cases where the method fails to deliver anything convincing, as we saw in the first example in this section, with a little art and imagination we may none the less succeed in demonstrating the invalidity of the sequent ourselves.

BOX 7.4

♦ A system of formal logic *s* is *Sound* if and only if every provable sequent is semantically valid, i.e. if and only if, for every and any sequent ***S***:

Soundness: $\vdash_s S \Rightarrow \models_s S$

♦ A system of formal logic *s* is *Complete* if and only if every semantically valid sequent is provable, i.e. if and only if, for every and any sequent A:

Completeness: $\models_s S \Rightarrow \vdash_s S$

♦ A system of formal logic is *Decidable* if and only if for each and every sequent of the system there exists a decision-procedure for validity and invalidity, i.e. an effective mechanical method in terms of which the validity or invalidity of any given sequent of the system can be determined in a finite number of steps.

PL is Sound, Complete and Decidable.

Monadic QL is Sound, Complete and Decidable.

Polyadic QL is Sound, Complete and Undecidable.

♦ More precisely, although there exists a decision-procedure for validity in polyadic QL there exists no decision-procedure for invalidity in polyadic QL. Consequently, the truth-tree method can generate *infinite trees*. An infinite tree contains at least one infinite branch. An infinite branch is a branch containing one or more formulas which, given the truth-tree rules, could be developed ad infinitum.

♦ The proof of the undecidability of polyadic QL is due to Alonzo Church [1936].

More detailed consideration of the undecidability problem and related issues lies beyond the scope of the present text. Interested parties can find the undecidability of quantificational logic beyond the monadic fragment discussed at length in Part Four of Geoffrey Hunter's *Metalogic*, especially pp. 219–51. Further, these issues are also considered at length and the undecidability of first order logic is proved (given Church's Thesis) in another text of seminal importance, namely, *Computability and Logic*, [1996], third edition, by George S. Boolos and Richard C. Jeffrey, Cambridge,

Cambridge University Press. In all honesty, the latter text is a goldmine of metatheory which includes formal proofs of all the important results at the level of QL as well as penetrating discussions of important issues beyond QL.

More to the point, readers of the present text will find that they can begin their consideration of the above text at p. 112, i.e. with Chapter 10 which is entitled 'First Order Logic is Undecidable'. The content of that chapter is precisely a proof of the result in question. Further, the reader will find that the notation for predicate logic used in that text is, in all important respects, identical to that which we have used here and that the formal method for demonstrating validity and invalidity of sequents of QL is exactly the truth-tree method which we have used. Moreover, the instantiation rules for quantified formulas of QL outlined on p. 123 of that text are in fact identical to the rules we have used here. As stated, this particular text is a genuine goldmine of metatheory and I warmly recommend it to you.

Now study Box 7.4 carefully.

IV
A Final Note on the Truth-Tree Method: Relations and Identity

Unusually, no exercise accompanied the previous section and you may well be disappointed at not having had the opportunity to construct truth-trees for fully polyadic QL sequents for yourself. At the end of the present section, I remedy this situation and provide you with just that opportunity. But before we proceed to Exercise 7.3, a few final points are worth a little more consideration.

First, recall the quantifier-equivalences we have exploited throughout this chapter:

1. $\sim\forall x\ [Fx] : \exists x\ [\sim Fx]$ 2. $\sim\exists x\ [Fx] : \forall x\ [\sim Fx]$

When constructing truth-trees for polyadic QL sequents it is crucial to appreciate that the same equivalences still hold, i.e. that these quantifier-equivalences hold quite generally. The point is perhaps easier to see if we now state the equivalences in a more general form. So, where 'v' represents any variable and '[. . .]' represents any matrix, we can restate the equivalences as follows:

1.′ $\sim\forall v\ [\ldots] : \exists v\ [\sim\ldots]$ 2.′ $\sim\exists v\ [\ldots] : \forall v\ [\sim\ldots]$

For example, consider the formula:

$\sim\exists x\ [\forall y\ [Rxy]]$

Given 2′ above, we can safely rewrite the formula as follows:

$\forall x\ [\sim\forall y\ [Rxy]]$

Hence, we might apply UIN to this formula and, for example, instantiate to the formula:

$\sim\forall y\ [Ray]$

Moreover, given 1′ above, we could in turn rewrite that formula as follows:

$\exists y\ [\sim Ray]$

Finally, we could apply EIN to this formula and instantiate to:

$\sim Rab$

In fact, we can give a still more general statement of the equivalences here which may make their general applicability clearer. Quite generally, then, where **Q** and **Q′** stand for any quantifier, 'v' stands for any variable and '[. . .]' for any matrix:

1″. ~**Q**v [. . .] : **Q′**v [~ . . .]

where **Q′** is ∃ if **Q** is ∀ and **Q′** is ∀ if **Q** is ∃

Consider the final statement of equivalence very carefully. Note that, in effect, given any QL formula with negation in initial position we can exploit the equivalence to, as it were, move the negation symbol to the right, i.e. into the appropriate matrix. Obviously, it is equally true to say that in effect we can exploit the equivalence to move a quantifier to the left. A little reflection quickly reveals that, if we liked, we could keep applying the equivalence until we had all the quantifiers on the left and all the other connectives behind them, i.e. on the right. A QL formula which has every quantifier on the left before any other connective is said to be in **prenex form**.

Further, for every QL formula there is an equivalent prenex formula. I will not attempt to prove this here but a little reflection on the points just made and a close look at the final equivalence statement above should reassure you that, for any QL formula which you give me, I can quickly give you back the same formula in prenex form, i.e. the appropriate equivalent. The general statement of equivalence and the notion of prenex form are very useful when constructing truth-trees of polyadic QL sequents, and prenex form is also very useful for other important formal purposes.[9]

Finally, as ever, we seem to have left considering identity to last. In this particular instance, however, that may be quite appropriate. In fact, the addition of identity to polyadic QL makes no difference whatsoever to the formal properties of the system. Monadic QL is Sound, Complete and decidable with or without identity and again, polyadic QL is Sound, Complete and undecidable (or, at best, semi-decidable) with or without identity. These facts show us very clearly then that it is the addition of relational expressions and only the addition of relational expressions which is momentous in terms of the formal properties of quantificational logic. As Church puts the point: 'The undecidability of quantification theory depends essentially on the presence of polyadic [relational expressions].'[10]

Formulas involving identity can easily be developed in the process of tree-construction simply by exploiting a familiar rule of inference for identity which we defined in Chapter 6, i.e. *identity-elimination* (again, because we are only concerned with instantiation here the introduction rule can be ignored). Hence, given an identity on a branch together with a formula which tells us something about one term in that identity, we may infer that the same something holds of the other term. Annotate any such line '=E'. Again, such an inference might involve a predicate but, equally, it might involve another identity, e.g. consider the tree which establishes the *transitivity of identity* in QL:

$$\models \forall x\,[\forall y\,[\forall z\,[((x=y)\,\&\,(y=z)) \rightarrow (x=z)]]]$$

1.	$\sim\forall x\,[\forall y\,[\forall z\,[((x=y)\,\&\,(y=z)) \rightarrow (x=z)]]]$	Negated conclusion
2.	$\exists x\,[\sim\forall y\,[\forall z\,[((x=y)\,\&\,(y=z)) \rightarrow (x=z)]]]$	1 Quantifier-equivalence
3.	$\exists x\,[\exists y\,[\sim\forall z\,[((x=y)\,\&\,(y=z)) \rightarrow (x=z)]]]$	2 Quantifier-equivalence
4.	$\exists x\,[\exists y\,[\exists z\,[\sim(((x=y)\,\&\,(y=z)) \rightarrow (x=z))]]]$	3 Quantifier-equivalence
5.	$\exists y\,[\exists z\,[\sim(((a=y)\,\&\,(y=z)) \rightarrow (a=z))]]]$	4 EIN
6.	$\exists z\,[\sim(((a=b)\,\&\,(b=z)) \rightarrow (a=z))]]$	5 EIN
7.	$\sim(((a=b)\,\&\,(b=c)) \rightarrow (a=c))$	6 EIN
8.	$(a=b)\,\&\,(b=c)$	From line 7
9.	$\sim(a=c)$	From line 7
10.	$(a=b)$	From line 8
11.	$(b=c)$	From line 8
12.	$(a=c)$	10,11 =E
	$\times$ 9,12	

In effect, having stated the negated conclusion, the first thing we do is to rewrite that formula in prenex form via the quantifier-equivalences over lines 2–4. Next, we simply instantiate using EIN before developing the resulting negated conditional. Finally, we apply =E to the formulas on lines 10 and 11 to derive $(a = c)$ on line 12 which contradicts $\sim(a = c)$ on line 9. Thus, we generate a contradiction in the trunk and so prove the validity of the original sequent.

Finally, where ρ is any proper name, any formula of the form: $\sim(\rho = \rho)$ may be treated as a contradiction for purposes of tree-construction. In other words, we supplement the existing rules by allowing that the presence of any formula of this form on a branch is sufficient to kill the branch. Given that final amendment, the rules are now able to deal perfectly well with sequents involving identity and, therefore, I include some such sequents in the following, final, exercise. Note carefully that this exercise consists of twenty sequents. These are a rather mixed bag. Further, the level of difficulty is generally on the increase and a number of the later sequents involve relations. Should you at any point find yourself trying to develop an infinite tree (and you should!) indicate which branch you believe to be responsible. Before you attempt Exercise 7.3, consider the following carefully: many promissory notes have been issued in this chapter. I have asserted that a full definition of satisfaction can be given for the well-formed formulas of QL and that *true under I* can be completely defined for each and every kind of QL formula. Moreover, I have asserted that polyadic QL is Sound, Complete and undecidable as regards invalidity. In all honesty, I have not given the relevant definitions here. Nor have I proved any of these results. However, the relevant definitions can be given and each of those results certainly can be proved. And so can a number of other important results. Each of the definitions and all of the relevant results (and more) are proved by George S. Boolos and Richard C. Jeffrey in their *Computability and Logic*, [1996], third edition, Cambridge, Cambridge University Press.

Again, I emphasise that this particular text is a goldmine of metatheory which includes formal proofs of all the important results at the level of QL as well as penetrating discussions of important issues beyond QL. Moreover, the present reader will find: (i) that the notation for quantificational logic used in that text is, in all important respects, identical to that which we have used here; (ii) that the formal method for demonstrating validity and invalidity of sequents of QL is exactly the truth-tree method we have been using here; and (iii) that the instantiation rules for quantified formulas of QL outlined on p. 123 there differ from the rules which we have exploited in this chapter only in being referred to (more simply) as 'UI' and 'EI' respectively. Therefore, you are now well placed to tackle the account of metatheory given in *Computability and Logic* and, again, I warmly recommend it to you.

EXERCISE 7.3

1 Test the following sequents for validity using the truth-tree method for QL. If you find any to be invalid make an Invalidating QL-Interpretation explicit. If you consider any tree or any branch of any tree to be infinite indicate the tree or branch in question:

1. $: \exists x\ [Fx \vee {\sim}Fx]$
2. $\forall x\ [Fx \rightarrow Gx] : \exists x\ [Fx\ \&\ Gx]$
3. $: \exists x\ [Fx] \vee \forall x\ [{\sim}Fx]$
4. $\forall x\ [Fx \vee Gx] : \forall x\ [Fx] \vee \forall x\ [Gx]$
5. $: (\forall x\ [Fx] \rightarrow P) \rightarrow (\exists x\ [Fx] \rightarrow P)$
6. $\exists x\ [{\sim}Fx], \forall x\ [Fx \vee {\sim}Gx] : \forall x\ [{\sim}Gx]$
7. $\forall x\ [Fx \leftrightarrow Gx] : \exists x\ [Fx] \leftrightarrow \exists x\ [Gx]$
8. $\forall x\ [\exists y\ [Rxy]] : \exists y\ [Ray]$
9. $\forall x\ [\forall y\ [Rxy]] : \forall x\ [Rxx]$
10. $\forall x\ [\exists y\ [Ryx]] : \exists y\ [\forall x\ [Ryx]]$
11. $\forall x\ [\exists y\ [Ryx]], \forall x\ [\forall y\ [Rxy \rightarrow Sxy] : \forall x\ [\exists y\ [Syx]]$
12. $\forall y\ [Fy \rightarrow Gy] : \forall x\ [\exists y\ [Fy\ \&\ Rxy] \rightarrow \exists y\ [Gy\ \&\ Rxy]]$
13. $: \exists x\ [\forall y\ [{\sim}Rxy]]$
14. $: \forall x\ [x = x]$
15. $\forall x\ [\forall y\ [(x = y) \rightarrow (y = x)]] : \exists x\ [Rxx]$
16. $\forall y\ [\exists x\ [Rxy\ \&\ Rxa] \rightarrow (a = y)], \exists x\ [Rxb\ \&\ Rxa] : (a = b)$
17. $\exists x\ [Fx\ \&\ \forall y\ [(Gy \rightarrow Rxy)]] : \forall y\ [Gy \rightarrow \exists x\ [Fx\ \&\ Rxy]]$
18. $\exists x\ [Fx], \forall x\ [\forall y\ [(Fx\ \&\ Fy) \rightarrow (x = y)]] :$
    $\exists x\ [Fx\ \&\ \forall y\ [Fy \rightarrow (x = y)]]$
19. $\forall x\ [{\sim}Rxx], \forall x\ [\forall y\ [\forall z\ [(Rxy\ \&\ Ryz) \rightarrow Rxz]]] :$
    $\forall x\ [\forall y\ [Rxy \rightarrow {\sim}Ryx]]$
20. $\exists x\ [\exists y\ [(Fx\ \&\ Fy)\ \&\ (Rxy \vee Ryx)]], \forall x\ [Fx \rightarrow {\sim}Rxx] :$
    $\exists x\ [\exists y\ [{\sim}(x = y)\ \&\ (Fx\ \&\ Fy)]]$

Notes

1 I personally am indebted to Stig Rassmussen for the following version of the truth-tree method for QL. However, the version is, in essence, that presented in Mendelson, Elliot, [1987], *Introduction to Mathematical Logic*, third edition, California, Wadsworth and Brooks. See especially pp. 110–16.

2 I mean in monadic QL.

3 Boolos, George S. and Jeffrey, Richard C., [1996], *Computability and Logic*, third edition, Cambridge, Cambridge University Press, pp. 96–102.

4 Hunter, Geoffrey, [1971], *Metalogic: An Introduction to the Metatheory of Standard First-Order Logic*, London and Basingstoke, Macmillan, pp. 137–95.

5 Ibid., p. 14. The use of italics on 'computing' is mine.

6 See Church, Alonzo, [1936], 'A Note on the Entscheidungsproblem' and 'Correction to "A Note on the Entscheidungsproblem"', *Journal of Symbolic Logic*, 1 (1 and 3), March and September.

7 This useful way of describing the situation is due (as far as I am aware) to Stephen Read and Crispin Wright. See Read, Stephen and Wright, Crispin, [1993], *Read and Wright: Formal Logic, An Introduction to First Order Logic*, fifth edition, revised, Departmental Publication, St Andrews, University of St Andrews, Ch. 16.

8 The sequent in question is an adaptation of an example due to Stig Rassmussen.

9 For example, in proving the Herbrand Gentzen Theorem. See Lyndon, Roger C., [1966], *Notes on Logic*, Van Nostrand Mathematical Studies #6, Princeton NJ, D. Van Nostrand, pp. 74–7.

10 From Church, Alonzo and Quine, W.V.O. [1952], 'Some Theorems on Definability and Decidability', *Journal of Symbolic Logic*, 17 (3), September.

Glossary

&E See **and-elimination**.

&I See **and-introduction**.

A See **assumption**.

adequate (set of **connectives**) Any set of connectives with enough expressive power to represent every possible **truth-function** in **PL**. This idea is discussed more fully in the final section of Chapter 4.

affirming the antecedent A **valid** pattern of inference from a **conditional** which involves asserting that the **antecedent** holds and then concluding that the **consequent** must hold, i.e. **modus ponens**.

affirming the consequent An **invalid** pattern of inference from a **conditional** which involves asserting that the **consequent** holds and then concluding that the **antecedent** must hold.

aggregative (sense of 'all') That sense of 'all' in which we think of the term as meaning everything in sum total *as one single thing*, e.g. 'all folk singers' understood as a *totality*, as distinct from the **distributive** sense.

and-elimination (rule of &E) One **conjunct** may be removed from a **conjunction** by one application of &E. The **line number** of the conjunction must be cited together with '&E'. The **dependency-numbers** of the new line are identical with those of the original line containing the conjunction.

and-introduction (rule of &I) Any two **well-formed formulas** may be conjoined. The relevant **line numbers** and '&I' must be cited. The **dependency-numbers** of the new line consist of all the dependency-numbers of both lines used.

antecedent See **conditional**.

anti-symmetrical A **relation** *R* is anti-symmetrical if whenever it is both the case that *a* bears *R* to *b* and that *b* bears *R* to *a* it follows that *a* and *b* are identical objects.

argument In logic an argument consists of a set of sentences (the **premises**) which may or may not establish another sentence (the **conclusion**) as a consequence. Not every set of sentences constitutes an argument in this sense. Premises should be given as reasons for their conclusion and a pattern of inference should be identifiable. Tips: words such as 'because', 'since' and 'for' indicate premises while words such as 'therefore', 'so', 'hence' or 'thus' indicate conclusions. Look for words such as these when trying to identify an argument.

argument-form See **logical form**.

argument-frames Schemas or patterns of inference with gaps which (in PL) are plugged by sentential **variables** or **sentence-letters**.

arrow-elimination An alternative name for **modus ponens**.

arrow-introduction An alternative name for **conditional proof**.

assumption (rule of A) Any **formula** may be assumed on any **line of proof**. The line must be annotated 'A' for 'assumption'. The **dependency-number** of the assumed formula is identical with the **line number** of the line on which it is assumed. Tip: assumptions are used not for **categorical reasoning** but for **hypothetical reasoning**. As such, they are essential to strategies for proof-construction which exploit, for example, **CP, RAA, vE** or **EE**.

asymmetrical A **relation** R is asymmetrical if and only if for every a and b, if a bears R to b then b does not bear R to a, i.e. iff $\forall x\ [\forall y\ [Rxy \rightarrow {\sim}Ryx]$.

atomic formula (of PL) The simplest **well-formed formulas** of PL, i.e. individual **sentence-letters**.

augmentation The practice of using the & rules to form a **conjunction** of **formulas**, one of which is subsequently inferred from the conjunction and exploited for purposes of applying a **discharge rule**. The practice enables applications of such rules by engineering the appropriateness of a formula's set of **dependency-numbers**.

autonym(s) Symbols which act as names for themselves and thus can be used within natural language sentences without use of quotation marks. Autonyms are useful because they help to minimise the proliferation of brackets while preserving the *use/mention distinction*. Ordinarily, when a word or symbol is mentioned rather than actually used it should be placed in quotation marks to indicate that we are talking about the word or symbol rather than using it to talk about something else. For example, in the sentence 'The name "Paul" has four letters', we mention rather than use the name. Hence, we place the term in quotation marks and so form a name for the name. As autonyms act as names of themselves we need not place autonyms in quotation marks.

biconditional (i) The symbol '↔'. (ii) Any formula whose **main connective** is '↔'.

biconditional-elimination/ ↔ elimination (rule of ↔ E) Given a **biconditional** on a line we may rewrite that **formula** as the **conjunction** of the relevant pair of **conditionals** on the next **line of proof**. The new line is annotated with the **line number** of the old line and takes as **dependency-numbers** all and only those of the old line.

biconditional-introduction/ ↔ introduction (rule of ↔ I) Given a pair of **conditionals** on two lines such that the **antecedent** of the first is the **consequent** of the second and the consequent of the first is the antecedent of the second you may write '↔' between the two **formulas**, antecedent and consequent, on a new line. The new line is annotated with the **line numbers** of both lines used and '↔ I'. The **dependency-numbers** of the new line are all of those of both lines used.

binary (connectives) Those **logical connectives** of PL which require at least two **formulas** to form a **well-formed formula**, i.e. '&', 'v', '→' and '↔'.

bivalent See **principle of bivalence**.

Boolean algebras A Boolean algebra is any algebraic system of a kind first described by George Boole. In effect, a Boolean algebra is a **formal system** for sets. By identifying the familiar **truth-functions** of the **logical connectives** with specific Boolean algebraic operations, propositional logic can itself be represented as a Boolean algebra.

bound variable See **variable**.

categorical reasoning Reasoning from **premises** which are considered to be asserted as true. This kind of reasoning contrasts with **hypothetical reasoning**, which is reasoning with **assumptions** which, in contrast, need not be asserted to be true.

classical logic The system of formal logic designed by Gottlob Frege and Bertrand Russell (as opposed to alternative or 'deviant' systems such as relevant logic or **intuitionist** logic).

classical negation **Negation** as understood in traditional or **classical logic**. In essence, negation of this kind is denial and has the effect of reversing the truth-value of the **formula** to which the negation is applied.

colon ':' The symbol in a **sequent** which represents 'therefore'.

comparative truth-table test A **sequent** is **invalid** if it is possible that its **premises** be true and its **conclusion** false. Quite generally, **truth-tables** allow us to calculate whether any **compound formula** is true or false. Hence, we can construct truth-tables for all of the **formulas** which compose a sequent and then *compare* overall values to establish whether the premises are true and the conclusion false under any of the assignments of truth-values. If so, the sequent is shown to be invalid. If not, the sequent is shown to be **valid**.

Completeness The (highly desirable) property of certain systems of formal logic that every **semantically valid sequent** is also **syntactically** valid in them, i.e. provable using the **rules of inference**.

compound formula (of PL) Any **well-formed formula** produced from the **sentence-letters** of PL using the **connectives** according to the rules given in the **recursive definition** of a **well-formed formula** in the final section of Chapter 3.

conclusion See **argument**.
conditional (i) The symbol '→'. (ii) Any **formula** whose **main connective** is '→'. Each such formula consists of a **sub-formula** occurring before the arrow (the **antecedent**) and a sub-formula occurring after the arrow (the **consequent**).
conditional proof (rule of CP) To prove a **conditional** as **conclusion** assume the **antecedent** and derive the **consequent**. Enter the conditional on a line along with 'CP'. The **line numbers** are both that where the antecedent is assumed and that where the consequent is derived. Tip: at the line where CP is applied the **dependency-number** of the assumed antecedent must be discharged (deleted) from the set of dependencies.
conjunct See **conjunction**.
conjunction (i) The symbol '&'. (ii) Any **formula** whose **main connective** is '&'. Each such formula has two **conjuncts** as **sub-formulas**.
connective See **logical connectives**.
consequent See **conditional**.
consistency A set of **well-formed formulas** is consistent if and only if every member of that set can be true simultaneously. Otherwise, that set is **inconsistent**.
consistency-tree See **truth-tree method**.
contingent formula A **formula** which, given the **truth-table** definitions of the **connectives**, is true under one **interpretation** but false under another interpretation.
contradiction The **conjunction** of any **formula** and its **negation**.
corresponding conditional Any **sequent** of the form: $\{A_1 \ldots\ldots A_n\} : B$ can always be rewritten by constructing a **conditional** of the form: $A_1 \rightarrow (A_2 \rightarrow (\ldots A_n \rightarrow B))$. Any such conditional is the conditional *corresponding* to the sequent. Tip: read the final section of Chapter 2 on the **deduction theorem for PL**, which describes the relationship between **entailment** and **material implication** in more detail.
counterexample (to a **logical form**) Any **substitution-instance** of that form which has true **premises** and a false **conclusion**.
counterexample set For any **sequent**, the set consisting of the **conjunction** of the **premises** together with the **negation** of the **conclusion**.
CP See **conditional proof**.
creativity problem The problem of explaining how, from a finite linguistic input, the language-learner comes to be able to recognise and produce a potentially infinite linguistic output which includes sentences the language-learner has never heard before.
D See **domain**.
decidability A given property is *decidable* if and only if for any and every **sequent** of that system there exists an **effective decision-procedure** by following which we can conclusively establish either that the sequent

has or that it lacks the property in question. A **proof** that there is no such procedure for a given system is a proof of the undecidability of that system. Alonzo Church has provided such a proof (on a certain assumption) for **polyadic quantificational logic** which is generally known as *Church's Theorem* (the assumption is generally known as *Church's Thesis*). The (un)decidability of **QL** is discussed and illustrated in Section III of Chapter 7.

deduction theorem for PL That metatheory which clarifies the relation between the **turnstile** and the arrow in PL. This is fully discussed in the final section of Chapter 2.

deductive apparatus The set of **rules of inference** belonging to a **formal system**.

deductive argument Here a deductive **argument** is a **valid** argument.

definite descriptions See **theory of descriptions**.

denying the antecedent An **invalid** pattern of inference from a **conditional** which involves denying that the **antecedent** holds and then concluding that the **negation** of the **consequent** must also hold.

denying the consequent A **valid** pattern of inference from a **conditional** which involves denying that the **consequent** holds and then concluding that the **negation** of the **antecedent** must also hold, i.e. **modus tollens**.

dependency-numbers That part of a **line of proof** which indicates which **formulas** (if any) the formula on that line depends upon. Tip: this is a useful and important part of a line of proof which you should be careful to complete. Note that every **rule of inference** contains a recipe for identifying the appropriate set.

derivation Here, a **proof**.

derived rules Any **rule of inference** whose addition to the set of rules of inference will not allow anything new to be proved in the **formal system**. The contrast here is with **primitive rules**, i.e. rules of inference whose addition to a system will allow new **sequents** to be proved in the system.

development rules (for **syntactical trees**) A set of syntactical rules for breaking down **well-formed formulas** into their most basic components in a way which makes the nature of a **formula** and its **subformulas** fully explicit.

development rules (for PL **truth-trees**) A set of syntactical rules for breaking down **well-formed formulas** into their most basic components. These rules enable us to construct **truth-tree tests** for **consistency**.

discharge rule Any **rule of inference** whose use integrally involves previously having made **assumptions** for **proof**. When successfully applied, such assumptions may subsequently be deleted ('discharged'). Hence, reasoning with such rules is **hypothetical reasoning**.

disjunct See **disjunction**.

disjunction (i) The symbol 'v'. (ii) Any **formula** whose **main connective** is 'v'. Each such formula has two **disjuncts** as **sub-formulas**.

distributive (sense of 'all') That sense of 'all' in which we think of it as *distributing* properties over each and every particular thing of the relevant kind. For example, the use of 'all' in 'All folk singers are groovy' is one which distributes grooviness over each and every folk singer.

distributive laws (PL) A name given to a certain kind of law found, for example, in arithmetic, logic and set theory. In PL, (P & (Q v R)) ⊢ ((P & Q) v (P & R)) is a **distributive** law because it shows us how '&' may be *distributed* over 'v' in the process of inference. Analogously, the arithmetical law $(x \times (y + z)) = ((x \times y) + (x \times z))$ is a distributive law because it shows us how multiplication may be distributed over addition.

DNE See **double negation-elimination.**

DNI See **double negation-introduction.**

domain (**D**) That set of things (**elements**) over which **variables** of quantification range, i.e. the possible values which a **quantifier**'s variable may take. Also known as the **universe of discourse**. When a particular set of things is specified, e.g. animals, people, fish fingers, the domain is said to be **restricted** (to that set). Otherwise, the domain is said to be **unrestricted**.

double negation Literally, the result of **negating** a negated **formula**.

double negation-elimination (rule of DNE) Given the **double negation** of a **formula** on any **line of proof** you may write the original unnegated formula on a new line. Annotate the new line 'DNE' together with the **line number** of the line containing the double negative. The **dependency-numbers** of the new line are identical with those of the old line. Tip: the correctness of DNE in **classical logic** rests on the classical account of negation, i.e. **classical negation**. DNE is rejected by **intuitionist** logicians, who uphold a distinct account of negation.

double negation-introduction (rule of DNI) Given an unnegated **formula** on any **line of proof** you may write the double negative of that formula on a new line. Annotate the new line 'DNI' together with the **line number** of the line containing the original formula. The **dependency-numbers** of the new line are identical with those of the old.

dyadic relation Literally, a *two*-place **relation**, i.e. an expression with two gaps which describes a relationship. For example, . . . loves ---, , . . . hates ---, and so on. The gaps in such expressions are *places*.

EE See **existential elimination.**

effective decision-procedure An *algorithm*, i.e. a mechanical procedure which, when followed correctly, always gives a definite answer to the question as to whether a particular property holds of something or not.

EI see **existential introduction.**
EIN See **existential instantiation.**
elements See **domain.**
elimination-rule A **rule of inference** which allows a step in a **proof** from a **formula** which contains a **connective** to a formula which does not contain that connective.
empirical A philosophical term meaning *relating to or derived from experience*. Empirical sentences concern experiential matters, worldly matters. An empirical sentence is one whose truth-value is **decidable** in terms of experience.
entail A set of **premises** entails a **conclusion** when the conclusion is a **logical consequence** of that set of premises.
equivalence classes See **partition.**
equivalence relation Any **relation** which is **reflexive, transitive** and **symmetrical.**
exclusive (sense of disjunction) See **inclusive** (sense of disjunction).
ex falso quodlibet Here a **derived rule of inference,** meaning literally, 'from the (logically) false anything follows'.
existential elimination (rule of EE) To infer a **conclusion** from an existentially quantified **formula**: first assume a genuinely typical disjunct and then derive the desired conclusion from that disjunct. Restate the conclusion on a new **line of proof**. Annotate the new line 'EE' together with three numbers: (i) the **line number** of the original existential formula; (ii) the line number of the typical disjunct; (iii) the line number of the conclusion derived from the typical disjunct. The **dependency-numbers** of the new line consist of all the dependencies belonging to the derived conclusion except that you may discharge the dependency-number of the typical disjunct and replace it with the dependency-number of the original existentially quantified formula. Tip: note carefully that EE cannot be legitimately applied if (i) the name in the typical disjunct is itself contained in the conclusion derived for EE, or (ii) the name in the typical disjunct is itself contained in any **premise** or **assumption** used to derive the conclusion for EE from the typical disjunct.
existential instantiation (rule of EIN) The **instantiation** rule for existentially quantified **formulas** given by the **truth-tree method** for **QL** formulas. Where ν is any variable, $\phi(\nu)$ is any expression in which ν may occur, ρ is any proper name and $\phi(\rho)$ is the result of substituting ρ for ν, existential instantiation is defined as follows:

EIN: $$\frac{\exists\nu\,[\phi(\nu)]}{\phi(\rho)}$$

Restriction: Provided ρ does not occur in any formula in the branch or trunk.

existential introduction (rule of EI) Given a **formula** containing a name on any **line of proof** you may replace one or all occurrences of that name with a **variable**. Introduce the **existential quantifier** to the resulting **matrix** and write the formula on a new line. Annotate the new line 'EI' together with the **line number** of the original line. The **dependency-numbers** of the new line are identical with those of the line of the original formula.

existentially quantified conjunction Any existentially quantified **QL formula** the **main connective** in whose **matrix** is the ampersand. Tip: any natural language sentence of the form: 'Some As are Bs' should be translated into QL as an existentially quantified conjunction.

existential quantifier See **quantifier**.

extension (of a predicate) The set of **elements** of the **domain** assigned to a predicate under a **QL-interpretation**.

false under I See **satisfaction**.

formal definition (of **validity**) According to the formal definition an **argument** is valid if, and only if, it is an instance of a logically valid form of argument.

formal language A symbolism or notation containing a set of grammatical rules, and into which natural language sentences and **arguments** can be translated in such a way that their forms can be clearly expressed, i.e. here, either **PL** or **QL**.

formal system A **formal language** plus a **deductive apparatus**.

formula Any well-formed formal sentence belonging to **PL** or **QL**.

free logic In effect, any system of formal logic which does not take it as given that the names involved in the system refer to actually existing entities. Hence, free logics generally include an operator on names which makes explicit that a name does refer. Tip: **classical logic** is not a free logic.

free variable See **variable**.

function A **relation** or correspondence between certain members of one set (the *arguments* of the function) and a specific member of another (the *value* of the function) such that a unique value results from application of the function to its arguments. The arguments of the function are drawn from a set known as the *range* of the function. The function is defined over the range and so is defined for any ordered set of arguments from that range.

general form of a sequent A **sequent** consists of a (possibly empty) set of **premises**, a **colon** symbol and a **conclusion**. Using **metalinguistic variables** '$A_1 \ldots\ldots A_n$' and 'B' we can represent the general form or shape of a sequent as $\{A_1 \ldots\ldots A_n\}$: B.

general sentences Sentences involving terms such as 'all' and 'some', 'most' and 'many'.

golden rule A rule of thumb for identifying strategies for **proof**-construction in (in the first instance) PL. The rule consists of two questions and a prescription. First, ask whether the **conclusion** to be proved is a **conditional**. If so, the strategy should be **conditional proof**. If not, ask whether any or all of the **premises** are **disjunctions**. If so, the strategy should be **vElimination**. If not, the strategy should be **reductio ad absurdum**. Tip: the golden rule is first discussed in Chapter 3 for PL **sequents**, but note that it is also useful in **QL** as discussed in the final section of Chapter 6.

government See **scope** (of a **quantifier**).

hypothetical reasoning See **categorical reasoning**.

identity In **QL**, a primitive relational symbol, '=', which allows us to translate not the natural language 'is' of predication but the natural language use of 'is' which means 'is one and the very same as'.

identity-elimination (rule of =E) Given an **identity** statement on a **line of proof** and a **formula** on another line of proof which asserts a predicate or **relation** of a **term** in the identity you may infer that the same predicate or relation applies to the other term in the identity. The new line should be annotated '=E' together with the **line number** of both formulas used. The **dependency-numbers** of the new line consist of all of the dependency-numbers belonging to the two formulas used.

identity-introduction (rule of =I) Any **formula** of the form ($\rho = \rho$) may be entered on any line of any **proof**. Any such **line of proof** should be annotated '=I' and should have no **dependency-numbers**.

imply A set of **premises** implies a **conclusion** when the conclusion is a **logical consequence** of that set of premises.

import-export law A rule of **proof**-construction which stipulates that although **premises** may be freely imported into contexts involving **hypothetical reasoning**, **assumptions** may not be exported outside contexts involving hypothetical reasoning.

inclusive (sense of **disjunction**) That sense of 'v' in which (P v Q) will be true if both P and Q are true, in contrast to the **exclusive** sense, in which it will not be true in this case.

inconsistent See **consistency**.

inconsistent formula (PL) A **formula** which, given the **truth-table** definitions of the **connectives**, is always false under every **interpretation** of its constituent formulas.

indefinite descriptions See **theory of descriptions**.

induction/inductive argument Both the **premises** and the **conclusion** of an inductive **argument** concern **empirical** matters of fact. The premises of an inductive argument at best only support their conclusion, i.e. the premises of such an argument do not **entail** their conclusion.

infinite tree A particular kind of **truth-tree**. In **monadic QL** every truth-tree ends in a finite amount of time in one of two ways. (i) Every branch ends in contradiction and so every branch is dead, the tree is dead and the **sequent** is shown to be **valid**. (ii) Some branch is live and the sequent is shown to be **invalid**. Infinite trees are found in **polyadic QL** and represent a third possible outcome. (iii) Some branch is live but we are forced to keep developing the branch ever further. Such a branch is an infinite branch. An infinite tree is a tree with an infinite branch.

instantiation Any inference from a quantified **formula** to one of its instances.

internal grammatical structure This phrase draws attention to an important contrast between the **well-formed formulas** of **PL** and those of **QL**. While the **compound formulas** of PL have an internal structure determined by which **connective** and **sub-formulas** compose that formula, the **atomic formulas** of PL (**sentence-letters**) have no internal structure and so can represent nothing of the internal structure of natural language sentences. In sharp contrast, even the most basic QL formulas, e.g. **subject–predicate sentences**, do have an internal grammatical structure and thus can represent the internal grammatical structure of natural language sentences of that form, i.e. sentences which predicate a property of a named subject or individual. Moreover, quantified QL formulas enable us to represent the logical character of **general sentences**. Hence, QL allows us to represent the internal grammatical structure of natural language sentences in ways which transcend the expressive power of PL formulas.

interpretation (PL) An assignment of truth-values to each of the **atomic formulas** which go to make up that **formula**.

interpretation (QL) Intuitively, a QL-interpretation is a way of understanding the language QL, i.e. an assignment of meanings to the **elements** of formal vocabulary out of which **well-formed formulas** are composed. In effect, a QL-interpretation consists of four parts: (i) the specification of the **domain** for the interpretation; (ii) the assignment to each *name* of an element of that domain (the **referent** of the name); (iii) the assignment to each predicate-letter of a subset of elements of that domain (the **extension** of the predicate); (iv) the assignment to each relational expression of all and only those **ordered pairs** (triples, quadruples etc.) of elements of that domain which stand in that relation (the **solution set**). Tips: when constructing an interpretation, you should never include **connectives**, **quantifiers** or **variables**.

intransitive A **relation** R is said to be intransitive if and only if $\forall x\ [\forall y\ [\forall z\ [(Rxy\ \&\ Ryz) \rightarrow {\sim}Rxz]]]$.

introduction-rule A **rule of inference** which brings a **connective** into a **line of proof**.

intuitionism An alternative or 'deviant' system of formal logic which, most famously, rejects the **law of excluded middle**.

invalid An **argument** is invalid if it is possible that its **premises** be true and its **conclusion** false.

invalidating PL interpretation (IPLI) A **comparative truth-table test** or **truth-tree test for validity** shows a **sequent** to be **invalid** only if it identifies an assignment of truth-values to the **sentence-letters** which compose the sequent under which the **premises** are true and the **conclusion** is false. An **interpretation** (PL) is an assignment of truth-values. Hence, that interpretation which shows true premises and a false conclusion is precisely an invalidating PL interpretation of the sequent. To make the IPLI explicit, identify the particular assignment, list the **sentence-letters** and show which truth-value is assigned to each letter under that interpretation.

invalidating QL interpretation (IQLI) An **interpretation** (**QL**) under which a **sequent** is shown to have true **premises** and a false **conclusion**.

IPLI See **invalidating PL interpretation**.

IQLI See **invalidating QL interpretation**.

irreflexive A **relation** *R* is irreflexive if and only if it is not the case that an **element** of the **domain** can bear that relation to itself, i.e. if, and only if, $\forall x\ [\sim Rxx]$.

joint denial A **truth-functional connective** identified by Henry Sheffer. Two senses of joint denial can be distinguished: (i) 'neither A nor B' – this **function** gives the value 'true' only when both **disjuncts** are false, and is most commonly referred to as **Sheffer's stroke**; (ii) 'not both A and B' – this function gives the value 'false' only when both **conjuncts** are true. Sets consisting of either of these **truth-functions** are **adequate** in themselves. Hence, every possible truth-function can be represented in terms of a single **logical connective**.

key A list of the **sentence-letters** and the natural language sentences they stand for in any case of translation into PL.

law of excluded middle The law of logic which asserts that for any sentence it is a logical truth that either it holds or its **negation** holds.

law of identity The law of logic which asserts that everything is identical with itself. In PL, $\vdash P \rightarrow P$ is a surrogate for this law, whose content is more clearly represented by the **QL sequent** $\vdash \forall x\ [x = x]$.

left-handed vIntroduction See **vIntroduction**.

line number The number belonging to a particular **line of proof**.

lines of proof Every **proof** consists of a number of lines of proof. Each line of proof consists of four parts, a **line number**, a **formula**, the rule annotation and the **dependency-numbers** of the formula.

logical analysis The application of formal logic to natural language sentences and the **arguments** they compose so as to clarify the logical status, nature and implications of such sentences and arguments.

logical connectives Literally, those symbols which are used to connect either **atomic formulas, compound formulas** or both. PL contains five connectives: '~', '&', 'v', '→' and '↔'.

logical consequence A sentence is a logical consequence of a set of sentences if the truth of the set of sentences guarantees the truth of the sentence in question.

logical force The logical force of a **valid argument** derives from the fact that to accept the **premises** of such an argument while rejecting its **conclusion** is to contradict oneself.

logical form The schematic pattern or inferential structure of an **argument**. Tips: logical form is not really a once-and-for-all matter. Particular arguments can be instances of more than one logical form. Our ability to represent the logical form of an argument is limited by the formal vocabulary available to us.

logically proper name A technical term of art in Bertrand Russell's philosophy of language. In Russell's view, a name is logically proper if it is meaningful solely in virtue of referring to an object. Eventually, Russell came to the view that the demonstratives 'this' and 'that' are the only true logically proper names in natural language (because they invariably refer to objects).

logical strength A **relation** between **formulas** measured in terms of a comparison of the **logical consequences** of those formulas. Given two formulas, if each formula **entails** the other the formulas are of equal logical strength. If the first entails the second but the second does not entail the first then the first formula is logically stronger than the second.

logicism The programme associated above all with Gottlob Frege and Bertrand Russell which attempted to demonstrate that the truths of mathematics were ultimately logical truths. Tip: see the reading list given near the end of Section XI in Chapter 5.

main column That column in a **truth-table** which records the overall truth-value of the whole **formula**. As such, the main column is the column under the **main connective**.

main connective The **connective** whose **scope** is the entire **formula**. Tips: though a **well-formed formula** may contain any number of connectives, every formula has one *and only one* main connective (the remainder are connectives belonging to the **sub-formulas** which compose that formula). When trying to identify the main connective: (i) If the main connective is '~', the formula should begin with the following shape '~(. . . '. Otherwise, read any **negation** in initial position as applying to the shortest subsequent formula. (ii) If the main connective is a **binary** connective look for the place at which brackets go in opposite directions, i.e. the shape ' . . .) . . . (. . . '. If the whole formula is not negated, the connective between these brackets is the main connective.

material implication The **relation** between **formulas** represented by the arrow symbol. If A materially implies B then B is not false when A is true. Tip: it is crucial to appreciate that this relationship is really a humble grammatical one between sentences or formulas, i.e. it is simply an analogue of some of our talk with English 'If . . . then - - - ' sentences and most certainly does not describe a causal relation between events.

matrix See **scope** (of a **quantifier**).

m.c. The abbreviation for **main connective** in **truth-tables**.

metalanguage That language in which we talk about another language. For example, the metalanguage in which we talk about PL is really English plus certain elements of formal vocabulary such as the **turnstile**. From the viewpoint of the metalanguage, PL is an **object language**, i.e. that language which is the object of our metalinguistic talk. This is an important distinction for logicians and philosophers; especially for those interested in the theory of truth.

metalinguistic variables These **variables** belong to the **metalanguage** and as such range over the **formulas** of the **object language**. For example, if PL is the object language, the variables stand for formulas of PL.

metalogical arrow A metalinguistic representation of 'If . . . then . . . '.

modal definition (of **validity**) According to the modal definition an **argument** is valid if it is such that if the **premises** are true then the **conclusion** must be true, on pain of contradiction.

model An **interpretation** of a **formula** which makes that whole formula true.

modus ponens (rule of MP) Given a **conditional** on one **line of proof** and its **antecedent** on another line, its **consequent** may be inferred. The **line numbers** of each must be cited together with MP. The **dependency-numbers** of the new line consist of the dependency-numbers of both cited lines.

modus tollens (rule of MT) Given a **conditional** on one **line of proof** and the **negation** of its **consequent** on another, you may infer the negation of the **antecedent**. Annotate the new line with the **line numbers** of both lines used and 'MT'. The **dependency-numbers** of the new line are all those of both lines used.

monadic quantificational logic See **monadic QL**.

monadic QL That fragment of **QL** which confines itself to monadic predicates and to the methods for establishing the **validity** and **invalidity** of **sequents** involving only monadic predicates.

monotonicity Here, the property of **classical logic** that if a **conclusion** follows from some set of **premises** then that conclusion follows from any augmentation of that set of premises.

MP See **modus ponens**.

MT See **modus tollens**.

multiple generality Generalisation of a level of complexity such that accurate translation into **QL** requires the use of more than one **quantifier**.

necessary truth A sentence is a necessary truth if it could not possibly be false. The contrast here is with **contingent truth**. A sentence is a contingent truth if it is true but could have been false.

negated existential generalisation Simply, any negated existentially quantified **QL formula**. Tip: given that we use positive existentially quantified formulas to express **ontological commitments** it is most natural to use negated existentially quantified formulas to translate natural language sentences which deny existence.

negation (i) The symbol '~'. (ii) Any **formula** whose **main connective** is '~'.

negation-introduction This is **reductio ad absurdum**.

nested conditionals Complex **conditionals** whose **sub-formulas** are also conditionals.

non-identity A sentence or **formula** which denies **identity**, i.e. literally, the result of applying **negation** to any identity statement.

non-reflexive A **relation** *R* is non-reflexive if and only if it is both the case that some **element** of the **domain** bears that relation to itself and that some other element does not bear that relation to itself, i.e. if and only if $\exists x\, [Rxx]\ \&\ \exists x\, [{\sim}Rxx]$.

non-symmetrical A **relation** *R* is non-symmetrical if it is both the case that there is some *a* that bears *R* to some *b* when that *b* bears *R* to *a* and that there is some *c* which bears *R* to some *d* while *d* does not bear *R* to *c*, i.e. iff $\exists x\, [\exists y\, [Rxy\ \&\ Ryx]]\ \&\ \exists x\, [\exists y\, [Rxy\ \&\ {\sim}Ryx]]$.

non-transitive A **relation** *R* is non-transitive if there are both cases in which *a* bears *R* to *b*, *b* bears *R* to *c* and *a* bears *R* to *c* and cases where *a* bears *R* to *b* and *b* bears *R* to *c* but *a* does not bear *R* to *c*, i.e. if and only if $\exists x\, [\exists y\, [\exists z\, [(Rxy\ \&\ Ryz)\ \&\ Rxz]]]\ \&\ \exists x\, [\exists y\, [\exists z\, [(Rxy\ \&\ Ryz)\ \&\ {\sim}Rxz]]]$.

numerically definite quantification Simply, quantifying in a way which specifies the exact quantity of things which satisfy the relevant predicate. Just on its own, the **existential quantifier** is inexact in that although it specifies that at least one **element** of the **domain** satisfies the predicate it leaves open the possibility that a larger number of things also satisfy that predicate. Given **negation** and **identity** we can in principle quantify in a numerically definite way for any number of elements.

object language See **metalanguage**.

ontological commitments *Ontology* is the philosopher's word for the study of what there is. So, one's ontological commitments are one's existential commitments, i.e. what one believes exists.

ordered pairs See **relation**.

overall (strategy for) **proof** See **sub-proof**.

P See **premise-introduction**.

partial ordering relations Any **relation** which is **reflexive**, **transitive** and **anti-symmetrical**.

partition (of a set) A separation of the members of the set into subsets such that each member of the set is in one and only one subset. Where the set is partitioned by an **equivalence relation** the subsets obtained are **equivalence classes**.

philosophical logic Philosophical logic is *logical theory*. In effect, the constructs of formal logic, i.e. **formal systems**, formal **semantics**, etc., and the notions fundamental to those constructs are objects of critical reflection for philosophical logic. However, the discipline also studies, for example, the **theory of definition**, scientific methodology, logical issues in ethics and philosophy of religion and, indeed, stands on its own legs as an object of study. Tip: interested parties might consult Sybil Wolfram's *Philosophical Logic: An Introduction*, [1994], London, Routledge.

PL (**propositional logic**) Intuitively, the logic of sentences formed from sentence-letters, by grammatical rules, using an **adequate** set of logical connectives.

place-markers Symbols which mark gaps in **argument-frames**.

polyadic quantificational logic See **polyadic QL**.

polyadic QL **QL** understood as containing formal vocabulary adequate to the task of representing many-place **relations**.

premise-introduction (rule of P) Any **well-formed formula** may be introduced as a **premise** on any **line of proof**. The **dependency-number** of that line is identical with the **line number** of that line of proof.

premises See **argument**.

prenex form Any **QL formula** which has every **quantifier** on the left before any other **connective** is said to be in **prenex form**. The **quantifier-equivalences** can always be exploited to put quantified QL formulas into prenex form.

primitive rule See **derived rule**.

principle of bivalence A semantic principle according to which every sentence of a language is either true or false but not both and not neither, i.e. each and every sentence has exactly one of the two truth-values *true* or *false*.

problem of non-being In essence, the problem of understanding what it is that we talk about when we talk about what is not the case. Does the very meaningfulness of my assertion 'Santa Claus does not exist' itself imply the existence of Mr Claus? Bertrand Russell thought not and offers an analysis of the logical character of ontological disagreements which may solve this problem. Russell's analysis is discussed in the final section of Chapter 5.

procedural rules (for PL **truth-trees**) A set of rules governing the order in which different kinds of **well-formed formula** must be developed so as to enable effective applications of the **truth-tree method**.

proof Intuitively, a step-by-step way of getting from the **premises** to the **conclusion** of a **sequent**, each step being justified by a **rule of inference**.

proof-in-PL A **proof** in PL is a finite sequence of consecutively numbered lines, each consisting of a **well-formed formula** of PL, together with a set of numbers known as the **dependency-numbers** of that line, the entire sequence being constructed using the **rules of inference** of PL.

proof-in-QL A **proof** in **QL** is a finite sequence of consecutively numbered lines, each consisting of a **well-formed formula** of QL, together with a set of numbers known as the **dependency-numbers** of that line, the entire sequence being constructed using the **rules of inference** of QL.

proof-theoretic consequence The notion of consequence represented by the **turnstile**.

proof-theory The study of the **relation** of **logical consequence** among the **formulas** of a **formal system** in purely syntactical terms, i.e. in terms of what can be proved using the **rules of inference** and with no regard to the **interpretation** of formulas within the system.

proposition Intuitively, the meaning or sense of a declarative sentence. Many philosophers believe that propositions are the real bearers of truth-values.

propositional calculus The **proof-theory** of PL.

Propositional logic See **PL**.

QL See **quantificational logic**.

QL-interpretation See **interpretation** (QL).

quantificational logic (QL) Intuitively, the logic of **general sentences** and the **arguments** they compose.

quantifier Intuitively, a quantifier is any expression which specifies a quantity of things as having (*satisfying*) some predicate or other. Here, two specific quantifiers are particularly important: (i) The **universal quantifier** '∀' indicates that each and every thing (**element** of the **domain**) satisfies the predicates in the **formula**'s **matrix**. (ii) The **existential quantifier** '∃' indicates that some (at least one) thing (element of the domain) satisfies the predicates in the formula's matrix. Tips: the quantifier in initial position indicates the kind of quantified formula in question.

quantifier-equivalences Given **negation**, the **quantifiers** can be shown to be equivalent in meaning. Where 'v' is any **variable** and '[. . .]' any **matrix**, the general forms of the quantifier-equivalences are as follows:

1′. $\sim\forall v$ [. . . .] : $\exists v$ [~ . . . .]   2′. $\sim\exists v$ [. . .] : $\forall v$ [~ . . . .]

Tip: the following pair of equivalences are notable given their role in the **truth-tree method** for **QL sequents**: (i) $\sim\exists x\ [Fx] \vDash \forall x\ [\sim Fx]$; (ii) $\sim\forall x\ [Fx] \vDash \exists x\ [\sim Fx]$.

quantifier switch/quantifier shift fallacy Fallacious reasoning which can be formally represented by the reversal of the order of **quantifiers** across the inference from **premises** to **conclusion**.

Quining The application of Bertrand Russell's analysis of **definite descriptions** to singular nouns as suggested by W.V.O Quine (hence, Quining). This practice (which originates in Russell's own thinking) is discussed in the final section of Chapter 5.

RAA See **reductio ad absurdum**.

recursive definition Definition of a kind which exploits procedures which are applied to generate a result to which those procedures can be reapplied indefinitely. The definition of grammatical well-formedness for PL **formulas** given in the final section of Chapter 3 illustrates such a definition.

reductio ad absurdum (rule of RAA) If a **contradiction** has been shown to be derivable from an **assumption** we may write the **negation** of that assumption on a new **line of proof**. Annotate the new line with the line number of the line of the contradiction, the line number of the assumption and 'RAA'. The **dependency-numbers** of the new line will consist of all those of the old lines except that of the assumption from which the contradiction has been derived.

referent (of a name or **term**) That **element** of the **domain** which a name stands for under a **QL-interpretation**.

reflexive That formal property of **relations** which are such that every **element** of the **domain** bears that relation to itself, i.e. a relation R is reflexive if and only if $\forall x\ [Rxx]$.

refutation by counterexample The method of proving **invalidity** by means of a **counterexample**.

relations In **QL** the symbols R, S and so on are used to translate assertions involving expressions which, rather than simply predicate a property of a subject, assert a connection between things. Hence, relational expressions involve at least two gaps or places (which are plugged or filled by **terms**), as opposed to predicates which involve only one. Elements of formal vocabulary with one place are said to be monadic. Two-place relations are **dyadic**, three-place relations are triadic and many-place relations are polyadic. In any relation, terms are said to *stand in* or bear that relation to one another. Tips: the order in which terms stand in a QL **formula** can be crucial to the sense of the translation. Hence, relational expressions are satisfied not simply by pairs of **elements** of the **domain** but rather by **ordered pairs** of elements.

restricted See **domain**.

right-handed vIntroduction See **vIntroduction**.

rule-annotation That part of a **line of proof** which makes clear which **rule of inference** was used to construct the **formula** on the line and which formulas (if any) that rule was applied to.

rule of term-introduction A procedural rule for the **truth-tree method** for **QL**. The rule is as follows: always develop each and every existentially quantified **formula** and every negated universally quantified formula before you develop any universally quantified formula or negated existentially quantified formula.

rules of inference The set of rules in terms of which a **proof** of a **sequent** may be constructed.

salva veritate A Latinism meaning 'truth-preserving' (literally, 'saving truth'). The expression refers to one kind of result of **uniform substitution** of **terms**, i.e. two terms are inter-substitutable *salva veritate* just in case the truth-value of sentences into which such substitutions are made always remains unchanged.

satisfaction A very useful and important notion in formal **semantics** for **QL** in terms of which we bridge the gap between the *assignment* of **elements** of the **domain** which a **QL-interpretation** makes to names, predicates and **relations** and the subsequent *evaluation* of QL **formulas** composed from that vocabulary. Intuitively, we can think of satisfaction as a matter of sharing of elements of the domain between, for example, a name and a predicate. Thus, if the **referent** of a name is included in the **extension** of a predicate that name is said to *satisfy* that predicate. Given that notion, we can then go on to say that the subject–predicate formula formed from just that name and just that predicate is **true under I**, i.e. that the formula is true under that **interpretation**. Conversely, if the name and the predicate do not share an element of the domain the resulting subject–predicate formula is **false under I**.

scope (of a **connective**) The scope of a connective consists of the connective itself together with the **formulas** it connects.

scope (of a **quantifier**) The scope of a quantifier is that part of a **QL formula** which that quantifier looks after, controls or **governs**. Tips: the scope of a quantifier is always indicated by a pair of square brackets the first of which appears immediately after the **variable** belonging to the quantifier. These square brackets and their contents are the quantifier's **matrix**. Therefore, the scope of a quantifier is identical with its matrix.

semantic consequence The notion of **consequence** represented by the **semantic double turnstile**.

semantic double turnstile '$\models$' (i) The symbol in a **valid sequent** which asserts that validity has been demonstrated by a semantic method, e.g. the **truth-tree method**. (ii) The notion of **logical consequence** represented semantically, i.e. in terms of **model**-theory.

semantic equivalence Semantic equivalence is **truth-functional** equivalence, i.e. two or more **compound formulas** of PL are semantically equivalent if and only if for each and every assignment of truth-values to their component **sentence-letters** the overall truth-values of those **formulas** are one and the same.

semantics That approach to formal logic which explicitly focuses on the meaning and **truth-conditions** of **formulas**.

semantic validity The notion of **logical consequence** as represented by the **semantic double turnstile**.

sentence-letters Symbols (here, upper-case capital letters from P onwards) which stand in place of specific sentences.

sentential constants Another term for **sentence-letters**.

sentential-functions Simply, predicates and **relations**, i.e. gappy functional expressions which take names as **arguments** and give sentences as values.

sentential variables Symbols which mark gaps which may be plugged by sentences.

sequent Any finite (possibly empty) set of **well-formed formulas** (the **premises**) together with a single well-formed formula (the **conclusion**) the two being separated by the **colon** symbol ':'.

sequent-introduction (rule of SI) A **rule of inference** which exploits previously proved **sequents** by allowing the **conclusions** of such sequents to be drawn immediately from their **premises** in the process of proof-construction without spelling out the intermediate steps required in the original **proof** of the sequent.

shallow analysis The practice of showing a **sequent** of **QL** to be an instance of a PL form.

Sheffer's stroke See **joint denial**.

short-cut method A method of testing **sequents** for **validity** by first assuming that the **premises** are true and the **conclusion** false and then assigning truth-values to the **sub-formulas** involved in an attempt to generate the desired overall values. Tips: the short-cut method is generally the quickest PL procedure for testing for validity semantically. However, effective use of the method presupposes a good working knowledge of the **truth-tables**.

SI See **sequent-introduction**.

singular conditionals Unquantified **QL formulas** whose **main connective** is the arrow.

singular sentences In effect, any sentence the grammatical subject of which is a single, specific individual.

solution set (for a **dyadic relation**) A specification of all the **ordered pairs** for which the **relation** holds and those for which it does not.

sound A sound **argument** is a **valid** argument with true **premises**.

Soundness The (highly desirable) property of certain systems of formal logic that every provable **sequent** is also a semantically **valid** sequent. Tips: do not confuse this metatheoretical result with the humble notion of a **sound argument**.

sub-formula(s) Those **well-formed formulas** which compose a **compound formula**.

subject–predicate sentences See **internal grammatical structure**.

sub-proof An intermediate part of a **proof** during which a **formula** necessary to proving the **conclusion** of the **sequent** is derived. In a sense, a sub-proof is a proof within a proof. In traditional logical terms a sub-proof is a *lemma*. Tips: a proof may contain a number of sub-proofs each of which make a contribution to deriving the conclusion. Moreover, each sub-proof may involve a distinct strategy for deriving its contribution to the proof. Use the **golden rule** to identify the right strategy not just for the **overall proof** of the conclusion from the **premises** but also to identify the right strategy for each sub-proof. Strategic considerations for **PL** proofs are discussed in Sections VI, VII and VIII of Chapter 3. Strategic considerations for **QL** proofs are discussed in the final section of Chapter 6.

substitutional criterion of validity (for logical forms of **argument**) The criterion that a form of argument is **valid** if, and only if, every **substitution-instance** of that form is itself a valid argument.

substitution-instance (of a particular **logical form**) Any particular natural language **argument** which **instantiates** or exemplifies the logical form in question.

syllogistic The system of logic first designed by Aristotle.

symmetry A **relation** R is symmetrical if and only if $\forall x\ [\forall y\ [Rxy \to Ryx]]$.

syntactical tree (PL) A diagrammatic representation which clearly displays the internal syntactic structure of a PL **formula** down to explicitly naming each constituent symbol.

syntactical trees for QL formulas A diagrammatic representation which clearly displays the internal syntactic structure of a **QL formula** down to explicitly naming each constituent symbol.

syntax That aspect of formal logic which approaches **formulas** and **sequents** as uninterpreted shapes, i.e. in terms of grammar and the notion of **proof**.

target set of truth-values (TST) That set of truth-values for the **QL formulas** composing a **sequent** generated by an **invalidating QL interpretation** of that sequent.

tautology A **formula** which, given the **truth-table** definitions of the **connectives**, is true under every **interpretation** of its constituent formulas.

terms (QL) Names, i.e. the symbols 'a', 'b', 'c', etc.

theorems Logical laws as represented in a **formal language**. In effect, these are **premise**less **sequents** whose **validity** is underpinned by the nature of the **formula** as **conclusion** and the **rules of inference** of the **formal system**.

theorem-introduction (rule of TI) A **rule of inference** which allows previously proved **theorems** generally (or a stipulated list thereof) to be entered on any line of any **proof** with an empty set of **dependency-numbers** and annotated 'TI'.

theory of definition See **philosophical logic**.

theory of descriptions A paradigm of **logical analysis** of a certain kind of natural language expression known as **definite descriptions**, due to Bertrand Russell. Such descriptions are prefixed by the definite article, e.g. '*The* present Prime Minister', '*The* man in the street', etc. As such, they are distinct from **indefinite descriptions**, which are prefixed by the indefinite article, e.g. '*A* man . . . ', '*A* philosopher . . . ', etc. In essence, the theory of descriptions makes explicit the logical character of definite descriptions by analysing their contribution to the meanings of sentences in which they occur. Here, the theory is more fully discussed in the final sections of Chapter 5.

TI See **theorem-introduction**.

transitivity A **relation** R is transitive if and only if $\forall x\ [\forall y\ [\forall z\ [(Rxy\ \&\ Ryz) \rightarrow Rxz]]]$

transitivity of identity As an **equivalence relation**, **identity** is **transitive**. This property is exemplified by the **sequent** $a = b, b = c \vdash a = c$.

true under I See **satisfaction**.

truth-conditions Intuitively, those conditions which must pertain if a sentence is to be true. To grasp the truth-conditions of a sentence is (arguably) to understand the meaning of that sentence.

truth-function A **function** whose **arguments** and values are truth-values.

truth-functional connective Any **connective** which takes truth-valued **formulas** as arguments to give other truth-valued formulas as values, as defined by a **truth-table**.

truth-table (i) A diagrammatic definition of a **truth-functional connective**. (ii) A diagrammatic representation of the overall truth-value of a **compound formula** shown as a function of the truth-values of its constituent **formulas** and logical **connectives** (as specified in (i)).

truth-tree A diagrammatic way of representing and breaking down sets of **formulas**.

truth-tree method (for PL) A way of testing sets of **well-formed formulas** of PL for **consistency**. The method is fully set out and discussed in Section XI of Chapter 4.

truth-tree method (for **QL**) A way of testing sets of **well-formed formulas** of QL for **consistency**. The method is fully set out and discussed in Chapter 7.

truth-tree test for validity A method of testing **sequents** for **validity** which exploits the **truth-tree method** by using it to test the **counterexample set** to a sequent for **consistency**. A sequent is valid if and only if its counterexample set is inconsistent. Therefore if the counterexample set is inconsistent then the original sequent is valid. Hence, tests for consistency to counterexample sets reveal the validity or **invalidity** of sequents.

TST See **target set of truth-values.**

turnstile '⊢' (i) The symbol in a **valid sequent** which asserts the existence of a **proof** of the sequent. (ii) The notion of **logical consequence** represented syntactically, i.e. in terms of **proof-theory**.

UE See **universal elimination.**

UI See **universal introduction.**

UIN See **universal instantiation.**

unary (connectives) Connectives which can be applied to a single **formula** to form a **well-formed formula.** '~' is the only unary connective in PL.

uniform substitution The practice of replacing each and every occurrence of one symbol with another specific symbol throughout a **formula.**

universal conditional A universally quantified **formula** the main **connective** in whose **matrix** is the arrow. Tip: every natural language sentence of the form 'All As are Bs' should be represented as a universal conditional in **QL.**

universal elimination (rule of UE) Given any universal **formula** on any **line of proof** you may always infer any particular instance of that formula on another line of proof. The new line should be annotated with the **line number** of the universal formula in question and 'UE'. The **dependency-numbers** of the new line are identical with those of the line of the original universal formula.

universal instantiation (rule of UIN) The **instantiation** rule for universally quantified **formulas** given by the **truth-tree method** for **QL** formulas. Where v is any variable, $\phi(v)$ is any expression in which v may occur, ρ is any proper name and $\phi(\rho)$ is the result of substituting ρ for v, universal instantiation is defined as follows:

UIN: $$\frac{\forall v\ [\phi(v)]}{\phi(\rho)}$$

Restriction: Provided ρ does occur in some formula on the branch or in the trunk (if no terms occur, instantiate to any term).

universal introduction (rule of UI) Given a **formula** containing a name on any **line of proof** you may replace each occurrence of that name with a **variable**, introduce the **universal quantifier** to that **matrix** and write the resulting formula on a new line provided that the original formula containing the name does not include among its dependencies any formula containing that name. Annotate the new line 'UI' together with the **line number** of the original line. The **dependency-numbers** of the new line are identical with those of the line of the original formula.

universal quantifier See **quantifier**.

universe of discourse Another name for **domain**.

unrestricted See **domain**.

valid Validity is that property belonging to **arguments** whereby if the **premises** are true then the **conclusion** must be true, on pain of contradiction. So, if an argument is valid the conclusion of that argument is a **logical consequence** of its premises.

variable (QL) Intuitively, a symbol which marks a place for an unnamed thing (an **element** of the **domain**). In QL, the symbols 'x', 'y', 'z' are variables. Tips: read each such symbol as meaning 'thing'. If more than one variable is involved in a formula as in $\forall x\ [\exists y\ [Rxy]]$ read the formula as meaning 'Consider every thing, x, there is some thing, y, such that . . . '. Any variable which occurs within the **scope** of a **quantifier** is said to be a **bound variable** (bound by the quantifier). A variable which does not occur within the scope of a quantifier is a **free variable**. Note very carefully that in QL no variable may occur free, i.e. without a quantifier to bind it.

vE See **vElimination**.

vElimination (rule of vE) To draw an inference from a **disjunction** you must derive the desired **formula** from each **disjunct** first, i.e. assume each disjunct in turn and derive the desired formula from each. Having done so, you may repeat the conclusion on a new **line of proof**. Annotate the new line with five numbers followed by 'vE'. The five numbers are: (i) the **line number** of the disjunction; (ii) the **dependency-number** of the first disjunct assumed; (iii) the line number of the **conclusion** derived from the first disjunct; (iv) the dependency-number of the second disjunct assumed; (v) the line number of the conclusion derived from the second disjunct. Note carefully that vE is a **discharge rule**. Hence, at the line annotated 'vE' you may discharge the dependency-numbers of each disjunct and replace them with the dependency-number of the original disjunction together with the dependency-number of any other formula you used to derive the conclusion.

vI See **vIntroduction**.

vIntroduction (rule of vI) Given a **formula** on a **line of proof** you may infer the **disjunction** of that formula with any other **well-formed formula** on a new line of proof. Annotate the new line with the **line number** of the old line and 'vI'. The **dependency-numbers** of the new line are identical with those of the old line. Tip: the disjunction of a formula may be inferred by introducing 'v' to the right of the formula and then completing the disjunction (**right-handed vIntroduction**). But equally the disjunction of a formula may be inferred by introducing 'v' to the left of the formula and completing the disjunction (**left-handed vIintroduction**).

well-formed formula (of PL) All and only those **formulas** sanctioned by the **recursive definition** given in the final section of Chapter 3.

Bibliography

Anscombe, Elizabeth, [1963], *An Introduction to Wittgenstein's Tractatus*, London, Hutchinson University Library.

Aristotle, *De Interpretatione*, in: J.L. Ackrill, [1963], *Aristotle's Categories and De Interpretatione*, Oxford, Clarendon Press.

—— *Metaphysics*, in: Jonathan Barnes (ed.) [1984], *The Complete Works of Aristotle*, Revised Oxford Translation, Bollingen Series LXXI, Princeton NJ, Princeton University Press.

Baker, G.P. and Hacker, P.M.S., [1984], *Language, Sense and Nonsense: A Critical Investigation into Modern Theories of Language*, Oxford, Blackwell.

Barker, Stephen F., [1957], *Induction and Hypothesis: A Study of the Logic of Confirmation*, Ithaca NY, Cornell University Press.

Benacerraf, Paul, and Putnam, Hilary (eds.), [1983], *Philosophy of Mathematics: Selected Readings*, second edition, Cambridge, Cambridge University Press.

Black, Max, [1964], *A Companion to Wittgenstein's Tractatus*, Cambridge, Cambridge University Press.

Boehner, Philotheus, [1952], *Medieval Logic: An Outline of its Development from 1250 to c.1400*, Manchester, Manchester University Press.

Boole, George, [1847], *Mathematical Analysis of Logic*, Cambridge; reprinted [1948], Oxford, Oxford University Press.

Boolos, George S., and Jeffrey, Richard C., [1996], *Computability and Logic*, third edition, Cambridge, Cambridge University Press.

Broadie, Alexander, [1987], *Introduction to Medieval Logic*, Oxford, Clarendon Press.

Carruthers, Peter, [1989], *Tractarian Semantics: Finding Sense in Wittgenstein's Tractatus*, Oxford, Blackwell.

—— [1990], *The Metaphysics of the Tractatus*, Cambridge, Cambridge University Press.

Chomsky, Noam, [1957], *Syntactical Structures*, The Hague, Mouton.

—— [1965], *Aspects of the Theory of Syntax*, Cambridge MA, MIT Press.

—— [1980], *Rules and Representations*, Oxford, Blackwell.

—— [1986], *Knowledge of Language: Its Nature, Origin and Use*, New York, Praeger.

—— [1988], *Language and Problems of Knowledge*, Cambridge MA, MIT Press.

—— [1995], 'Language as Natural Object', *Mind*, 104, pp. 1–63.

Church, Alonzo, [1936a], 'A Note on the Entscheidungsproblem', *Journal of Symbolic Logic*, 1 (1), March, pp. 40–1.

—— [1936b], 'Correction to "A Note on the Entscheidungsproblem"', *Journal of Symbolic Logic*, 1(3), September, pp. 101–2.

—— and Quine, W.V.O., [1952], 'Some Theorems on Definability and Decidability', *Journal of Symbolic Logic*, 17(3), September, pp. 179–87.

Cook, V.J. and Newsom, Mark, [1996], *Chomsky's Universal Grammar: An Introduction*, second edition, Oxford, Blackwell.

Curry, Haskell B., [1976], *Foundations of Mathematical Logic*, Dover Edition, New York, Dover Publications.
De Morgan, Augustus, [1847], *Formal Logic: The Calculus of Inference, Necessary and Probable*, London, Taylor and Walton.
Dummett, Michael, [1977], *Elements of Intuitionism*, Oxford, Clarendon Press.
—— [1978], 'The Philosophical Basis of Intuitionist Logic', in *Truth and other Enigmas*, London, Duckworth.
Faris, J.A., [1964], *Quantification Theory: Monographs in Modern Logic*, London, Routledge & Kegan Paul.
Frege, Gottlob, [1879], *Begriffsschrift, eine der arithmetischen nachgebildete Formelsprache des reinen Denkens*, Halle, L. Nebert.
—— [1892], 'On Sense and Reference', in: A.W.Y. Moore (ed.), [1993], *Meaning and Reference*, Oxford, Oxford University Press.
Hamilton, A.G., [1978], *Logic for Mathematicians*, revised edition, Cambridge, Cambridge University Press.
Hookway, Christopher, [1988], *Quine*, Oxford, Polity Press.
Hunter, Geoffrey, [1971], *Metalogic: An Introduction to the Metatheory of Standard First-Order Logic*, London and Basingstoke, Macmillan.
Jeffrey, Richard C., [1967], *Formal Logic: Its Scope and Limits*, New York, McGraw-Hill.
Kneale, William, and Kneale, Martha, [1962], *The Development of Logic*, Oxford, Clarendon Press.
Korner, Stephan, [1960], *The Philosophy of Mathematics: An Introductory Essay*, London, Hutchinson University Library.
Lear, Jonathan, [1980], *Aristotle and Logical Theory*, Cambridge, Cambridge University Press.
Lemmon, E.J., [1965], *Beginning Logic*, London, Thomas Nelson and Sons.
Luce, A.A., [1958], *Logic*, London, English Universities Press.
Lukasiewicz, Jan, [1951], *Aristotle's Syllogistic from the Standpoint of Modern Formal Logic*, Oxford, Clarendon Press.
Lyons, John, [1970], *Chomsky*, Fontana Modern Masters, London, Fontana/Collins.
Lyndon, Roger C., [1966], *Notes on Logic*, Van Nostrand Mathematical Studies #6, Princeton NJ, D. Van Nostrand.
Marsh, Robert C. (ed.), [1984], *Logic and Knowledge Essays 1901–1950*, London, George Allen & Unwin.
Mates, Benson, [1953], *Stoic Logic*, University of California Publications in Philosophy, Vol. 26, Berkeley and Los Angeles, University of California Press.
—— [1972], *Elementary Logic*, second edition, New York, Oxford University Press.
Mendelson, Elliot, [1987], *Introduction to Mathematical Logic*, third edition, Monterey CA, Wadsworth and Brooks.
Moore, Adrian (ed.), [1993], *Meaning and Reference*, Oxford, Oxford University Press.
Morse, Warner, [1973], *Study Guide for Logic and Philosophy*, second edition, Belmont CA, Wadsworth.
Mounce, H.O., [1981], *Wittgenstein's Tractatus: An Introduction*, Oxford, Blackwell.
Neale, Stephen, [1990], *Descriptions*, Cambridge MA, MIT Press.
Newton-Smith, W.H., [1985], *Logic: An Introductory Course*, London, Routledge.
Phillips, Calbert (ed.), [1995], *Logic in Medicine*, London, British Medical Journal Publishing Group.
Popper, K.R., [1972], *Conjectures and Refutations*, fourth edition, London & Henley, Routledge and Kegan Paul.
Quine, W.V.O., [1963], *From a Logical Point of View*, New York and Evanston, Harper & Row.
—— [1986], *Philosophy of Logic*, second edition, Cambridge MA and London, Harvard University Press.

Read, Stephen, [1980], '"Exists" is a Predicate', *Mind*, 89, pp. 412–17.
—— [1988], *Relevant Logic*, Oxford, Blackwell.
—— [1995], *Thinking about Logic: An Introduction to the Philosophy of Logic*, Oxford, Oxford University Press.
—— and Wright, Crispin, [1993], *Read and Wright: Formal Logic, An Introduction to First Order Logic*, fifth edition, revised, Departmental Publication, St Andrews, University of St Andrews.
Russell, Bertrand, [1905], 'On Denoting' in: Robert C. Marsh (ed.), [1984], *Logic and Knowledge: Essays 1901–1950*, London, George Allen & Unwin.
—— [1918], 'Lectures on the Philosophy of Logical Atomism', in: Robert C. Marsh (ed.), [1984], *Logic and Knowledge: Essays 1901–1950*, London, George Allen & Unwin.
—— and Whitehead, Alfred North, [1910–13], *Principia Mathematica*, Cambridge, Cambridge University Press.
Sheffer, Henry M., [1913], 'A Set of Five Independent Postulates for Boolean Algebras, with Applications to Logical Constants', *Transactions of the American Mathematical Society*, XIV, pp. 481–8.
Strawson, Sir Peter, [1950], 'On Referring', in: P.F. Strawson, [1971], *Logico-Linguistic Papers*, London, Methuen.
Tennant, Neil, [1978], *Natural Logic*, Edinburgh, Edinburgh University Press.
Wittgenstein, Ludwig, [1953], *Philosophical Investigations*, Oxford, Blackwell.
—— [1961], *Tractatus Logico-Philosophicus*, London, Routledge & Kegan Paul.
Wright, Crispin, [1983], *Frege's Conception of Numbers as Objects*, Aberdeen, Aberdeen University Press.

Index

Note:

- **Words in bold have entries in the glossary.**
- **Page references from 1–188 will concern general concepts and applications of the formal language PL. Thereafter they will refer to applications of the formal language QL.**
- **A search for information on any subject should begin with its earliest page reference, constituting its introduction to the text. Subsequent page references may give more information, repeat earlier details, illustrate applications, or link the subject to other concepts, etc.**